Bioinspired Inorganic Materials
Structure and Function

Inorganic Materials Series

Series editors:
Duncan W. Bruce, *University of York, UK*
Dermot O'Hare, *University of Oxford, UK*
Richard I. Walton, *University of Warwick, UK*

Titles in the Series:
1: Pre-combustion Carbon Dioxide Capture Materials
2: Post-combustion Carbon Dioxide Capture Materials
3: Solar Energy Capture Materials
4: Bioinspired Inorganic Materials: Structure and Function

How to obtain future titles on publication:
A standing order plan is available for this series. A standing order will bring delivery of each new volume immediately on publication.

For further information please contact:
Book Sales Department, Royal Society of Chemistry, Thomas Graham House, Science Park, Milton Road, Cambridge, CB4 0WF, UK
Telephone: +44 (0)1223 420066, Fax: +44 (0)1223 420247,
Email: booksales@rsc.org
Visit our website at www.rsc.org/books

Bioinspired Inorganic Materials
Structure and Function

Edited by

Simon R. Hall
University of Bristol, UK
Email: simon.hall@bristol.ac.uk

ROYAL SOCIETY
OF **CHEMISTRY**

Inorganic Materials Series No. 4

Print ISBN: 978-1-78801-146-4
PDF ISBN: 978-1-78801-580-6
EPUB ISBN: 978-1-78801-873-9
Print ISSN: 2472-3819
Electronic ISSN: 2472-3827

A catalogue record for this book is available from the British Library

The Royal Society of Chemistry is a charity, registered in England and Wales, Number 207890, and a company incorporated in England by Royal Charter (Registered No. RC000524), registered office: Burlington House, Piccadilly, London W1J 0BA, UK, Telephone: +44 (0) 20 7437 8656.

For further information see our web site at www.rsc.org

Printed in the United Kingdom by CPI Group (UK) Ltd, Croydon, CR0 4YY, UK

Series Preface

Materials chemistry has grown hugely in the past few decades and continues to evolve rapidly. This is especially the case since the subject now finds itself central to many of the major challenges that face global society, such as those concerning energy and the environment; but applications of materials are also found in many everyday situations, with ongoing developments in technology around electronics and the miniaturisation of devices. Although the study of materials is a highly interdisciplinary topic, the role of innovative, fundamental chemistry continues to lie at the heart of the discovery of new materials, the understanding of their atomic-scale structure and the relationship of structure to a useful property.

With the *Inorganic Materials Series*, our aim is to provide a series of themed volumes, each reflecting the diversity of the topic as it continues to develop over the coming years, with many chosen to relate closely to emerging practical applications. The use of the title *Inorganic Materials* is by no means limiting, but instead emphasises the study of the chemistry of all elements in the Periodic Table and how they impact the development of new materials. The chapters in each book will have a pedagogical flavour with a target audience from first-year postgraduate student upwards, but with wide subject coverage ranging from continuous inorganic solids, to molecular and soft matter systems. We will use expert, active researchers as editors and contributors to provide an up-to-date perspective of their field. At the same time our aim is to provide an international perspective, so to reflect the diversity and interdisciplinarity of the now very broad area of the subject.

Duncan W. Bruce, York, UK
Dermot O'Hare, Oxford, UK
Richard I. Walton, Warwick, UK

Inorganic Materials Series No. 4
Bioinspired Inorganic Materials: Structure and Function
Edited by Simon R. Hall

Published by the Royal Society of Chemistry, www.rsc.org

Preface

There is a richness to life on earth that manifests itself in many ways. A particularly lush source of complexity and beauty can be found in the multifarious structures that are produced by flora and fauna. These serve a utilitarian function, as organised organic scaffolds give rise to organic and/or inorganic structures that protect, augment and enhance the survivability of the organism. Over the course of four billion years, since the emergence of prokaryotic life, evolution has continually shaped and re-shaped biological structures in response to a bewildering range of stimuli. We, therefore, have at our fingertips a world full of organic and inorganic structures that are at the apex of functionality, formed with the utmost efficiency.

Since the Victorian-era, scientists have intentionally turned to the natural world for inspiration whenever there was a need to produce materials that were structurally sound yet functionally capable. The gargantuan load-bearing capability of leaves of the Guyanese lily *Victoria amazonica*, for example, inspired the British architect Joseph Paxton to design and construct the Crystal Palace for the Great Exhibition of 1851. The structural motifs that enabled the lily leaves to withstand enormous stresses were translated into glass and steel and the world's largest greenhouse was the result. From then on, a deeper understanding of the underlying mechanisms involved in *how* natural structures grew enabled scientists to produce ever more complex functional materials. There is virtually no field of scientific endeavour that has not felt the 'bioinspired' ethos of products, created *via* an understanding of the mechanisms of compartmentalisation, boundary-organised biomineralisation, organic/inorganic templating and epitaxial growth that are a feature of the natural world.

This book comes, therefore, at a most propitious time. Great advances are being made in energy storage, smart surfaces and regenerative medicine by adopting the principles and practices found in the natural world.

Inorganic Materials Series No. 4
Bioinspired Inorganic Materials: Structure and Function
Edited by Simon R. Hall

Published by the Royal Society of Chemistry, www.rsc.org

In addition, there are materials now being produced *via* a bioinspired approach that are iridescent, multi-coloured and even invisible to certain wavelengths of radiation. We aim to capture all of these incredible advances in this book and show how, by being bioinspired, we can work smarter, cleaner and, most importantly, greener.

Putting you, the reader, right at the cutting edge of this exciting field, the scientists themselves describe in detail the work currently being done and their ideas on how bioinspiration will transform the materials of the future.

Simon R. Hall
Bristol, UK

Contents

Inorganic Materials Series No. 4
Bioinspired Inorganic Materials: Structure and Function
Edited by Simon R. Hall

Published by the Royal Society of Chemistry, www.rsc.org

CHAPTER 1

Bioinspired Synthesis: History, Fundamentals and Outlook

R. BOSTON

University of Sheffield, Materials Science and Engineering, Sir Robert Hadfield Building, Mappin Street, Sheffield S1 3JD, UK
Email: r.boston@sheffield.ac.uk

1.1 Introduction

Times throughout human history have often been characterised by the materials that define them, with the peoples of each era using their most up-to-date technologies to shape their world. This continues today, albeit in a subtler way, through the ubiquity and continuous development of new functional inorganic materials, often with increasingly complex function, composition, or fabrication methods. The development and manufacture of the materials of today and of the future are taking a much more holistic approach. The emphasis is no longer solely on function, but also on the impact that production might have on the environment (either through mining, energy use, harmful emissions or similar).[1] Taking inspiration from nature is then a practical and efficient way to solve some of these increasingly important issues.[2] Our ever-increasing demand for materials with ever-better function is pushing beyond the known capabilities of inorganic chemistry. Increasingly, researchers are turning to nature for inspiration: biology, having had a few billion years' head start, has invariably already found a solution, in terms of either structure or process.

The family of inorganic materials covers an incredibly diverse range, including (but not limited to) those that are covered herein: ceramics,

Inorganic Materials Series No. 4
Bioinspired Inorganic Materials: Structure and Function
Edited by Simon R. Hall

Published by the Royal Society of Chemistry, www.rsc.org

glasses, and metals. Across these families, almost every conceivable function, form, and structure can be found, making the field of inorganic materials both highly versatile and ubiquitous in every civilisation throughout history. And yet, despite 15 million years of evolution, the sum of human development is still in its infancy compared to the control of materials in the natural world.

Bioinspired work can take many forms, including copying a natural structure to replicate a particular macroscopic effect, recreating natural chemical processes in the laboratory to produce a particular structure or product, and use of the naturally occurring materials themselves to create the functional materials required.

1.1.1 Early Human Use of Nature

Natural resources have always been central to human success and expansion across the globe.[3] As we learn more about early human history, we find more and more evidence of use of the resources available, rolling back the date at which the idea of "man-made technology" emerged.[4] Some of the earliest form of bioinspiration can be seen in early art; people of the time were inspired by the world around them, taking ideas from nature for subjects, colours or media.[5] These earliest known artworks, on the walls of caves, use natural materials (muds/clays) as the media and depict the natural world as the artist knew it (Figure 1.1).[6] Moving forward through time, natural forms can be seen in every type of art from every culture, illustrating the importance of the influence of the natural world: the original bioinspired work.

It is well-known that early science emerged as a distinct subject from natural philosophy, and so pioneers of the subject were well acquainted with

Figure 1.1 Example of prehistoric cave paintings at the Lascaux Cave.[7] Reprinted with permission from P. M. Martin-Sanches, A. Nováková, F. Bastian, C. Alabouvette and C. Saiz-Jimenez, *Environmental Science and Technology*, **46**, 3762. Copyright 2012, American Chemical Society.

the workings of the natural world.[8] Indeed, it could be argued that science as a distinct subject emerged as a result of the desire of early scientists to better understand nature and its functions (*i.e.* biology), with the other core areas of science (chemistry and physics) stemming from this as the next step to a deeper level of understanding. It is perhaps surprising then that it is only in the last half century that science has returned to complex natural materials or processes for inspiration to solve problems on the leading edge of technology. The reason for this lies in the very complexity required by modern technology. The types of structure or process of interest are so intricate that we only now have the tools to start unlocking their secrets. This knowledge, however, has the potential to be transformative in multiple fields, although the risks are high, the rewards have the potential to be immense.

1.1.2 Structures in Nature

Nature, and more specifically, biology, produces some of the most beautiful and intricate structures known, all without the use of harsh conditions such as high temperatures or pressures, of both organic/biological and classically inorganic substances. The creation and control of inorganic materials by biology is one of the most elegant and efficient examples of necessity being the mother of invention. Nature has harnessed the power of inorganic materials for a huge range of purposes. They not only serve specific purposes for the life forms that create them, they give a tantalising glimpse of what is possible under relatively benign conditions, such as the wide variety of seashell structures, a complex example of which is given in Figure 1.2.[9]

Much of the great success of natural structures is the ability to control morphology across multiple length scales, often simultaneously, the true definition of a "bottom-up" method. The properties, structure and function of materials are controlled, fundamentally, by the type and arrangement of

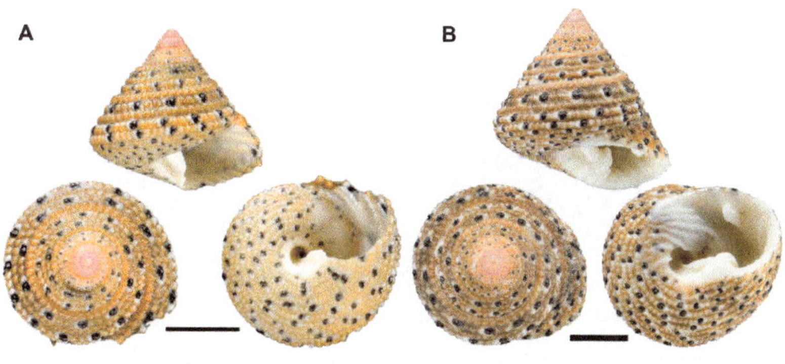

Figure 1.2 Seashell structures giving an example of the complexity of structures accessible.[9] Scale bars, 5 mm.
Reproduced from ref. 9, https://doi.org/10.1371/journal.pone.0156664, under the terms of the CC BY 0 licence, https://creativecommons.org/publicdomain/zero/1.0/.

atoms. Composition aside, the level of order (or disorder) in a material is one of the most important and defining aspects, ranging from entirely disordered or amorphous structures (*e.g.* glasses), structures with a mixture of short- and long-range order (*e.g.* cements and metals), through to highly ordered structures (*e.g.* ceramics) where order within the structure can reach the macroscale. As this is such a defining feature, we will examine each in turn within the context of nature's direct control and use, followed by how these methods might be used in the laboratory, either through replication of structures by various means, harnessing the same chemical processes, or through using the naturally-created structures as templates.

1.2 Materials with Short-range Order: Glasses

1.2.1 Introduction

Of all the inorganic materials in everyday use, glass is both one of the most commonly used and important, and yet the least well understood. It has an infinitely variable structure, and a high degree of disorder, which gives rise to glass's familiar properties: hard but brittle, transparent but (relatively) resistant to corrosion.

Glasses such as obsidian exist in nature, and are formed in rapidly cooling lava flows, where the rate of cooling is high enough to prevent crystallisation of the molten rock (Figure 1.3).[10]

Obsidian has been used from the earliest eras in history[11] as blades, axes, other tools, and even mirrors. It can be cut to a fine edge and polished to a high

Figure 1.3 Obsidian: natural glass formed in volcanic eruptions.[10] Reprinted from *Journal of Volcanology and Geothermal Research*, **310**, J. K. Shields, H. M. Mader, L. Caricchi, H. Tuffen, S. Mueller, M. Pistone and L. Baumgartner, Unravelling textural heterogeneity in obsidian: Shear-induced outgassing in the Rocche Rosse flow, 22, Copyright 2016, with permission from Elsevier.

shine thanks to its small granular structure. Although not bioinspired or controlled, it is an excellent example of early humans using the resources available to them. The relative scarcity of obsidian meant that it was generally reserved for ceremonial or decorative purposes, leading to the development of more widely available stone or metal alternatives as technology progressed.

Eventually, however, human ingenuity lead to the development of glass-making processes. Glass has been identified as being made by the ancient Mesopotamians as far back as 3000 BC, possibly even earlier. The ancient Egyptians perfected the technique as early as 1500 BC,[12] with examples of some artefacts being shown in Figure 1.4.[13]

Evidence suggests that the frequency of glass making significantly increased in the 1st century BC, when the people of the time learnt to increase the heat of their furnaces using blown air. Glass artefacts became more common, but were still highly prized due to the relative difficulty of production. Although rare, glass from throughout the Roman period is still found on a regular basis (Figure 1.5)[14] and shows a degree of sophistication and variety in production techniques (pressed beads and blown vessels). Throughout the next 2000 years, glass remained an expensive item, made with a high degree of artisanal skill and used mainly decoratively (Figure 1.6).[15] When mass production of glass become common in the late 1800s, it quickly became one of the most commonly used materials for storage of food, decoration and construction.[16]

1.2.2 What are Glasses?

Glasses are highly disordered amorphous structures, often containing high levels of silica (although it is possible to make glasses out of a large number of different materials). The high level of disorder is often achieved through rapid

Figure 1.4 Examples of Egyptian glass artefacts dated to 1178 BC, during the reign of Rameses III.

Figure 1.5 Examples of Roman glass fragments from 1st century AD.
Reproduced from ref. 14, https://doi.org/10.1016/j.jas.2015.05.004, under the terms of the CC BY 4.0 license, https://creativecommons.org/licenses/by/4.0/.

Figure 1.6 Decorative stained glass windows in the cathedral of Segovia, Spain, created in the 1540s.
Reproduced from ref. 15 with permission from John Wiley and Sons, © 2015, The American Ceramic Society and Wiley Periodicals, Inc.

cooling, "freezing" atoms into their disordered positions and preventing the formation of long-range order. The most commonly manufactured glasses are based on a mixture of silica, sodium carbonate, sodium oxide, and calcium carbonate (lime), earning this type of glass the name "soda-lime".[17] Soda-lime glass is found most commonly in everyday life, in the glass used for glazing, mirrors, food containers, and screens for electronic devices. When molten, glass can be highly viscous which means it can be carefully controlled and manipulated into a wide variety of forms. Air can be used to blow the glass, creating hollow structures, or it can be flowed around a former or into a press to create a specific shape. The large flat sheets of glass required for glazing are made using the float-glass process pioneered in the 1950s, which uses gravity to flow the glass onto a bed of molten metal, polishing the surfaces with blown nitrogen gas. The disordered nature of the structure means that glass formulations can easily incorporate many metallic ions, the careful selection of which can create colour that is evenly distributed throughout the structure. There are a number of mechanisms that govern this, including: inclusion of nanoparticles (*e.g.* gold nanoparticles as in Figure 1.7a), which forms a red glass through surface plasmon resonance; ions in solution in the glass (*e.g.* cobalt, Figure 1.7b, and iron oxide, Figure 1.7c); and colour centres within the glass (*e.g.* cadmium, when used in combination with selenium and sulfur, Figure 1.7d), which forms opaque red.

Greenish uranium glass was very popular in the early 20th century, and has an almost luminescent quality under normal illumination. When exposed to UV light, the glass fluoresces bright green (Figure 1.8), as the UV

Figure 1.7 A variety of coloured glasses: (a) nano-sized gold causes the colour in this mid-18th century glass dish, (b) an inkwell coloured with cobalt, (c) 19th century iron oxide-coloured glass doorstop and (d) a cadmium sulfide red vase.

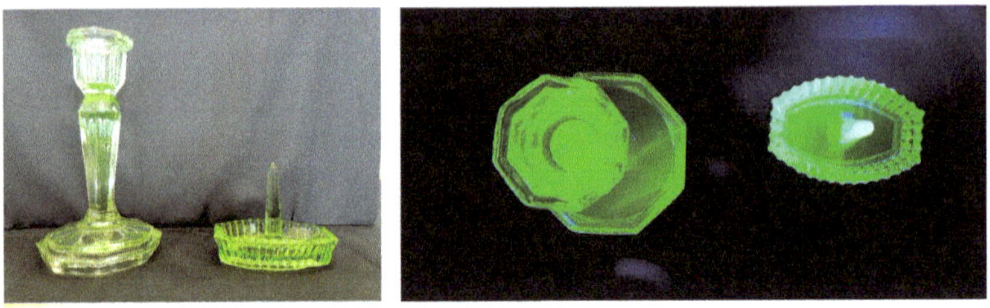

Figure 1.8 Uranium-containing glass candlestick and ring holder from the early 20th century, with and without UV irradiation.

photons interact with the uranium ions within the glass. The quantity of uranium is not high enough to be dangerous (around 1%);[18] however, it is enough to register on a Geiger counter. Although no longer produced, the fact that uranium has been so successfully incorporated in glass is still providing inspiration for the next generation of nuclear waste storage.[19]

Given the incredibly long half-lives of many waste nucleotides, long-term, stable storage solutions are a source of sizeable current research efforts. Decaying nucleotides tend to destroy highly ordered materials, making the majority of crystalline solids, including metals, cements, and ceramics, unsuitable for long-term (geological time-scale) storage.[20] As glasses are already highly disordered, any incident nucleotides would not create a huge increase in entropy in a glass system.[21] This, coupled with the extremely high tolerance of glasses to the incorporation of differently sized atoms, means that radioactive materials can be incorporated into glasses, potentially without the dangers of degradation (and therefore leeching of the nucleotides into the environment) over time.

1.2.3 Biological Glasses

There are also biological examples of silica formation in nature. Perhaps the most beautiful and intricate are the almost impossibly complex structures formed by diatoms, a group of microalgae that are found in both saline and freshwater environments all over the planet (Figure 1.9).[22,23] They have a common structure in that they produce delicate silica shells in myriad forms, which act as a protecting and containing layer for the cells of the diatom.

This unusual cell wall is known as a frustrule[24] and is synthesised inside the diatom itself. As with many protein-mediated natural processes, the transfer of silica taken up from the environment into the cell wall of the diatom is not well-characterised; however, at least three genes have been identified that are responsible for silica transport.[25] The silica in the shell is amorphous;[26] however, upon heating it forms cristobalite.[24] Studies have also indicated that diatom silica is well hydrated, and is likely to be dominated by a six-membered ring structure.[24,26] The wide variety of structures formed by diatoms are passed on from one generation to the next, indicating

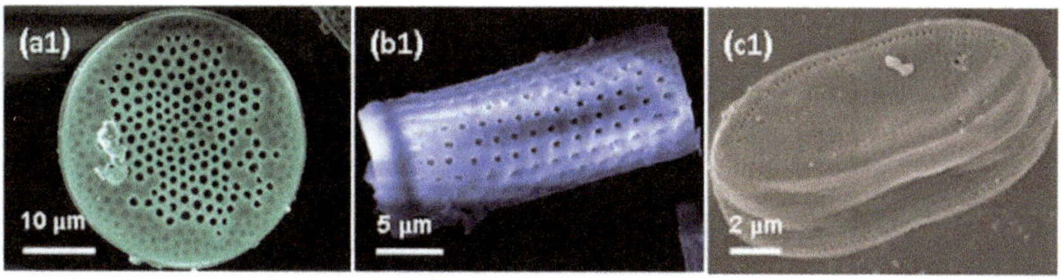

Figure 1.9 A selection of diatoms: (a1) *Coscinodiscus* diatom, (b1) *Melosira* diatom and (c1) *Navicula* diatom.
Adapted from ref. 23 with permission from the Royal Society of Chemistry.

that the processes used to form the silica shells are controlled and passed on genetically,[27] which gives some indication of the types of molecules likely to be involved in controlling silica synthesis.

Diatom silica is, in fact, a composite material, containing long-chain polyamines (specifically (*N*-methylated) poly(propyleneimine) attached to putrescine)[28] and polypeptides (silafins), although this varies from species to species. In the presence of monosilicic acid, silafins promote rapid precipitation of silica, forming the shell of the diatom.[27] Any proteins present at this stage also precipitate into the silica, forming a composite material.

Given the apparent and comparative simplicity of the diatom, it is unsurprising that there have been attempts to recreate the process of silica formation in the laboratory. Early experiments used synthetic peptides to control precipitation of silica,[29] and this was later expanded into a large library of synthetic materials that can be used to control precipitation,[30] although none of these can recreate the intricate shapes found in nature. We still have a way to go in our understanding of the process before it can be replicated in the laboratory. That is not to say, however, that the research performed so far has not found application; for example, peptide-precipitated silica has been shown to be useful as a high surface area drug delivery vector[31] and in the creation of thin silica films.[32]

Diatoms themselves have also been used as templates: by introducing the silica shells to gaseous magnesium it is possible to convert the silica to MgO/silica composites whilst retaining the macrostructure.[33] MnO_2 replicas have also been created using the diatoms as templates in hydrothermal reaction, which resulted in pure, hollow, microstructured MnO_2 replicas after removal of the diatoms using NaOH.[23]

1.2.4 Bioinspired Glasses

Of all the classes of materials, bioinspired glasses have perhaps been the most successful so far in terms of commercialisation. In industry, glasses are not generally bioinspired, but there are some notable examples where nature has been used to either direct or inspire structures to improve the already-useful properties of glasses.

1.2.4.1 Self-cleaning Technology

As a building material, glass has been used for more than two thousand years. Despite this, nature is still proving to be a source of inspiration to improve the properties for construction materials. With the growth of high-rise buildings, for example, self-cleaning windows become increasingly necessary, and, as is often the case, nature has already found an elegant solution to this, which can be observed in numerous types of leaf (Figure 1.10).[34] The best-known of these is arguably the superhydrophobic properties exhibited by the lotus leaf, which has both a physical structure and a chemical coating designed to shed water easily. The lotus leaf uses micrometre-scale surface structures to increase the contact angle of water droplets on the surface. This works by utilising the surface tension of the water to minimise the contact area (and therefore wetting), causing any

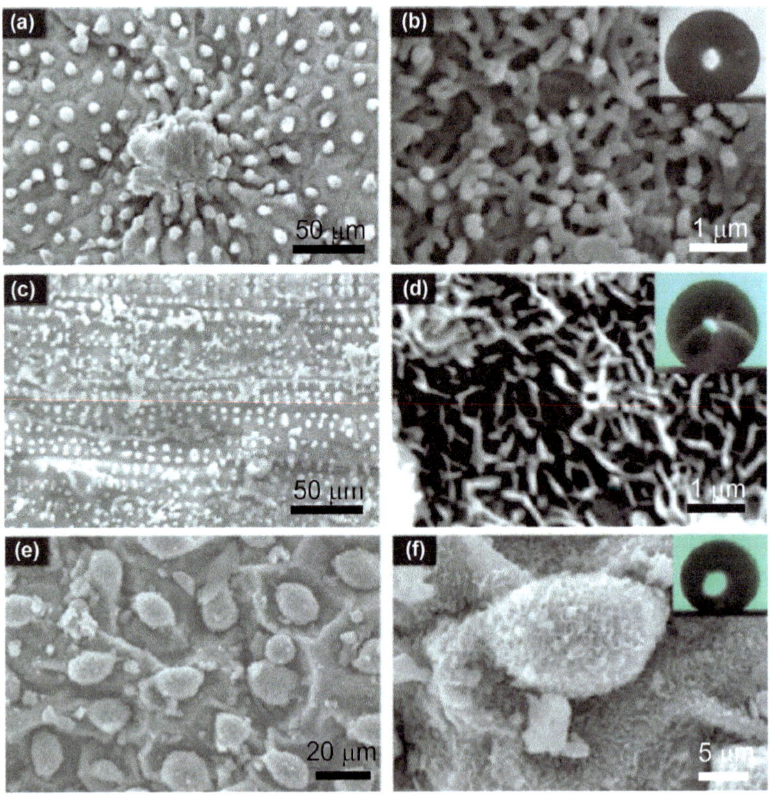

Figure 1.10 Scanning electron microscope (SEM) images of three varieties of hydrophobic leaf: (a), (b) The well-known lotus leaf, (c), (d) a rice plant leaf, and (e), (f) a taro plant leaf.
Reproduced from ref. 34 with permission from the Royal Society of Chemistry.

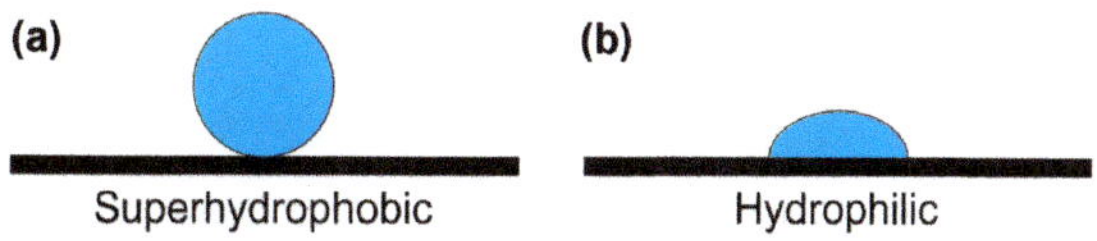

Figure 1.11 Examples of differences in contact angle on (a) a superhydrophobic and (b) a hydrophilic surface.

incident water to form almost spherical droplets (see Figure 1.11) which can easily move about on the surface of the leaf under gravity.

As they roll, the water droplets can pick up any surface particles such as dust or mud, carrying them away and cleaning the surface of the leaf. This is a highly desirable function, particularly on glass, where dust or contamination can be easily seen. By using these natural structures as a starting point, similar forms can be patterned onto the surface of window glass,[35] meaning that the windows can self-clean every time it rains. There are numerous considerations here,[36] however, including durability and effect on transparency; nobody wants a self-cleaning window that you cannot see through!

Hydrophobic leaves are not the only source of inspiration for self-cleaning materials. Many natural structures exhibit similar properties, including butterfly wings (which also have interesting optical properties, as we will see later). Also of interest are hydrophilic surfaces, which can similarly be used to self-clean, and can reduce drag,[37] particularly in water, and repel oil.[35] As such they are often found in naturally wet environments. Fish scales,[38] the shells of snails,[39] and shark skin[40] are all hydrophilic and have all been used as biotemplates for hydrophilic materials.

1.2.4.2 Biomedical Applications

Glasses are well suited to use in the human body, as they are biologically inert and generally composed of non-toxic or biocompatible materials. Glasses for biomedical use tend to be composed of SiO_2, CaO, and P_2O_5,[41] and are able to bond with host tissues, making them commonly used in the repair of tissues such as bone[42] and skin (and epidermal layers),[43] as well as a host of other tissues.[41,42] These, however, do not generally take inspiration directly from nature. To find this we must look at applications that require micro- and nano-scale structures. These types of structures have inherently high surface areas which, when combined with the biocompatibility of certain glass compositions, are useful for drug delivery. Taking inspiration from the structure of pine-cones, it has been possible to create bioactive glass nanoparticles with high surface areas (280 $m^2 g^{-1}$) using a sol–gel method. The pine-cone structure was created using hexadecyl trimethyl ammonium bromide and ethylacetate to form colloidal particles. These were then hydrolysed using ammonia, before addition of tetraethyl orthosilicate, triethyl phosphate, and calcium nitrate, followed by autoclaving at 100 °C and calcination at 650 °C.[44] This resulted in high surface area pine-cone-like morphologies (Figure 1.12)[44] which showed improved delivery of

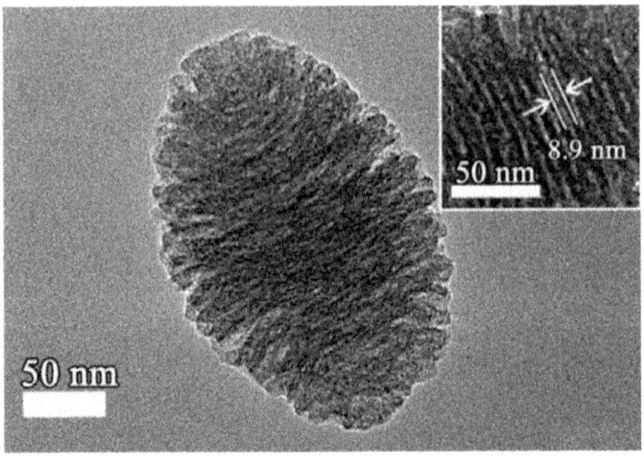

Figure 1.12 Pine-cone-like structures in bioglass.
Reprinted from *Materials Letters*, **148**, Q. Liang, Q. Hu, G. Miao, B. Yuanand X. Chen, A facile synthesis of novel mesoporous bioactive glass nanoparticles with various morphologies and tunable mesostructure by sacrificial liquid template method, 44, Copyright 2015, with permission from Elsevier.

doxorubicin, an anti-cancer medication, that was shown to release over four days.[45]

1.2.5 Summary

In terms of synthesis, glasses are challenging to make using bioinspired methods as these tend to promote crystallisation over an amorphous state; however, nature has and continues to provide inspiration for the shapes and nanostructures in glasses. This enables the recreation of functional and structural properties found in nature, focused mainly on surface structural features. As a result, bioinspiration can now be found in construction materials and may eventually form a key component in drug delivery. Natural glass synthesis in organisms such as diatoms is also providing new insights into the biological control of silica on the nanoscale.

1.3 Materials with Long-range Order: Metals and Ceramics

1.3.1 Introduction

After glasses, the often highly ordered metals and ceramics are some of the next most common materials used in modern life. They both generally have long-range order and are composed of crystals, the composition and arrangement of which determines the structural and functional properties.

Metals are found in a multitude of applications due to their high electrical and thermal conductivity, ductility, (generally) high melting point, strength and lustrous appearance, including in construction, electrical wires, aerospace and automotive parts, and in almost every other aspect of modern technology. They also feature heavily throughout history, defining eras such as the Iron or Bronze ages, demonstrating the development of metals through technology (see Figure 1.13). They can be used alone or as alloys with multiple components, but all, generally, have a degree of long-range order, metals are usually crystalline. It might not be immediately obvious how metals can be synthesised using bioinspired methods: the techniques we will examine generally use noble metals (*i.e.* those that do not readily oxidise) and are concerned with macroscopic control for producing high surface area materials; for instance, nanoscale metallic structures such as foams.

When metallic elements are joined by a non-metal (often oxygen, carbon or nitrogen) the materials that form are termed ceramic. Ceramic materials, of one form or another, make up a significant portion of materials used in everyday life and in nature. Ceramic is a word that covers a vast range of materials (examples shown in Figure 1.14), including clay-based materials as might be found in pottery or earthenware ceramics most usually found on the dinner table or as a container for your morning coffee. It also covers the majority of oxides, carbides and nitrides with long-ranger order. Generally, these have interesting structural or functional properties and, as such, play a vital, although often unseen, role in everyday life. This includes functional oxides in electronic circuitry such as dielectric materials for capacitors, piezoelectrics as might be found in headphones, speakers or accelerometers, and, perhaps most pervasively, as the cathode in Li-ion batteries. Structural and technical ceramics are also key to modern technology, appearing in applications such as ceramic brakes, in dental or medical implants, and in temperature-resistant coatings in stoves and boilers.

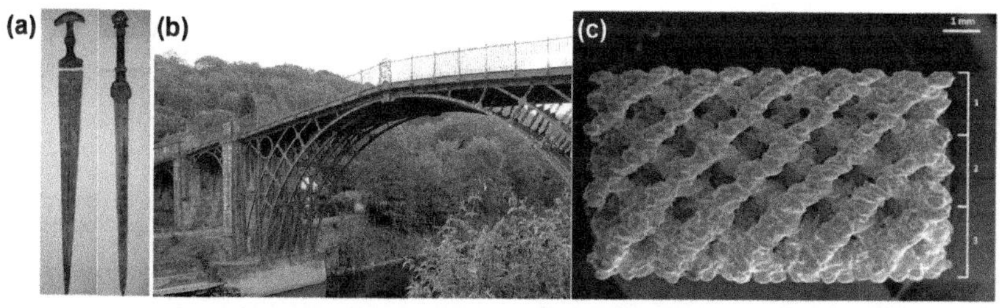

Figure 1.13 Metalwork throughout history. (a) Japanese Bronze Age swords, from around 3000–1200 BC,[46] (b) the bridge over the River Severn in Shropshire, built in 1781, the first large bridge to be made of cast iron,[47] and (c) 3D-printed steel structures.[48]
(a) Reprinted from *Nuclear Instruments and Methods in Physics Research B*, **407**, S. Shizuma, T. Kajimoto, S. Endo, K. Matsugi, Y. Arimatsu and H. Nojima, Non-destructive analysis of ancient bimetal swords from western Asia by γ-ray radiography and X-ray fluorescence, 244, Copyright 2017, with permission from Elsevier. (b) Source: https://commons.wikimedia.org/wiki/Ironbridge#/media/File:Iron_Bridge.JPG. (c) Reproduced from ref. 48, https://doi.org/10.1016/j.scriptamat.2016.10.030, under the terms of the CC BY 4.0 license, https://creativecommons.org/licenses/by/4.0/.

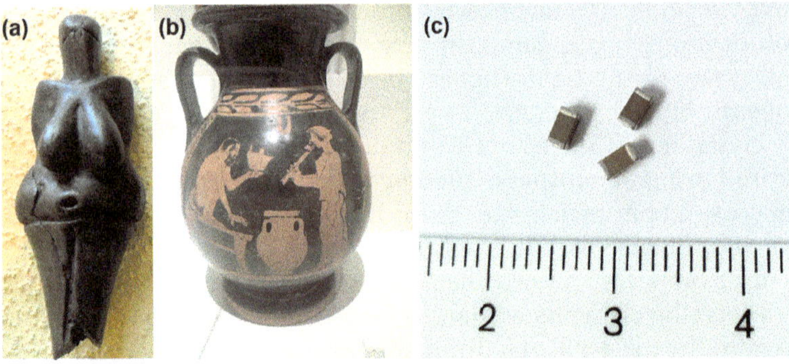

Figure 1.14 Examples of ceramic materials: (a) one of the earliest known ceramics, the Dolni Věstonice Venus,[49] (b) an ancient Greek ceramic vase, and (c) a modern multilayer capacitor.
(a) Reproduced from https://commons.wikimedia.org/wiki/File:Doln%C3%AD_V%C4%9Bstonice_Venus_-_Fossils_in_the_Arppeanum_-_DSC05513.JPG.

Each of these (and the myriad other applications) requires specific structural or functional properties that are usually determined by the composition and crystalline structure, or a combination of both, and, as such, there are multiple techniques where nature and biology can provide a source of ideas. This might be in the form of replication of a process or structure (as was seen in the case of self-cleaning window glass), or it might be by using a natural material directly to control structure as crystallites form.

1.3.2 Natural Processes

Nature makes use of the functionality in ceramic materials. For example, magnetotactic bacteria (Figure 1.15) create magnetic iron oxide nanoparticles to use as tiny internal compasses to navigate towards new food sources following magnetic field lines. These bacteria fall into two categories, one producing Fe_3O_4 (magnetite) nanoparticles, the other Fe_3S_4 (greigite).[50] The crystallisation of the iron-containing species is controlled by the concentration of iron and the pH within the nanoparticle-producing cells (magnetosomes), which are compartments within the bacteria.[51] The genomes of some species of these types of bacteria have been sequenced[52] and it was found that the bacteria use cation diffusion facilitators to transport the iron to the correct place within the cell, and to actively increase concentration.[53]

1.3.3 Natural Calcium Carbonate

Perhaps a little more well-known are the intricate exoskeletons found on beaches and in gardens all over the world: molluscs have been controlling the crystallisation of calcium carbonate for well over 500 million years. Their first appearance in the fossil record is in the Cambrian period;[54] however, the

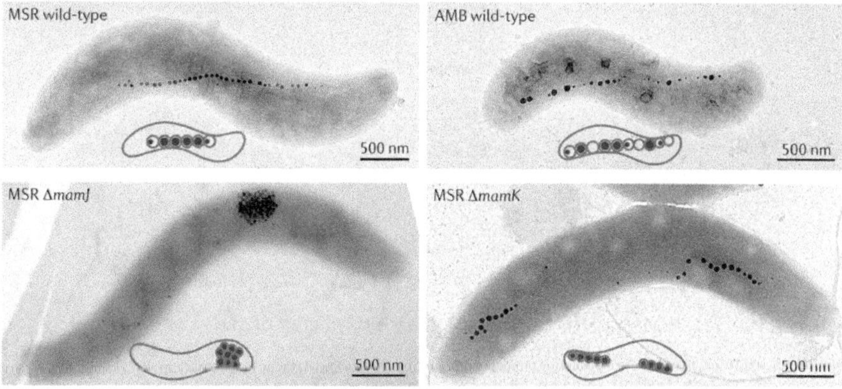

Figure 1.15 Magnetotactic bacteria of various forms showing the magnetic nanoparticles distributed within the cell.[53]
Reprinted by permission from Springer Nature: Macmillan Publishers, *Nature Reviews Microbiology*, Magnetosome biogenesis in magnetotactic bacteria, R. Uebe and D. Schüler, **14**, 621, Copyright 2016.

evolutionary origins are not known and, indeed, by the late Cambrian the fossil record contains primitive forms of the majority of modern mollusc species.

Molluscs are a tremendously diverse and successful class of animals. They have adapted to a huge range of environments from the deep ocean to the suburban garden, thanks in large part to their protective shells (examples can be seen in Figure 1.2). These shells are primarily composed of calcium carbonate, making this one of the most widespread ceramic materials produced in nature.

The controlled deposition of calcium carbonate by molluscs has been widely studied and yet is still not fully understood. Shells are made up of three layers; the outer protein-rich periosteum, the prismatic layer, and an inner layer of nacre.[55] The outermost layer is a protein upon which crystallisation can occur to create subsequent layers of shell[56] and is formed by specialised cells found in the organism's mantle. The prismatic layer, also formed by the cells within the mantle, is the middle, thicker layer of calcium carbonate in the form of aragonite.[57] The inner surface of the shell is perhaps the best-known layer, nacre (or mother of pearl), and is composed of hexagonal platelets of aragonite, interlayered with a range of elastic biopolymers (for example, chitin or lustrin).[58] It is hard and shiny and displays structural colour due to the arrangement and size of the platelets, as can be seen in Figure 1.16.[59] Nacre is secreted by a different type of cell within the mantle, and is primarily designed to protect the soft tissues of the mollusc. It is also used as defence, depositing successive layers on any invading parasites or grit, forming a pearl.[60]

Of course, molluscs are not the only creatures to form shells, there are many others, including crabs, oysters, barnacles, brachiopods, annelid worms, sea urchins, and, perhaps most ubiquitous of all, the eggs produced

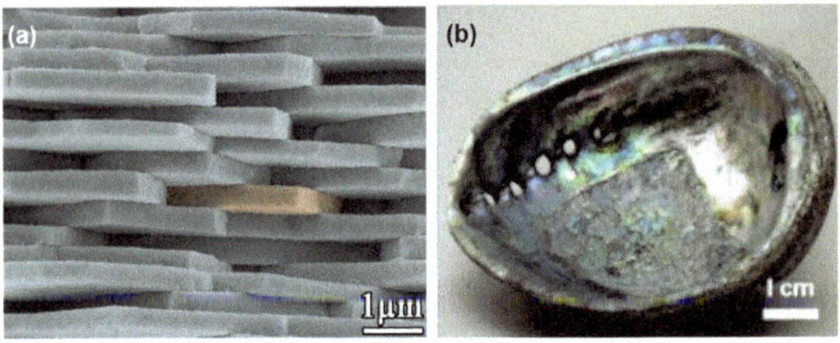

Figure 1.16 (a) SEM and (b) optical microscope images of nacre. (a) Reproduced from ref. 59 with permission from The Royal Society of Chemistry; (b) reproduced from ref. 61 with permission from the Royal Society of Chemistry.

by birds and reptiles. These all use calcium carbonate in some form and each has their own formation mechanism. For now, however, we will continue to consider the structure of nacre as a source of bioinspiration for directing the structure of ceramics.

1.3.4 Use of the Nacre Structure in Ceramics

The intricate layered nacreous macrostructures formed in seashells have long provided a source of inspiration to materials scientists. Rather than the more familiar calcite, nacre is made up of a different polymorph: aragonite. This has an orthorhombic crystal structure (as opposed to the trigonal structure found in calcite), and in nacre the crystals exist as pseudo-hexagonal plates separated by organic phases.[62] These plates, which make up around 95% of the structure by volume,[63] are separated by biopolymers such as β-chitin, which provide a scaffold and separation of the plates.[64] By creating plates of random sizes, they can be randomly interdigitated, which is where much of the strength arises. This also has the beautiful effect of creating structural colour, due to the interaction of light with the regularly spaced plates. The slight macroscopic variations in layer separation across the surface of the shell are what give rise to the different colours observed.

The structure of nacreous shells (Figure 1.17) often includes a trigonal/prismatic (calcite) layer covering the internal nacreous layer, with the two layers often being separated by a "green sheet", a thick layer of organic material that mediates the transition from prismatic calcite to nacre.[61] The growth of the nacreous plates themselves is currently under investigation, but is thought to occur due to aggregation of nanoparticulate calcium carbonate in a fibrous organic matrix;[65] once a critical number of particles is reached, these coalesce into nacre platelets.[61] Insights into this process have been gained through the use of polarisation-dependent imaging contrast and X-ray photoemission spectroscopy.[63] The work on red abalone (*Haliotis rufescens*) demonstrated that the kinetics of formation were driven by certain growth directions being more favourable than others, and that this

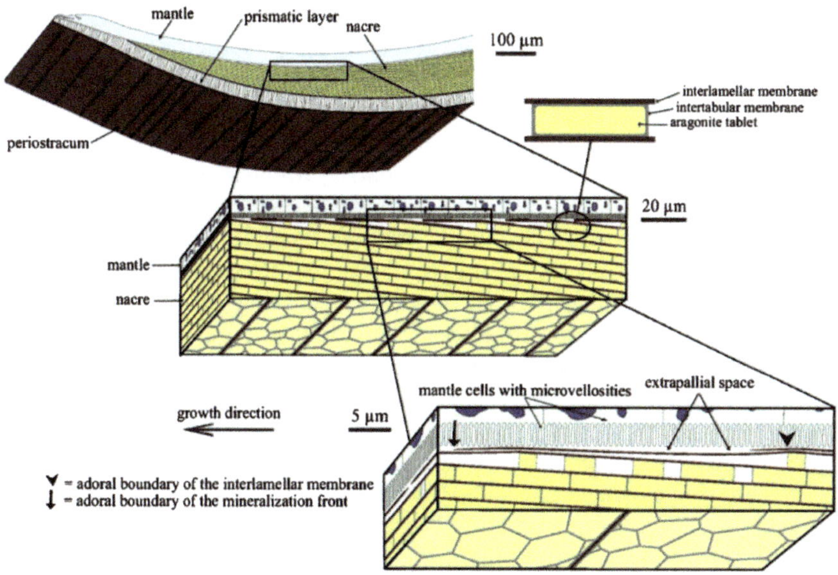

Figure 1.17 The structural features of a sea shell.
Reproduced from ref. 61 with permission from the Royal Society of Chemistry.

dependence was linked to the distance between the forming nacre plate and the green sheet, resulting in the well-orientated plates eventually winning the competition for space inside the shell.[63] In particular, it is thought that the organic matrix suppresses the growth of the crystallographic *c*-axis (perpendicular to the surface of the shell), in contrast to the mineral form of aragonite, which generally shows preferential growth on the *c*-axis. It is from here that some of the favourable mechanical properties may arise, particularly when found as an organic/inorganic composite material.

The mechanical properties of nacre are well characterised[66] and thought to arise from a combination of its organic/inorganic composite structure and the hierarchical arrangement of structures over several length scales.[67] Nacre has been demonstrated to have fracture toughness that is several orders of magnitude higher than the mineral aragonite, and this is known to rely heavily on the organic component, as wet and dry nacre show very different results. This also holds true for compressive strength, which is highly unusual, as strength and toughness are often mutually exclusive properties. Attributing these properties to specific structural features within the nacre structure, however, remains a challenge.[66]

The lack of understanding of the formation of nacre makes creating artificial nacre in the laboratory using natural growth processes impossible currently. However, the structural form (and therefore properties) of nacre can be replicated in the laboratory using templated[68] or self-assembly techniques,[69] offering a route to create materials with highly sought after properties, such as ceramics with both high strength and toughness. Often, this is achieved using composites,[70] introducing a ductile or flexible phase; however, using a nacre-like structure, it has been demonstrated that these physical properties in ceramic structures can also be greatly improved.[71]

Replicas of nacre-like structures have, however, been successfully created by freeze-casting a laminated chitosan matrix, which was then mineralised by precipitating calcium carbonate onto the layers. Silk fibroin was then infiltrated between the layers, mimicking the organic component of nacre.[72] This work showed that the rising crack-extension resistance of synthetic nacre was of a similar order to naturally occurring nacre, and that the synthetic version was able to disperse incident energy in the same way, through trapping of cracks by the organic layers.[72]

Similar demonstrations have been made in alumina/graphene oxide and poly(vinyl alcohol) structures (Figure 1.18), which were shown to resolve some of the tension between strength and toughness, and in particular it was found that this was aided by using a range of platelet sizes concurrently.[73] All-ceramic materials have also been created. By freezing alumina platelets under flow, hierarchically structured ceramics have been created with a nacre-like appearance, albeit without the presence of any organic phase.[74] Instead, a mixed nanoparticulate alumina and silica/calcia-based glassy phase was used to mimic the proteinaceous organic layer in nacre, giving rise to excellent mechanical properties when compared with standard polycrystalline alumina.[74] This is an elegant bioinspired solution to the brittleness of ceramic materials; however, systems currently have to be developed on a material-by-material basis.

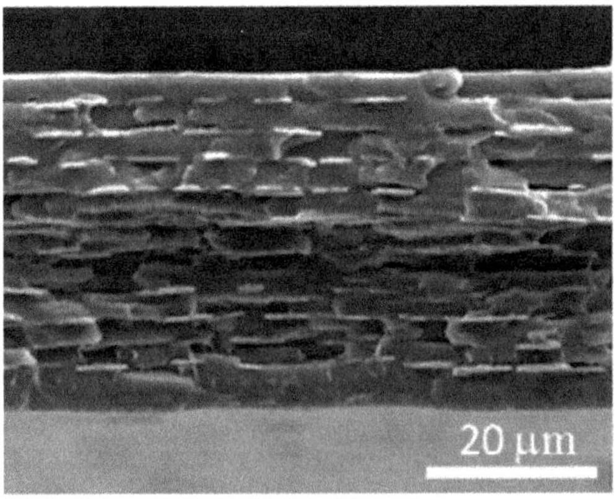

Figure 1.18 Alumina, graphene oxide and polyvinyl alcohol composite, replicating the nacre structure.[73]
Reprinted with permission from Bioinspired hierarchical alumina-graphene oxide- poly(vinyl alcohol) artificial nacre with optimised strength and toughness, J. Wang, J. Qiao, J. Wang, Y. Zhu and L. Jiang, *ACS Appl. Mater. Interfaces*, 2015, 7, 9281. Copyright 2015, American Chemical Society.

1.3.5 Artificial Calcium Carbonate

Nature does not have a monopoly on control of crystallisation in calcium carbonate. In the laboratory, $CaCO_3$ has been considered from multiple viewpoints, including fundamental crystallisation, control of functional properties and interaction with organic and biological agents. As such, it forms a suitable starting point for our consideration of control of bioinspired inorganic materials in the laboratory.

The fundamental crystallisation of $CaCO_3$ may appear straightforward,[75] however, it has been the source of much controversy, as investigation of the precise nature of the earliest nucleation has proved extremely difficult, mainly due to the extremely small length and time-scales over which it occurs.[76] A deep discussion of this topic in both calcium carbonate and other related materials is beyond the scope of this book.[77,78] Whatever the exact process is, here we are concerned with the process immediately following the emergence of structures on the length scale that could be termed long range.

Understanding the role of biological molecules in the crystallisation of calcium carbonate is critical to determining both how nature creates the morphologies observed and how we can utilise this in the laboratory, in calcium carbonate and in other materials. For example, work has been performed to understand the role of amino acids in the morphology and control of functionality. During crystallisation, simple amino acids such as aspartic acid can be incorporated into the crystal structure of calcium carbonate, fundamentally changing the way that the crystals form and resulting in obvious changes to the shape of the crystal.[79] Interestingly, some amino acids can act as "chaperones" for other functional molecules such as dyes, resulting in the ability to produce calcite crystals with specific functionalities (*e.g.* fluorescence[79] or colour,[80] Figure 1.19). It is worth noting that the molecules would not incorporate into the crystal structure on their own, affinity with the amino acid is key. Modelling suggests that this is due to the particular size and shape of the amino acids, and may well hold the key to why calcium carbonate is seen so often in nature, and in the incredible variety of morphologies.[79] These techniques may have significant application in the formation of composite or tuneable materials in the future.

Of course, all formation of ceramics, particularly biotemplating, requires some form of crystallisation and, in all cases, this occurs as a supersaturation, or creation of particles that are larger than the critical size for the nucleus of the crystallising material to be stable and not to disintegrate. The material may crystallise directly into the product phase or it may form intermediate materials, which form the product phase through later solid–solid or solid–liquid reactions. These are often governed by the same rules as classical solid-state reactions; however, they may occur at reduced temperatures due to better mixing or smaller particle size.

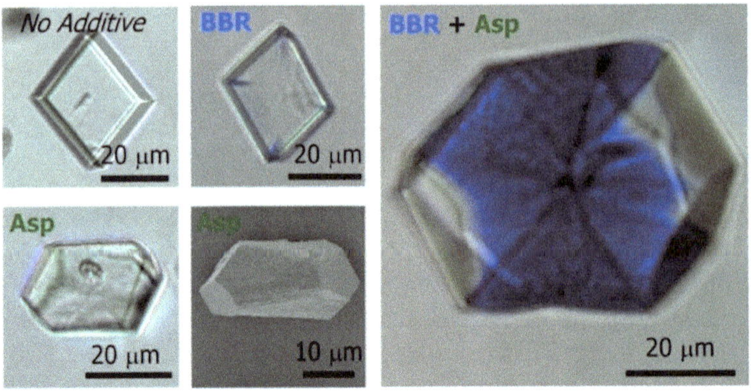

Figure 1.19 Laboratory-grown calcium carbonate (top, left) showing the effect of incorporating aspartic acid (Asp) to change the morphology of the crystallites (bottom, left and middle). Furthermore, organic molecules can also be incorporated (in this case Brilliant Blue R dye, BBR), but only when chaperoned by the aspartic acid (right). Without the chaperone, the dye does not incorporate fully (top, middle). Reproduced from ref. 80, https://doi.org/10.1002/anie.201804365, under the terms of the CC BY 4.0 license, https://creativecommons.org/licenses/by/4.0/.

1.3.6 Chemistry and Control in Bioinspired Synthesis

Standard synthesis processes in oxide ceramic materials rely on solid-state reactions, which generally occur at high temperatures and/or over long periods of time. The process uses ionic migration from one compound into another (or into several depending on the complexity of the product), which requires large amounts of energy to overcome the energetic barriers. This often results in a significant change in crystal structure or tilt, or gives rise to lattice distortions, potentially limiting the number of accessible polymorphs. Despite all of this, solid-state processing is still the method of choice for most large- or industrial scale manufacture of oxide powders, and is highly reliable and easily scalable.[81]

Solid-state processing has a number of limitations, however. The high temperatures require tremendous amounts of energy, often accounting for a significant proportion of the energy used in production.[82] This is compounded by the long durations required, often in excess of 12 hours, with repeated regrinding and reheating to redistribute unreacted elements. These requirements mean that, very often, volatile elements are lost during processing and that the materials undergo uncontrolled grain growth, resulting in particles with random shapes and sizes (Figure 1.20). As such, many modern ceramics rely solely on increasingly complex chemistries to create the desired functionality, with the common $BaTiO_3$ multilayer ceramic capacitors or lead-free piezoelectrics often containing multiple fractional dopants of rare or often toxic elements.[83–85] Of course, their fractional nature and small component size means that recovery of these materials is currently next

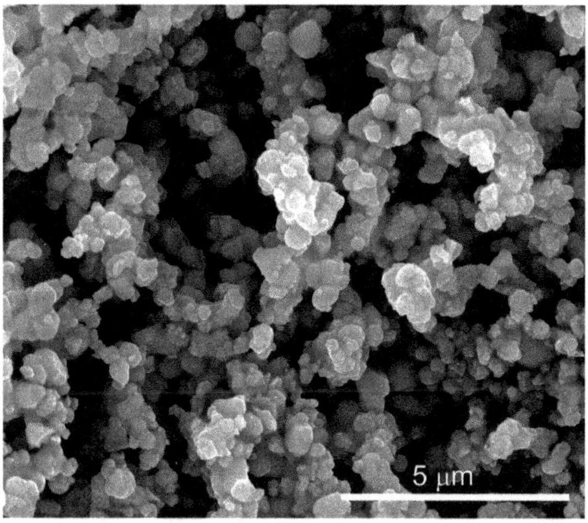

Figure 1.20 Typical particle morphologies and sizes found in solid-state processed materials, in this case, barium titanate ($BaTiO_3$).

to impossible and means that we will be facing a future crisis of sustainability within the device industry unless alternative solutions can be found.

Control of morphology through bioinspiration may, in some cases, offer a solution to this. Certain functionality, including superconductivity and cathodic materials, has known physical benefits when morphology is controlled. The critical current that can be carried by a superconductor can be significantly influenced by crystallographic alignment,[86] and so it might be favourable to create similarly shaped crystallites that show the same crystallographic alignment for use in devices. Similarly, many cathodic materials (for example in Li-ion batteries) can be improved by increasing the surface area of the material, *i.e.* by creating nanoscale crystallites,[87] something that is not easily achieved using traditional processing techniques.

Biotemplating, the use of naturally occurring molecules to control and direct crystallisation, offers an elegant bioinspired solution to both the need to control crystallite size and morphology and to reducing the energy cost of production by enabling the use of reduced processing temperatures and times.

1.3.7 Biotemplating in Oxides

The process for creating biotemplated oxides is relatively simple. A solution of metal ions is created and mixed with the chosen biotemplate, either as a second solution or as a direct additive, which then dissolves. The positive metal cations interact with the negatively charged regions that form along the biotemplate through deprotonation/dissociation of the functional groups along the chain, forming a large, disordered metal–organic complex, modelled by an egg-box-type structure (Figure 1.21).[88] This effectively holds

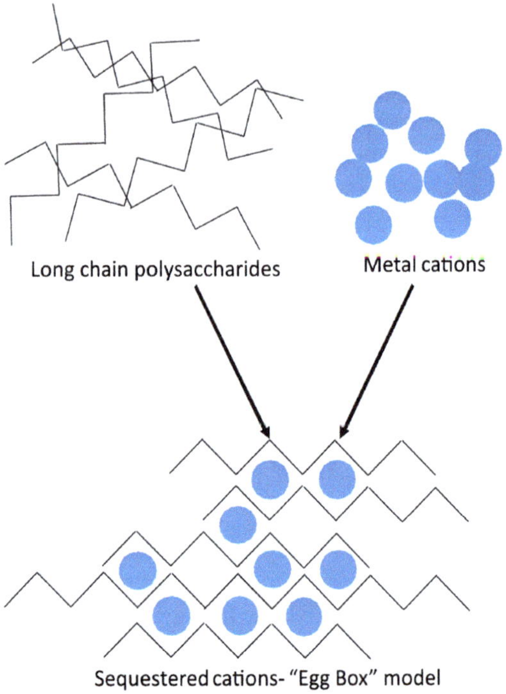

Figure 1.21 Schematic of the egg-box model:[88] ions chelated into a matrix of biopolymer.

the metal ions in place and, upon drying, prevents the recrystallisation of the starting metal salts and enables the production of atomically homogeneous composites.

Upon heating, the template starts to combust, removing the carbon and hydrogen (as CO_2 and H_2O, respectively) as well as scavenging some of the oxygen from the template or metal salts. At this point, the first crystalline phases emerge. In some cases, this might be the product phase, as in the case of binary oxides or certain spinels that have been demonstrated to crystallise without any intermediate phases.[89] In more complex oxides, however, these initial crystallisation events are often a binary oxide or carbonate phase of one of the reactants. These initial crystallisations are usually driven by the increasing concentration of material as the template outgasses. The ions are released from chelation and coalesce as the template combusts, driving the nucleation of ions above the critical size for stability. Herein is the key to successful structural replication, as is observed in many biotemplated systems. These nuclei are small (of the order of a few nm) and are extremely well-mixed and, as the temperature is increased further, they are able to undergo solid-state reactions, just as would be observed in traditional solid-state synthesis, except on the nanoscale and with a far higher degree of mixing than would normally be accessible. This results in the product forming more rapidly and at lower temperatures than in the solid state in many materials.

1.3.8 Types of Templating

Templating, biological or otherwise, can be broadly split into two forms: hard and soft.[90] Hard templating uses, as the name suggests, a solid template with particular form or structure, which is then either replicated by the final product or acts as a former, which is later removed to leave the inverse structure.[91–93] Soft templating, on the other hand, uses materials that have no defined macrostructure (for example, proteins) and which produce materials solely using bottom-up control.[94] We will examine both of these in the context of materials with long-range order (metals and ceramics) and how they can be employed to change structure and function; often demonstrating improvements in properties. We will not consider every type of template here as there are simply too many;[95,96] however, we will examine some key examples in order to study the significant morphologies that can be formed, their underpinning mechanisms, the main concepts, key considerations and limitations.

1.3.9 Soft Biotemplating

Soft biotemplating relies on the solubility of the template to create chelation sites. With life fundamentally being water-based, many biological processes, and therefore the materials that control them, are water-soluble, making them excellent candidates for soft templating applications.

1.3.9.1 Polysaccharides

Of all the molecules found in nature, the six-membered ring structure of glucose is perhaps the most common, and arguably the most important, forming as it does most of the fuel for life. As such, it is a common biogenic molecule, either as a monomer or in its more common polymeric form. These polymers can be many thousands of Daltons in mass and form linear or branched structures, or, more commonly, a combination of the two. Most importantly, glucose contains up to four hydroxyl functional groups, the number of which decreases as the glucose monomer polymerises or becomes branched. These species can be substituted for other groups (for example, carboxylates), forming a wide-ranging family of molecules based around glucose – the polysaccharides, some of which we will examine here.

Functional groups are fundamental to the biotemplating process: their number, density, and species are all important factors to be considered, as they have a direct impact on the product.

Many biological polymers are based on the polysaccharide structure. These include sugars such as dextran, more complex materials such as chitin and sodium alginate, and insoluble polymers such as starch or cellulose (Figure 1.22). These types of material therefore make up the majority of natural materials and are a key source of material for biotemplates.

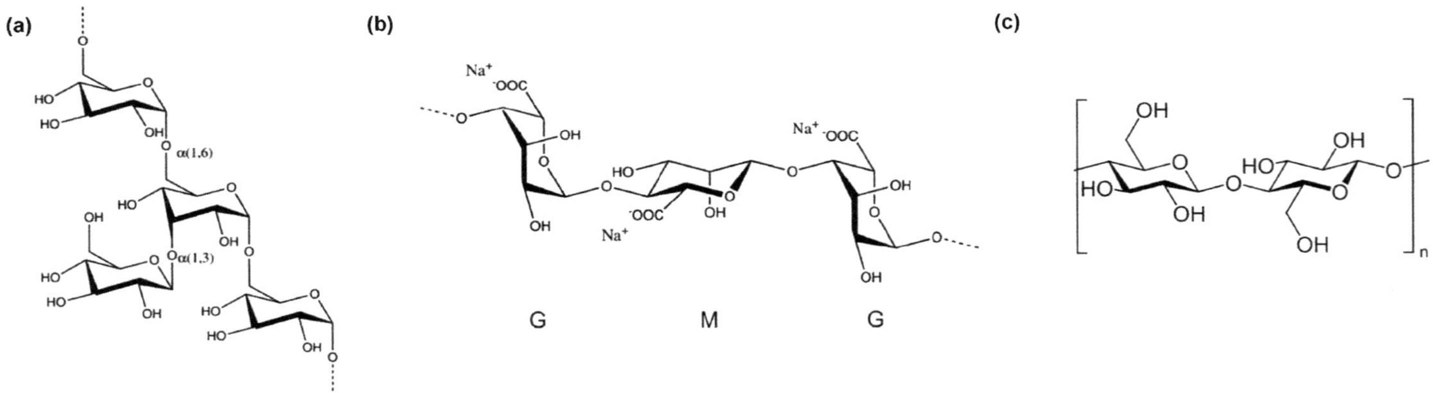

Figure 1.22 Example polysaccharide structures (a) dextran, (b) sodium alginate containing G (guluronate) and M (mannuronate) groups and (c) cellulose.

1.3.9.1.1 Simple Polysaccharides. The glucose ring structure is incredibly flexible in terms of chain length and degree of branching. Of these, dextran is one of the more successful templates for metallic and ceramic materials. It has been widely used across a number of materials and functionalities, often with the resultant physical properties of the product being improved over the solid-state synthesised counterparts.

One of the first uses of dextran as a template was for the formation of metallic sponges.[97] Noble metals were used (specifically silver and gold) as these exist preferentially as metals rather than oxides and so would resist oxidation during heating in ambient atmosphere. Dextran was dissolved into solutions of either silver nitrate or gold chloride, dried and calcined in air. The resulting sponges were highly reticulated, being formed of micrometre-scale crystallites joined in a porous structure. Surface area measurements demonstrated the scale of the texturing, with surfaces areas being measured at around 1 $m^2 g^{-1}$, significantly higher than foils or bulk materials. The work also investigated copper oxide, which was equally successful, forming self-supporting monolithic high surface area foams.[97] The work also detailed a method by which copper oxide particles could be embedded within a metallic silver sponge, simply by adding a small amount of copper nitrate solution to the original silver sponge preparation. The oxide particles were discrete and well distributed throughout the sponge; this is interesting given that the starting solution will have been atomically homogeneous, and that the entire reaction will have occurred in the solid state. It is likely, therefore, that the copper ions have been mobile in the early stages of reaction (as the template is removed), forming copper oxide nuclei that have been able to migrate through the metallic component, agglomerating into the observed micrometre-scale particles.

The next obvious step from a metallic sponge is to investigate the formation of sponges of metal oxides. It was soon found that dextran could indeed be used to form high surface area oxide materials, such as binary materials (*e.g.* Fe_2O_3,[98] ZnO[99]). With relatively limited numbers of potential impurity phases, however, there was still scope to explore the flexibility of dextran to create higher-order materials, examining the ease with which more complex materials could be created, whilst avoiding the usual kinetically stable intermediate phases.

The high temperature superconductor $YBa_2Cu_3O_{7-\delta}$ (YBCO) is one such higher-order material. With a highly complex phase diagram and a wide range of stable phases (mainly binary and ternary oxides), YBCO is an ideal test for a template that has to deal with three different types of ion with a variety of valence states, sizes and chemical affinities. The superconducting properties of YBCO are also extremely sensitive to oxygen content and so any deviation from the ideal composition due to synthesis method could be easily identified. Dextran, as with many templates, is more than a match for these challenges. The early work with dextran-templated YBCO was highly successful, replicating the same reticulated sponge structure as had been observed in the metallic sponges.[100] Later investigation of the intermediate phases formed indicated that nanoscale binary oxides/carbonates were formed (Y_2O_3, $BaCO_3$ and CuO),[101] exactly the same as would be seen in a traditional solid-state reaction. What was surprising here, and in all dextran-templated YBCO studies, however, is that unlike in a solid-state reaction, the phase formed was highly oxygenated.[100–102] In solid-state processing, there is not enough oxygen available to attain the desired $YBa_2Cu_3O_{6.9}$ stoichiometry, with the molar ratio of oxygen often being significantly lower, to the detriment of the superconducting transition temperature, and in extreme cases changing the desired orthorhombic crystal structure to an entirely non-superconducting tetragonal structure with $O_{6.5}$ or less.[103] The solid-state method then usually requires a post-anneal step under flowing oxygen to suitably re-oxygenate the sample. In dextran-templated YBCO, the $O_{6.9}$ phase has been found to emerge preferentially to the less oxygenated form. This is likely to be due in part to the higher surface area and in part to the shorter processing times required, as significantly more of the material will be exposed to the air. This enables easier uptake of oxygen from the surrounding atmosphere at intermediate temperatures and the shorter times held at high temperatures mean that less oxygen will be lost from the structure. It is likely that the former is mainly responsible for the high levels of oxygenation.

Investigation of the intermediate phases also reveals how the final YBCO phase forms (Figure 1.23). A common intermediate phase observed, particularly close to the final calcination temperature (and indeed as a minor impurity phase after the end of the calcination step) is the Y211 "green" phase, or Y_2BaCuO_5. This indicates that the likely mechanism of formation is migration of the Ba and Cu ions into the Y_2O_3 crystal structure. Given that CuO is highly stable and is also often seen as a minor impurity after calcination,[101] the Y211 phase is an unsurprising secondary phase, as, after forming CuO at low temperatures during heating, there will be less copper available for reaction, thus promoting the formation of a phase that is proportionally richer in yttrium. Ideally, of course, the biotemplating method would produce a well-mixed intermediate preventing this kind of phase separation; however, this is at least partly dependent on the materials and their stable intermediates, and partly on the nature of their interaction with the template. In the case of YBCO, copper oxide is one of the first crystalline products to form upon heating which means that it is has more time for smaller particles to agglomerate as the heating programme progresses, so the copper ions are locked away in larger crystallites and hence are less available for reaction.

As was seen in the metallic sponge work, dextran-templated oxides form reticulated, sponge-like structures with small particles and high surface area (Figure 1.24). It has also been shown that a flux of sodium ions (in a molten salt style synthesis) can lead to alignment of the grains, which in YBCO lead to an improved critical current density.[104] The reticulated structure is not just observed in YBCO; other materials have shown this type of structuring (Figure 1.24 c), which suggests that it is common to all dextran-templated materials.

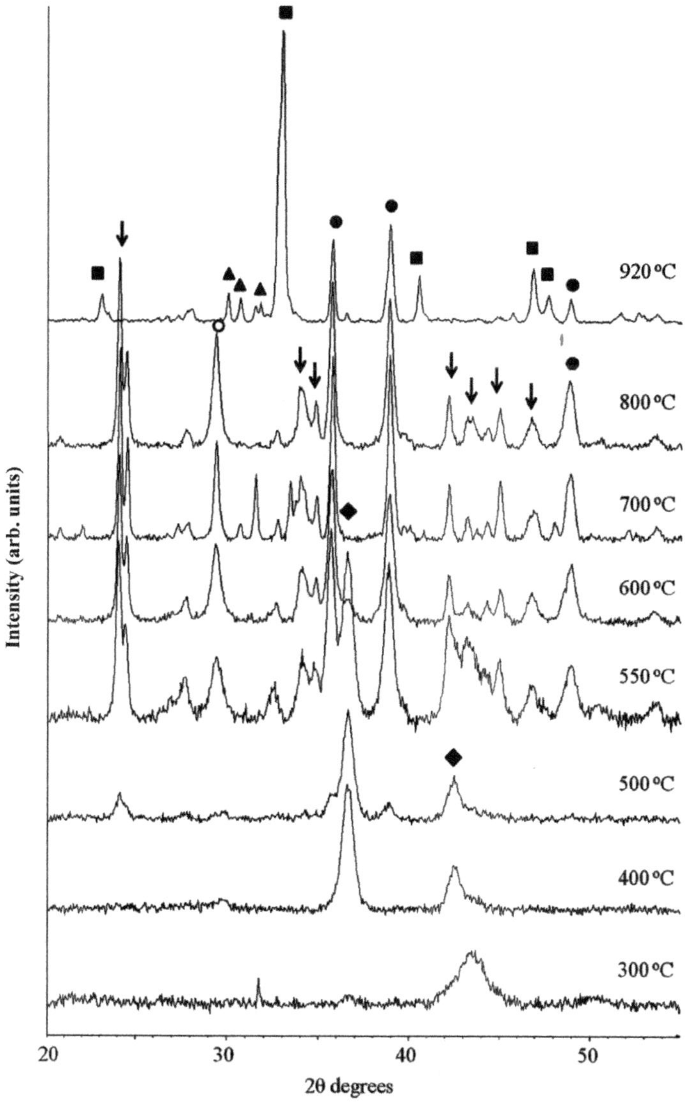

Figure 1.23 X-ray diffraction patterns showing phase formation with temperature in dextran-template YBCO: ■ YBCO, ● CuO, ↓ $BaCO_3$, ◆ Cu_2O, O Y_2O_3 and ▲ Y_2BaCuO_5, using CuKα radiation with a wavelength of 1.5406 Å. Reproduced from ref. 101 with permission from the Royal Society of Chemistry.

Laboratory-grade dextran often has an extremely high molecular weight (*e.g.* 10–150 kDa), and so there will naturally be variation in chain length and branching within these samples. It is therefore important to understand the role (if any) of these variables in the product formed. Work using oligo-saccharides of known chain length and degree of branching demonstrated

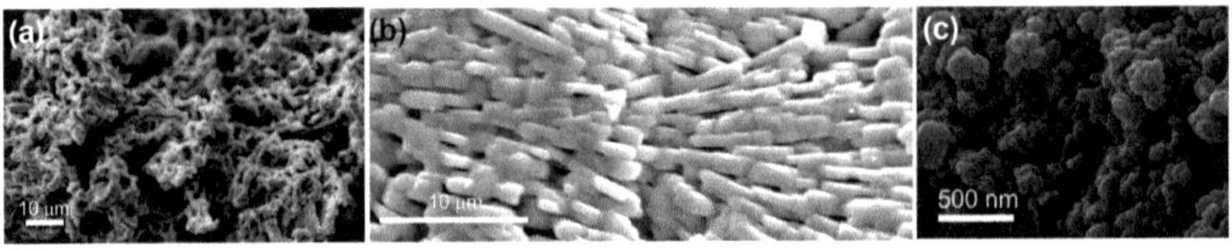

Figure 1.24 Dextran templating in (a) YBCO,[100] (b) YBCO with added sodium chloride to provide a flux[104] and (c) sodium nickel manganate.[89]
(a) Reprinted from D. Walsh, S. C. Wimbush and S. R. Hall, *Chem. Mater.*, 2007, **19**, 647. Copyright 2007, American Chemical Society. (b) Reproduced from ref. 104 with permission from The Royal Society of Chemistry. (c) Reproduced from ref. 89 with permission from the Royal Society of Chemistry.

subtle control of crystal morphology in YBCO,[105] with longer chains producing more plate-like crystals, which also showed a corresponding increase in critical current. Although a relatively small effect, it demonstrates that even within the same family of template there are parameters that can be optimised for highly targeted crystal morphology.

Since the initial work on YBCO, there have been numerous uses of dextran as a biotemplate, including sodium nickel manganese oxide (a potential sodium ion battery cathode material) (Figure 1.24c),[89] other materials such as praseodymium barium copper iron oxide,[106] the high temperature superconductor bismuth strontium calcium copper oxide,[107] and simple binary oxides such as iron oxide.[98] And there is nothing to suggest an upper limit on complexity of composition: materials with five or more elements should be equally amenable to dextran or other simple polysaccharide templating.

1.3.9.1.2 Complex Polysaccharides. Complex soluble polysaccharides have also been shown to direct structure through indirect means, most notably in the production of nanowires in a range of different materials.[108–110] All of the templates that have been used successfully to create nanowires have a range of similar features; namely, the presence of glucose-like ring structures and several different types of functional group on the ring structure.

Sodium alginate (seaweed)[111] and chitosan (from chitin found in crab shells)[108] are most regularly used for nanowire synthesis. Sodium alginate is particularly interesting in its composition, being made up of two acidic building blocks, guluronic and mannuronic acid. Both contain hydroxyl and carboxylate groups that are able to chelate metal ions. By having two types of functional group, sodium alginate can chelate more ions across a wider range of pH than, for example, dextran, which only has hydroxyl groups. This results in strong gelation when the metal ion solution is mixed with the dissolved alginate, to such an extent that gelled material can form enclosed "bubbles" of solution inside a skin of alginate/metal complex within the alginate solution. In fact, this technique has been used by the food industry to create spheres or pearls of sauce that are stable on the plate (known as spherification), such is the efficacy and stability of the skin formed. In the laboratory, this presents an issue since separated phases, particularly large regions of metal ion solution, are highly undesirable as, if allowed to persist into drying, they would result in large recrystallised regions. As such, full mixing of the metal ions and alginate solutions is highly important for the overall phase of the product. Sodium alginate also has, as the name suggests, a sodium counter ion and it appears that the presence of sodium ions is important to catalyse nanowire growth. Strangely, the Na ions are not usually observed in the final product, and debate is still ongoing as to the exact role of the sodium in these cases.

Once chelated and dried there are no crystalline phases present (the same as in most biotemplated intermediates); however, this changes rapidly upon heating, where well-mixed nanoparticles of intermediate oxides or carbonates form. In the case of YBCO nanowire preparation, the intermediate compounds formed are CuO, Y_2O_3 and $BaCO_3$, which remain as unreactive nanoparticles until the decomposition temperature of $BaCO_3$ (811 °C, Figure 1.25).[108,112]

Instead of simply decomposing, however, the $BaCO_3$ has been found to melt, almost certainly as a result of the presence of sodium. This partial melting of the structure results in mobile, molten, barium-rich nanoparticles inside a solid matrix rich in yttrium and copper ions.[113] Due to outgassing of the template at lower temperatures, this matrix is porous,[113] allowing the molten droplets to move around inside the reaction mixture. As they do so, they collect yttrium and copper ions from the surrounding bulk and, once the concentration of the ions inside the droplet reaches the critical saturation required to allow crystallisation, the YBCO product precipitates out as a solid phase. If this occurs when the droplet is at the surface, then crystallisation occurs at the liquid–solid interface, resulting in anisotropic growth, or nanowires. The precise nature of the nanowire growth that occurs is known to depend on the porosity of the bulk material. In a highly porous

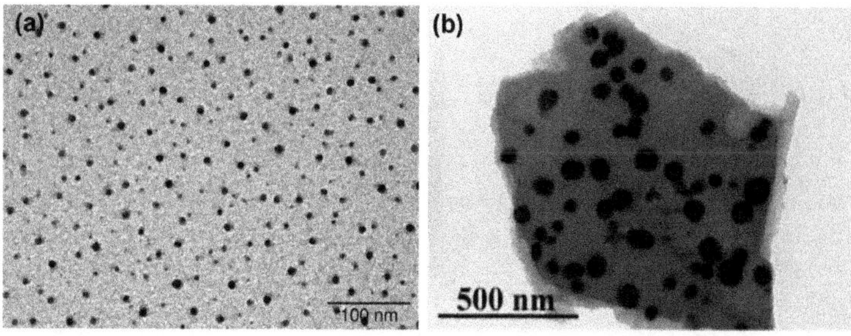

Figure 1.25 Barium carbonate nanoparticles in a Y_2O_3 and CuO matrix formed using (a) chitosan and (b) sodium alginate as templates.
(a) Reproduced from ref. 108 with permission from John Wiley and Sons, © 2006 WILEY-VCH Verlag GmbH & Co. KGaA, Weinheim. (b) Reproduced from ref. 112 with permission from John Wiley and Sons, © 2008 WILEY-VCH Verlag GmbH & Co. KGaA, Weinheim.

or open structure, the droplet can sit proud of the surface, promoting crystallisation between the droplet and the bulk. This initial crystallisation acts as the nucleation site for more YBCO to form, and the nanowire grows out from the surface, with the molten droplet located at the free end, in classic vapour–liquid–solid (or in this case liquid–solid) growth.[112] This results in single crystalline, tapered nanowires that end in a point as the droplet is exhausted (Figure 1.26).

There is another case of growth that is observed where, instead of escaping the surface, the droplet is contained within a pore on the solid surface. Once initial crystallisation has occurred on the external surface, crystallisation continues between the liquid–solid interfaces, with the solid portion of the newly forming wire being fed from below by new molten material. This is known as the microcrucible mechanism (Figure 1.27a)[113] and results again in a single crystalline nanowire. This time the nanowire generally has a uniform cross-section, as the droplet can be continually replenished by more molten droplets that arrive at the surface.

This can result in some interesting morphologies, however. To create the correct stoichiometry in our YBCO example, the arriving Ba-rich droplets require more Y or Cu ions to drive crystallisation: these are scavenged out of

Figure 1.26 Vapour–liquid–solid-like growth observed in sodium alginate-templated YBCO.
Reproduced from ref. 112 with permission from John Wiley and Sons, © 2008 WILEY-VCH Verlag GmbH & Co. KGaA, Weinheim.

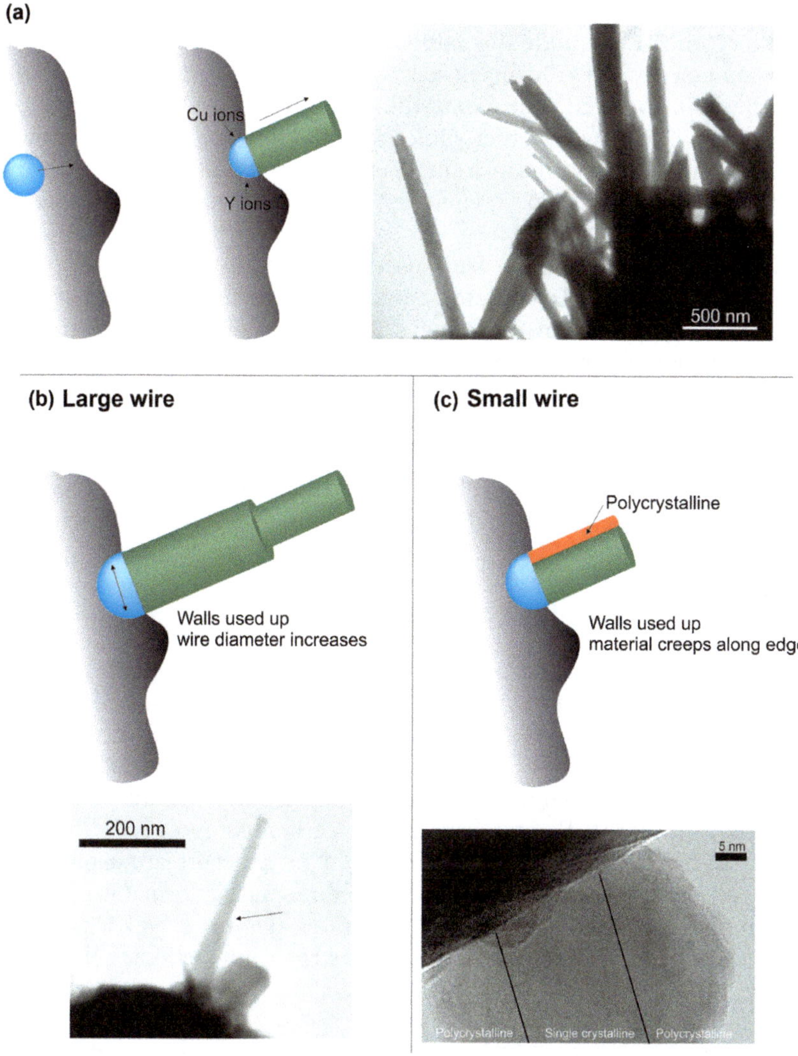

Figure 1.27 (a) Schematic of the microcrucible mechanism demonstrating the action of the molten nanoparticles (in this case Ba-rich through a Cu/Y oxide matrix) with examples of the resulting morphologies; (b) an example of telescopic growth observed in larger (>100 nm) wires where the walls of the microcrucible are consumed and the growing wire increases in diameter; and (c) creep growth observed in shorter nanoscale wires where, as the microcrucible breaks down, additional material flows along the edge of the nanowire, forming a polycrystalline edge or surface.[113]

the bulk material in the immediate vicinity. As these ions are used up, the walls of the microcrucible break down. In larger nanowires, this can result in "telescopic" style morphologies (Figure 1.27b). If the nanowire is short enough, this enables molten material to flow along the edge of the crystal, resulting in a core–shell style nanowire (Figure 1.27c) composed of a single crystalline centre surrounded by additional polycrystalline material.

If two nanowires are growing close to one another, it is likely that at some point the bulk material that forms the dividing wall will be entirely consumed, allowing the two microcrucibles to join. This means that the growing nanowires will now continue as a single entity, albeit often with two separate lengths of end, resulting in "crenellated" or stepped morphologies, as can be seen in Figure 1.28.

Sodium alginate has also been used to produce nanowires in other materials, including lead zirconate titanate,[109] lanthanum gallium silicate,[110] and lanthanum strontium manganese oxide.[111] There is still much to be understood about the precise role of sodium alginate in controlling nanowire growth, and there is now a drive to replicate the types of growth observed without the template and in other materials. Certainly, nanowires have been one of the more unexpected results in biotemplated syntheses, and they continue to be a focus for research into understanding the wider concepts of the technique, as well as providing an intriguing morphology to create in functional oxides.

1.3.9.2 *Proteins*

In theory, proteins should be excellent candidates for soft templating: they are often soluble and have a large number of functional groups, which should result in good levels of chelation. Of all the many proteins and protein-rich available, gelatine has been used successfully to template ceramic materials. These encompass a range of different materials, including oxides,[114] carbides,[115] nitrides,[116,117] and oxide/nitride/carbon composites.[118] Materials without oxygen are calcined under nitrogen or an inert atmosphere (depending on the desired product) in order to direct phase formation. Gelatine as a template often creates foamed structures (Figure 1.29a) which form spontaneously during drying as a result of strong interactions between the

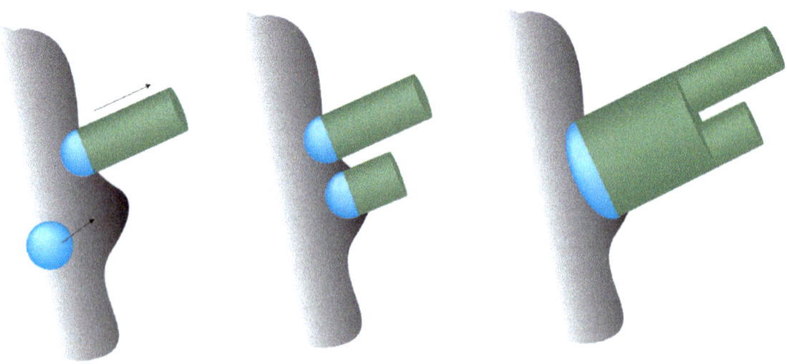

Figure 1.28 Schematic of two nanowires growing *via* the microcrucible mechanism in close proximity. As the wall between the two regions of molten material is consumed, the wires are able to join together resulting in stepped morphologies.[113]

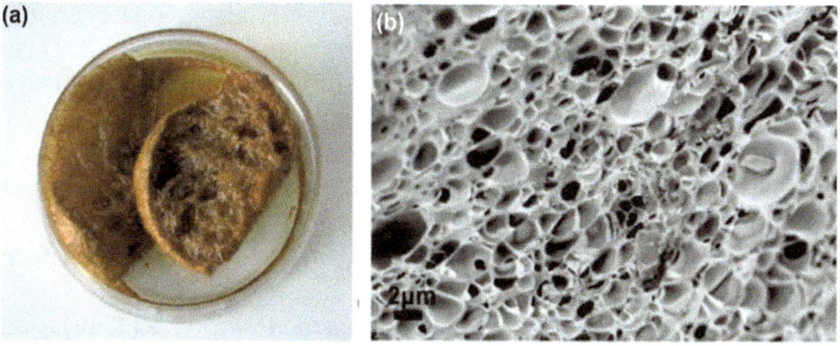

Figure 1.29 (a) Foam pre-calcination and (b) retention of the foam structure in iron carbide.
Adapted from ref. 117 with permission from the Royal Society of Chemistry.

chelated metal cations and the functional groups present on the gelatine molecule.[117] The bubbles of gas that form the foam result from interaction of the precursor anion (nitrate in this case). Upon calcination this foam structure is retained, resulting in hierarchically textured iron nitride (Figure 1.29b), which demonstrated increased catalytic activity.

1.3.10 Hard Biotemplating

Of course, biotemplating is not limited to bottom-up soft templating techniques. The macroscopic natural world is filled with intricate structures that can be used as hard templates to replicate levels of complexity that are difficult to achieve by any other method. Hard templating relies on the stability of the template, enabling deposition of the product whilst also offering easy removal. Template removal can be achieved in a number of ways including thermally or using dissolution with water or weak acid. This is heavily material/template dependent of course, as whatever is used for template removal must not adversely interfere with the product phase or morphology. This can present a significant challenge to hard-templating techniques, which often have to be considered on a case-by-case basis for each template/product combination. Despite these challenges, hard templating has been widely used across a wide range of length scales. At the nanoscale, some of the smallest biological structures, viruses, have been successfully used to template materials, through to some of the most well-known and readily available macrostructures, including wool, wood, and feathers.

1.3.10.1 Viruses

Viruses are some of the smallest and most successful self-contained organisms in the natural world. They have remarkable resistance to extreme environments, which is one of the reasons for their prevalence around the globe, exhibiting stability in extremes of pH and temperature. Both of these characteristics are attractive and useful for biotemplating, as they enable the use of a wide range of precursors and avoid many of the pitfalls faced by polysaccharide biotemplating. Viruses are nanoscale structures, and so their use for hard templating invariably results in a nanoscale product. As the product or precursor is deposited onto the surface and the virus later removed, this almost always produces hollow nanostructures, something that is traditionally challenging to do using top-down methods. This means that templating using viruses was an early and highly successful form of biotemplating.

Of all the viruses used in research, the *tobacco mosaic virus* (TMV) is one of the most common. The virus is roughly cylindrical, around 300 nm long and with a diameter of 18 nm, and it has a number of charged residues on the surface.[119] These charged residues mean that deposition of materials onto the surface is relatively straightforward. In 1999, it was demonstrated that the charged residues could be used to drive nucleation of a variety of different materials (iron oxide, lead sulfide, cadmium sulfide) to form 22 nm diameter nanotubes.[119] Use as a hard template has also been successfully achieved (Figure 1.30). For example, silica to form nanoscale hollow silica shells,[119] for metal nanowires,[120] and for alloy nanoparticles.[121] Of course, viruses are rarely found in isolation and so tend to self-assemble. Indeed, even once coated, the silica-coated versions demonstrated end-to-end self-assembly, forming long chains of silica shells resembling fibres.

This self-assembly has been further exploited to form mesoporous silica, as the virus is known to form nematic liquid crystal structures when prepared at high concentrations. By infiltrating the surrounding solution with silicate precursors and carefully controlling the rate of hydrolysis to match the rate of self-assembly of the virus particles, it was found that a continuous silica framework could be created, albeit with inclusions of the virus particles themselves.[123]

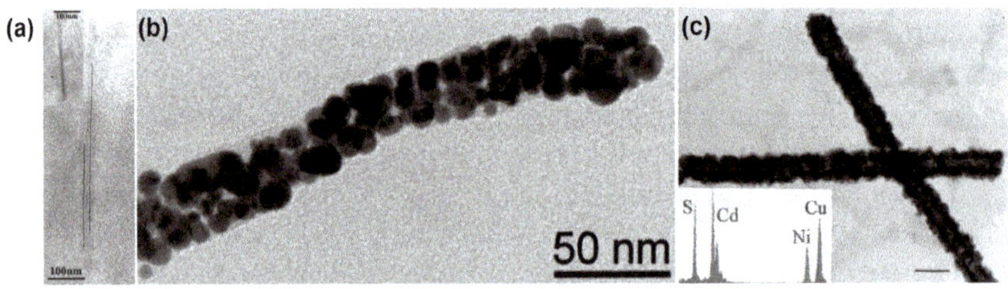

Figure 1.30 Structures in a variety of materials that can be accessed using TMV as a template: (a) a Ni nanowire,[122] (b) gold nanoparticles,[120] and (c) cadmium sulfide; scale bar 10 nm.[119]
(a) Reprinted with permission from M. Knez, A. M. Bittner, F. Boes, C. Wege, H. Jeske, E. Maiβ and K. Kern, *Nano Letters*, 2003, **3** (8), 1079. Copyright 2003, American Chemical Society. (b) Reproduced from ref. 120 with permission from the Royal Society of Chemistry. (c) Reproduced from ref. 119 with permission from John Wiley and Sons, © 1999 WILEY-VCH Verlag GmbH, Weinheim, Fed. Rep. of Germany.

One of the advantages of using a biologically well-characterised organism is that genetic manipulation of TMV is possible and more recent work on biotemplating using this structure has focused on using such manipulation to control the surface morphology or charges to direct deposition to bespoke regions on the surface.[124] This has been highly successful in the production of Co and Ni metal nanowires, this time using the internal structure of the virus as the template (Figure 1.30a).[122]

Most excitingly, genetically modified protein capsids on M13 bacteriophage has been used to control the structure of ZnS and CdS.[125] This peptide engineering, achieved using phage-display techniques, enabled the chalgonenides to form at highly specific sites and in specific, defined patterns. Interestingly, the mineralisation was further influenced by the specific orientation of the surface peptides, allowing engineering of the crystallographic orientation of the nanocrystals. It was found that upon further processing that the materials formed could be assembled into nanowires that retained the preferred orientation. This opens the door for genuinely bottom-up designed structures; perhaps in the future, genetically modified organisms will be used to produce bespoke nanomaterials.

1.3.10.2 DNA

DNA is a ubiquitous structure in the natural world, found in almost every cell of every organism on the planet. The master of self-assembly, and biologically well understood, these types of structures are already widely manipulated, and so their use as templates, both in their natural forms and as designed structures, is an obvious step. The size and morphology of DNA in particular makes it ideally suited to nanotechnological applications; for example, for producing wires for nanocircuitry, and, as such, much attention has been given to the metallisation of DNA to form metal nanowires.[126]

1.3.10.3 Dextran

Dextran may seem an odd inclusion as a hard template as it is generally known to be very soluble. With additional cross-linking, however, dextran can act as a hard template. Carboxymethylated dextran, commonly found in gel filtration matrix materials such as Sephadex, has enough cross-linking for it to be much less soluble and retain its structure in water. This additional degree of cross-linking reduces the number of available chelation sites, but enough remain to uptake sufficient ions for the product phase to replicate the shape. Commercially available Sephadex has a hollow spherical form, and so can be used to create hollow, polycrystalline spheres. Interestingly, when used to template YBCO,[101] the additional carbon content *versus* standard dextran has a small impact on the intermediate phase formation. Instead of CuO emerging as the first crystalline phase, Cu_2O is observed. This is a more reduced phase and indicates that there is less oxygen available in the early stages, as it is lost through additional CO_2 formation.[100] This appears not to

Figure 1.31 Hollow YBCO spheres formed using Sephadex.[101]

impact the final phase, however, as well-oxygenated YBCO is still formed, albeit with a hollow sphere morphology (Figure 1.31).

Of course, in order to preserve the structure, enough ions are required to form a contiguous structure upon heating. It is entirely possible to chelate too few ions, resulting in collapse of the macrostructure during heating. Conversely, adding a surfeit of ions simply results in the extra material remaining in solution rather than being chelated by the template. This may cause issues in some systems, however: if the solution is allowed to evaporate (rather than being actively removed) then large crystals of unchelated material will form within the system, affecting stoichiometry, as materials are not well distributed. For the Sephadex method, this is not an issue as the chelated spheres are filtered and excess solution is rinsed off, leaving only chelated ions within the system.

1.3.10.4 Proteins: Wool and Feathers

Both wool and feathers are principally made of the same protein: keratin. This protein forms fibres and is ubiquitous in many mammals, birds, reptiles and amphibians, who make use of it in numerous ways, and is often thought of as equivalent to chitin in arthropods (indeed, it has similar toughness).

As with many structural proteins, there are no obvious functional groups available for chelation, and, as such, any bioinspired templating occurring here is a true hard-templating method, with the template acting as a scaffold onto which material can be precipitated. This is followed by template removal, forming, as might be expected, hollow fibrous metal oxides in a variety of binary oxide materials (Al_2O_3, NiO, *etc.*).[127] The level of replication achieved varies across the different materials, and is dependent on the degree of adsorption of metal ions onto the surface, in line with previous results.[127] In the case of Al_2O_3, for example, rather than relying on the weak intermolecular adsorption forces to bind the cations, they were precipitated onto the surface using an ethyl acetate solution (in which the precursor, aluminium nitrate, is insoluble). After calcination, this resulted in the formation of nanocrystalline

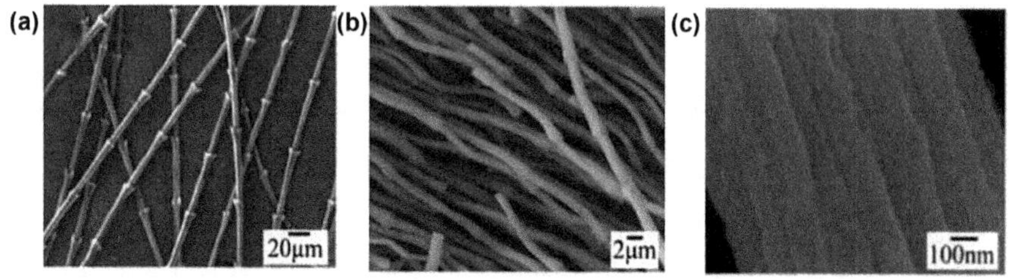

Figure 1.32 Feathers as templates: (a) the original feather material and (b) and (c) with the Al_2O_3 replicated structures.
Adapted from ref. 127 with permission from The Royal Society of Chemistry.

Al_2O_3 on the surface of the feather fibres (Figure 1.32) as the template was burned away. This approach was also successful for SiO_2 and TiO_2 using parrot feathers, resulting in photonic structures.[128]

1.3.10.5 *Wood and Cellulose*

Wood is primarily composed of cellulose, and is one of the most abundant natural materials on the planet. It is insoluble and has high directional strength and flexibility. The reasons for these properties can be seen on the microscale, as wood has a chambered structure made up of cellulose and lignin and it is this structure that can be replicated in ceramic materials to produce high surface area inorganic materials. Wood is also an excellent candidate for non-oxide ceramic synthesis, as it is known to retain its structure on anaerobic combustion. (For example, in the production of charcoal, if the structure was lost charcoal lumps would not exist, a rather disastrous outcome for the summer barbeque season!)

Lithium cobalt oxide cathode material was successfully produced using wood as a template, replicating the directional, hierarchical porous structure of the starting material.[129] This technique has been used widely across a multitude of oxide and carbide materials (Figure 1.33), including chromium oxide,[130] manganese oxide,[131] yttria-stabilised zirconia,[132] and silicon carbide,[133] to name but a few.

Cellulose is also found in leaves and these have been used to create iron carbide structures that are not only magnetic, but also retain the delicate and intricate structure of leaf veins (Figure 1.34).[134]

1.3.10.6 *Butterfly Wings*

Some of the most colourful structures in nature are not all that they appear. Structural colour is found in many different places in the natural world, and it is a source of wonder and intrigue to learn that some of the brightest and most iridescent colours in nature are formed not by pigments but by structure and the interaction of light therewith. These structures are perhaps

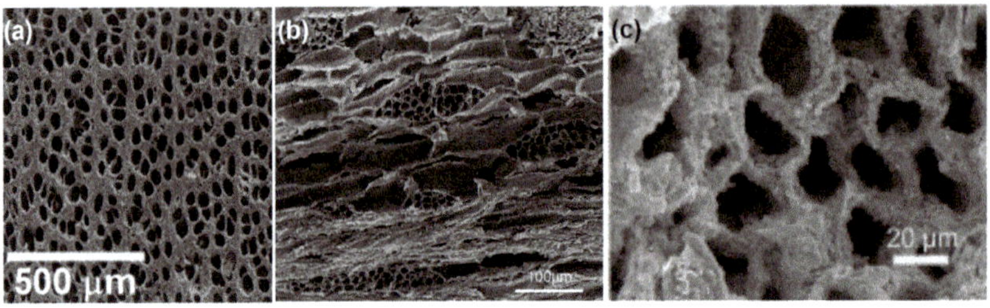

Figure 1.33 Typical porous structures formed when using wood as a template. (a) SiC,[133] (b) Mn oxide[131] and (c) Li cobalt oxide.[129]
(a) Reprinted from *Scripta Materialia*, **55**, T. E. Wilkes, M. L. Young, R. E. Sepulveda, D. C. Dunand, and K. T. Faber, Composites by aluminum infiltration of porous silicon carbide derived from wood precursor, 1083, Copyright 2006, with permission from Elsevier. (b) Reprinted from *Journal of the European Ceramic Society*, **26**, X. Li, T. Fan, Z. Lui, J. Ding, Q. Guo and D. Zhang, Synthesis of hierarchical pore structure of biomorphic manganese oxide derived from woods, 3657, Copyright 2006, with permission from Elsevier. (c) Reproduced from ref. 129 with permission from John Wiley and Sons, © 2018 WILEY-VCH Verlag GmbH & Co. KGaA, Weinheim.

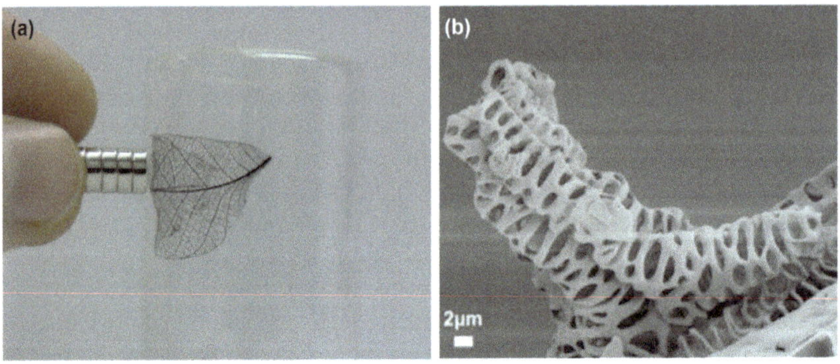

Figure 1.34 (a) Magnetic iron carbide structures replicating the veins inside a leaf and (b) the microscopic morphology retained during templating.
Adapted from ref. 134 with permission from John Wiley and Sons, © 2010 WILEY-VCH Verlag GmbH & Co. KGaA, Weinheim.

best-known on the *Morpho* butterfly, whose almost luminescent blue wings grace the rainforests of Central and South America (Figure 1.35).

The wings are made primarily of chitin, the same as is found in crab or crustacean shells, comprising innumerable overlapping scales (setae).[135] The size of these setae is comparable with the wavelength of light, so that they interact with incident light, acting in the same way as a photonic crystal.[137] As such their structures have been widely studied, both as a direct template to create photonic crystals in a number of materials[138] and as inspiration for similar structures created by different means.[139]

Figure 1.35 The wing structure of the blue *Morpho* butterfly on (a) the macroscale, (b) using optical microscopy to highlight the shifting blue/green colours and (c) and (d) SEM of the scales demonstrating the intricacy and scale of the chitin structures.
(a), (b) Reproduced from ref. 135 with permission from John Wiley and Sons, © 2012 WILEY-VCH Verlag GmbH & Co. KGaA, Weinheim. (c), (d) Reproduced from ref. 136 with permission from John Wiley and Sons, © 2008 WILEY-VCH Verlag GmbH & Co. KGaA, Weinheim.

A wide variety of materials have been made using butterfly wings as the template, including metals such as silver[140] and oxides, including ZnO,[141] TiO_2,[136] Al_2O_3[142] and many others.[143]

1.3.10.7 Cuttlebone

For some materials, there is limited or no affinity between the metal ions and the proposed template. Here it is advantageous to give the metal ions a "helping hand" to take on the shape of the template. Cuttlebone, for example, is a stable composite of calcium carbonate and β-chitin, arranged in a cell-like structure (Figure 1.36) with high porosity (around 93% by volume).[144] Early attempts to retain this structure in silica were partially successful, however much of the intricacy of the cellular structure was lost during processing.[144]

In later attempts using cuttlebone-templated YBCO,[145] ethylene glycol was added to the metal ion solution in order to increase the viscosity of the solution and to provide chelation sites for the ions. This provided the required

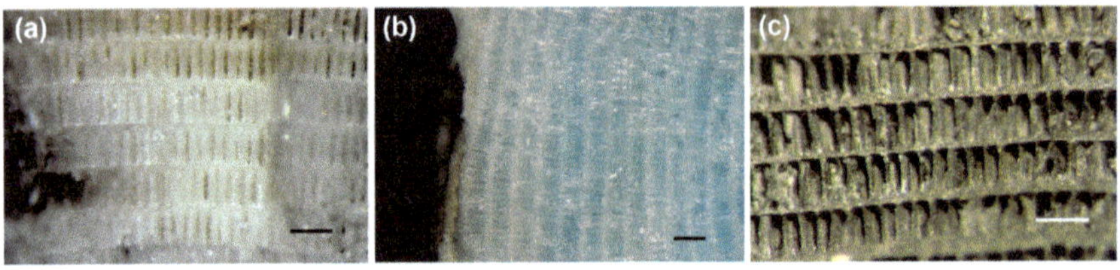

Figure 1.36 Optical microscopy of (a) native cuttlebone (scale bar, 500 μm), (b) cuttlebone infiltrated with an YBCO precursor solution (scale bar, 1 mm) and (c) the cuttlebone structure replicated in YBCO after calcination (scale bar, 500 μm). Significant calcium from the template remains in the structure after heating, and only about 12% by volume of the structure is superconducting.
Adapted from ref. 145 with permission from the Royal Society of Chemistry.

number of binding sites for the Y, Ba and Cu ions, preventing recrystallisation upon drying. This also enabled a thicker layer of coating material to be deposited each time the template was soaked in the solution, leading to better replication of the structure. In this case, the template was not fully removed during calcination, leaving behind a cuttlebone structure free of organic material, but composed of a mixture of YBCO and calcium oxide. This "thickened" solution method could be used to great effect for templates for which it may not be possible to "biotemplate" in the traditional sense.

1.3.10.8 Graphene Oxide

Like biotemplates, graphene oxide has many functional groups decorating the surface. These can be used as a direct analogue of a naturally occurring material, and are able to chelate metal ions from solution in exactly the same bioinspired way. Unlike many naturally occurring materials, graphene oxide can be manipulated into different structures across multiple length scales simultaneously. The individual sheets can be arranged in layers, or randomly, to create high surface area, sponge-like monoliths, and the monoliths themselves can be made to take on the shape of the container in which they are created.[146–148] Crystallisation during heating means that the inorganic phases only form where the template once was, making it possible to create direct replication of both the macroscale monolith and microscale internal structure.

This has been performed successfully in YBCO, resulting in high surface area monoliths with control of ordering across a number of length scales (Figure 1.37).[149]

1.4 Biotemplating Considerations

Biotemplating for ceramics (and indeed other materials) faces a number of issues, and indeed the use of every biotemplate for every material is not possible. Among the key considerations are the pH of the solution of metal ions, solubility, and the presence or introduction of any counter ions.

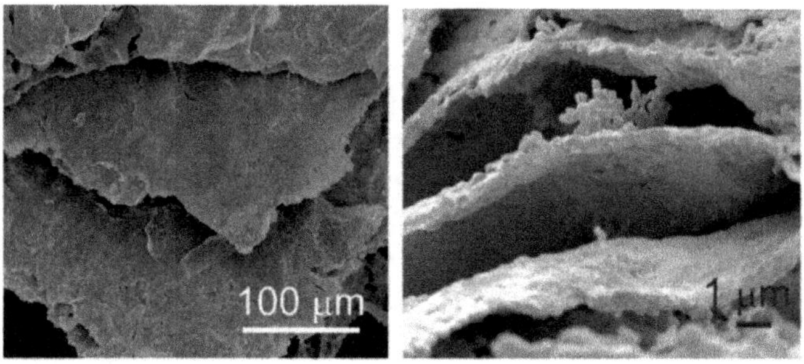

Figure 1.37 Graphene oxide-templated YBCO.
Adapted from ref. 149 with permission from the Royal Society of Chemistry.

1.4.1 Solubility

Every bioinspired scheme considered so far has used highly soluble metal salts such as acetates or nitrates as the source of ions. This requirement is dictated by the biotemplates themselves. In order to deprotonate/dissociate the chelating sites, the biotemplates must be dissolved in water, and hence any metal ions must also be free in an aqueous solution. Herein lies a significant challenge: not all metals have soluble salts, which limits the application of biotemplating and means that many of the technologically useful oxide products (titanates, *etc.*) are extremely challenging if not impossible to synthesise using these types of methods. There are solutions to this, however, which are covered in Section 1.5.

1.4.2 pH/pK_a

It not enough that the chosen metal salts can be dissolved, as the ions, counter ions and template are at some point necessarily in the same solution and so the interaction of all three together must be taken into consideration. The success or failure of a reaction at this stage hinges on the pH of the solution, which can often result in unexpected or undesirable effects such as precipitation of the solution or the formation of a "skin" of chelated material before the mixture has been fully mixed.

Before the template can even be considered, the solution of ions must be stable. Many metal salts are only sparingly soluble, and therefore highly sensitive to pH or temperature. In many cases, the addition of more than one salt to a solution can result in the precipitation of one or more of the metal ions. This obviously prevents one or more of the metal ions from taking part in chelation. Bismuth salts are notorious for this, often requiring a weakly acid solution or indeed a secondary small-molecule chelating agent such as EDTA before a stable solution can be formed and templating can occur.[107]

Every functional group has a pK_a value, the acid dissociation constant, and this becomes important when considering the pH of the metal ion solutions. Most functional groups commonly found in biotemplates (hydroxyls, carboxylates, *etc.*) have pK_a values around 4–5, which means that solutions with pH above this are desirable for full deprotonation (or dissociation) of the template. This is a key requirement, as deprotonation creates the chelating sites required. Many of the soluble metal salts, however, are weakly acidic in water, owing to the acetate or nitrate anion, and so the pH of the solution must be taken into account when designing experiments. This is often best done practically, and can be controlled by the concentration of the solution. This can present other issues, however, concerning precipitation of certain salts, which can sometimes be rather unexpected (depending on the anions used, especially if more than one moiety is involved) and/or difficult to circumvent. Solution design is therefore an important and fundamental consideration when developing a synthesis.

1.4.3 Counter Ions

The biotemplate/metal ion solution does not exist in isolation, other ions are present; namely the anion component of the metal salts used and of the biotemplate itself. This can be a very wide variety of species, although is most commonly nitrate, acetate, or carbonate, and less commonly oxide, sulfate or chloride.

1.4.3.1 Ions from the Template

Some templates do not use hydrogen as the counter ion for the functional groups. Carboxylates in particular use a wide range of monovalent ions (*e.g.* Na, K, *etc.*),[112] which must be considered in the reaction. Indeed, they are often central to the morphologies, for example, as in the Na-catalysed nanowire growth promoted by sodium alginate (see Section 1.3.9.1.2), but many times they simply result in contamination of the sample. Attempts to control the growth of YBCO with carrageenan, for example, were beset with issues: the high sulfur content of the template promoted the growth of a highly stable barium sulfate phase, which locked away the barium ions, making them unavailable to form the YBCO phase sought.[150] In theory, this should rule out most proteins, but the effect is material-specific and can be avoided where no stable sulfate phase forms, or can be ameliorated by controlling the quantity of template used. For example, a sulfur-rich protein was successfully used to template YBCO simply by lowering the quantity of template.[151]

1.4.3.2 Anions from the Salts

The anionic portion of the metal salts can cause difficulties for full dissociation of the template, limiting the available chelating sites. These counter ions remain in the system in the early stages of synthesis, which means that they are likely to play a role in the chemical pathway. This has, however, been exploited in a limited number of syntheses. For example, the choice of reagent has been shown to play a crucial role in nanowire synthesis (see Section 1.3.9.1.2) where a solution of yttrium oxide, barium acetate and copper acetate resulted in nanowires of a different shape compared with those produced when all-nitrate precursors were used. The difference was attributed to the porosity of the intermediates formed: the nitrate anion was lost at a lower temperature than the carbon-rich acetate form, resulting in more densely packed nanoparticle intermediate phases, which were better able to trap the molten particles at the surface, promoting growth *via* the microcrucible mechanism rather than the more usual liquid–solid nanowire growth.[113]

1.5 Non-aqueous (Bio)templating

All of the materials and methods considered so far have started from an aqueous solution. As discussed previously, such a solution cannot always be successfully created, and is often hindered by the lack of a suitable soluble ion. This is of particular relevance for titanate or niobate materials, for which there are few or no suitable soluble options (particularly when considering safety and scalability). Thus, there has been a recent drive to find means to harness the technique without the need for water, which opens up the possibility of using alkoxide precursors, which are more commonly found in sol–gel synthesis.[152]

1.5.1 Ionic Liquids

Another method that can be used to obviate the need for water as a solvent is to use ionic liquids. Ionic liquids are, as the name suggests, liquids composed of (usually) two small, organic, charged species that are stable over a wide range of temperatures. Generally, at least one of the components has a delocalised charge, which is required to prevent a stable crystalline lattice forming, resulting in a liquid that can act as a solvent, but with almost endlessly tuneable characteristics. This means that ionic liquids can be specifically developed to have the required chemical behaviour and interactions with metal ions. As yet, however, the trend is to use readily available, or off-the-shelf, ionic liquids that already have many of the required characteristics.

There are two main approaches that have been used with ionic liquid-based syntheses to date: as a template in their own right and as a replacement solvent for water with the specific aim of dissolving water-insoluble biopolymers.[153] The general synthetic method follows the same procedure for both. Any materials with soluble salts are dissolved in water as usual, and mixed with the ionic liquid. This mixture is then heated until all of the water has evaporated, a process that can take anything from 30 minutes to several hours depending on the size of sample. Once dehydrated, alkoxide precursors can be added. At this point in the synthesis the exact nature of the interaction between the alkoxides and the ionic liquid is not known. Evolution of gas is observed (thought to be, for example, isopropyl groups leaving the titanium isopropoxide precursor) but there is no water to hydrolyse the alkoxide, and so no oxide phase is observed (again, taking the titanium isopropoxide example, this would be titanium dioxide). Indeed, there is often some gelation or thickening of the mixture, which suggests some complexation between the ionic liquid and the added metal ion; but, certainly, there is no crystallisation at this stage. Thus, the normally recalcitrant elements can be kept in an atomically dispersed state, throughout the reaction mixture, a direct analogue of the solvent/solute situation in aqueous systems. At this point, the mixture can be calcined, enabling the high surface area solid-state reactions characteristic of biotemplating to occur.

Interestingly, studies have revealed that rather than combusting (as would be seen in standard biotemplating), ionic liquids tend to evaporate. This can be observed by using a combination of thermal gravimetric analysis and differential thermal analysis, which pinpoints the evaporation at between 200 and 300 °C.[154] This is interesting for two reasons. First, the ionic liquid is recoverable during heating and therefore possibly recyclable (although it remains to be seen if this can be done practically, and whether there is any degradation of the ionic liquid during heating). Second, the lack of a combusting phase may change the chemical pathway during synthesis.

1.5.1.1 *As a Solvent for Biopolymers*

Many biotemplates were excluded from the early investigations, as they are insoluble in water. Most notably this includes cellulose, one of the most abundant biopolymers on the planet as it is the main constituent in many plants, particularly trees. Since trees necessarily retain their structure when it rains, it is intuitive that cellulose is insoluble in water; however, it is soluble in, for example, 1-ethyl 3-methylimidazolium acetate, the ionic liquid that has been used for the majority of studies of this type so far.

Work conducted by Green *et al.*[153] demonstrated the efficacy of this method, by using an ionic liquid into which cellulose was dissolved and undertaking the synthesis of a wide range of complex oxides. Across all of the materials created, phase purity was achieved at lower temperatures than solid state (as would be expected in the high surface area systems), and was generally found to follow similar chemical pathways to standard biotemplating. This was particularly evident in the study conducted on YBCO: the same intermediate phases were observed in the cellulose/ionic liquid synthesis as were observed in, for example, dextran or sodium alginate (*i.e.* barium carbonate, yttrium oxide and copper oxide) and indeed are the same as would be used in the solid state. This indicates that the cellulose is playing the "normal" role of a biotemplate in the reaction mixture, providing carbon for the carbonate phase formation and spatial separation for all of the ions involved.[154]

All of the materials synthesised using this combinatorial technique (ionic liquid plus biotemplate) have resulted in nanoscale crystallites of product (Figure 1.38). Of the materials tested so far, generally functional properties have been retained, although no enhancement of properties was observed.

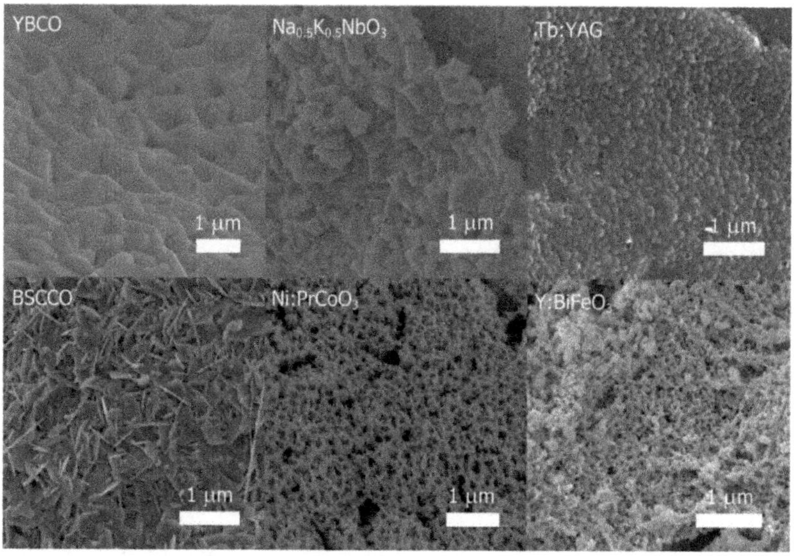

Figure 1.38 Examples of product phases formed using cellulose and ionic liquids. Reproduced from ref. 153 with permission from John Wiley and Sons, © 2012 WILEY-VCH Verlag GmbH & Co. KGaA, Weinheim.

1.5.1.2 As the Sole Template

Many ionic liquids contain components that exhibit the same types of functional groups as are seen on biotemplates, for example, hydroxyls and carboxylates. These can be used as chelating groups in exactly the same way, with metal ions interacting with these groups upon addition to the reaction mixture.

This approach has been used with success for the synthesis of low temperature lanthanum strontium titanate.[155] A 1-ethyl 3-methylimidazolium acetate ionic liquid was used as the "template" alone, relying on the carboxylate group present to act as a chelating agent. Here again, the readily soluble salts (lanthanum and strontium acetates in this case) were dissolved in water and mixed with the ionic liquid, and, after dehydration, titanium isopropoxide added. Upon heating, the ionic liquid simply evaporates, promoting crystallisation of the reagents at very low temperatures. In the case of 1-ethyl 3-methylimidazolium acetate, this is as low as 400 °C. With no carbon dioxide being produced, less reduction of the intermediate phases occurs than would be observed with a carbon-rich template, and so highly oxidised phases can form at much lower temperatures than would normally be observed. In the lanthanum strontium titanate materials, the formation of titanium dioxide, or strontium/lanthanum carbonates, is heavily suppressed, making the ions much more available for reaction into the final perovskite phase. This results in the product phase crystallisation occurring at around 420 °C, as opposed to 1100 °C as would be required for solid-state synthesis.[155] This low temperature also means that the crystallites produced are extremely small (approximately 30 nm) and the size can be carefully controlled using a heating protocol. Given the flexibility of the synthesis for other materials, this route, although not strictly biological, may represent a highly exciting and versatile new method for nanomaterial synthesis. Of course, the crystallisation dynamics of each material will need to be considered on a case-by-case basis, but if some of the more stable intermediates can be circumvented, nanoscale control becomes simply a matter of heating time or temperature.

1.5.2 Deep Eutectic Solvents

For all their promise, ionic liquids (particularly 1-ethyl 3-methylimidazolium acetate) have a major drawback: cost. This, coupled with their (current) non-recoverability, make them an unattractive proposition for industrial use. There are plenty of similar alternatives, however, in the form of deep eutectic solvents (DESs). These are similar in construction to an ionic liquid, being composed of small organic molecules that, when brought together, have a depressed melting point, enabling them to remain liquid down to room temperature or even slightly below. This, of course, makes for easy incorporation of ions, either from solution or as an alkoxide, directly following the method used for ionic liquids.

The first synthesis of an oxide material using a deep eutectic solvent was proof-of-concept work using barium titanate.[156] A deep eutectic solvent of malonic acid and choline chloride, both readily and cheaply available, was used, along with aqueous barium acetate and titanium isopropoxide, following the standard route. Whilst barium titanate formed readily at 950 °C (a 400 °C reduction in temperature over solid-state routes), the chemical pathway given the intermediate materials was not immediately clear. The work conducted looked closely at the intermediate phase formation, and concluded that barium chloride and titanium dioxide were the main phases formed at intermediate temperature, which then reacted together above 900 °C to form the titanate phase.[154] Whilst this is an effective route for nanoscale barium titanate formation, there remain some questions over the fate of the chloride ions: either they are retained on the oxygen sites in the highly adaptable perovskite structure (as is similarly observed with OH groups), or they are lost as HCl gas during processing. Neither of these are desirable outcomes. Retention of anything that is not oxygen in the perovskite structure in barium titanate is known to adversely affect both the transition temperature and magnitude of the relative permittivity, and, indeed, in the proof-of-concept work, this was found to be the case.

Given the limitations of chloride-containing DESs, similar but halide-free DESs exist, and generally fall into the emergent category of natural deep eutectic solvents or NADES. These are also made up of small organic molecules, many of which occur in nature. Work is therefore ongoing to explore the use of halide-free deep eutectics and the effect that these have on synthesis pathway and temperatures. They are a highly promising area of research, offering the adaptability of and ionic liquid (water-free) style synthesis, whilst being able to be incorporated with biotemplates and cheap and readily available, all of which make DES syntheses a potential step forward for bioinspired laboratory and industrial synthesis of materials.

1.6 Summary

As we have seen, nature, with its millions of years' head start, provides answers to some of the most challenging materials questions faced by modern society. Evolution over millions of years has solved many of the materials challenges needed by organisms to survive, overcoming the energy barriers presented by an ambient temperature existence through ingenious control of ions. Every example of nature controlling inorganic materials is for a specific end and advantageous to the organism. We have only just started to understand the processes involved, but, as we do so, it is natural that we will start to apply them to the materials required in modern life. Of course, our advantage is that we can closely control both temperature and composition. Where organisms are limited to ambient temperatures and the elements in their immediate environment, we can choose our materials with almost total freedom to make any composition required and can change the thermal conditions to suit. So, whilst we are limited in one sense, having an incomplete understanding of biological control of inorganic materials, we have already started to overtake nature in the control of complex and functional materials, which simply would not exist in nature and would be of little use to the organisms themselves.

Whilst by no means an exhaustive list, this chapter has highlighted some of the most important and widely used biotemplates. We have covered the mechanisms and chemical pathways for formation, focusing on complex materials such as YBCO as proof-of-concept. The ideas behind bioinspiration and biotemplating have been expanded back into the non-biological realm, enabling reliable creation of structures and polymorphs in a manner that has been all but impossible with traditional methods (and indeed with the natural variation which comes with biologically derived materials), and we are on the brink of seeing these types of techniques moving into mainstream industrial processes. Although some materials and processes are becoming well understood, and represent significant improvements over the current state of the art in terms of energy usage or production of waste, there are still significant challenges to overcome to successfully scale up these types of technique from the laboratory to full industrial production. Nature, as a source of inspiration, will therefore continue to offer a starting point as we move into the future of materials and manufacturing.

References

1. E. A. Olivetti and J. M. Cullen, *Science*, 2018, **360**, 1396.
2. R. R. Naik and S. Singamaneni, *Chem. Rev.*, 2017, **117**, 12581.

3. J. Lyle, *Design for Human Ecosystems: Landscape, Land Use, and Natural Resources*, Island Press, Washington D.C., USA, 1999.
4. C. W. Marean, M. Bar-Matthews, J. Bernatchez, E. Fisher, P. Goldberg, A. I. R. Herries, Z. Jacobs, A. Jerardino, P. Karkanas, T. Minichillo, P. J. Nilssen, E. Thompson, I. Watts and H. M. Williams, *Nature*, 2007, **449**, 905.
5. G. Sauvet, R. Bourrillon, M. Conkey, C. Fritz, D. Gárate-Maidagan, O. Rivero Vilá, G. Tosello and R. White, *Quat. Int.*, 2017, **432**, 86.
6. M. Aubert, A. Brumm, M. Ramli, T. Sutikna, E. W. Saptomo, B. Hakim, M. J. Morwood, G. D. Van Den Bergh, L. Kinsley and A. Dosseto, *Nature*, 2014, **514**, 223.
7. P. M. Martin-Sanchez, A. Nováková, F. Bastian, C. Alabouvette and C. Saiz-Jimenez, *Environ. Sci. Tech.*, 2012, **46**, 3762.
8. J. D. Bernal, *Science in History: Volume 1: The Emergence of Science*, Faber and Faber, London, 3rd edn, 2012.
9. S. T. Williams, S. Ito, K. Wakamatsu, T. Goral, N. P. Edwards, R. A. Wogelius, T. Henkel, L. F. C. de Oliveira, L. F. Maia, S. Strkopytov, T. Jeffries, D. I. Speiser and J. T. Marsden, *PLoS One*, 2016, **11**, 0156664.
10. J. K. Shields, H. M. Mader, L. Caricchi, H. Tuffen, S. Mueller, M. Pistone and L. Baumgartner, *J. Volcanol. Geotherm. Res.*, 2016, **310**, 137.
11. E. Frahm and C. A. Tryon, *J. Archaeol. Sci. Rep.*, 2018, **21**, 472.
12. G. C. Righini, *Proc. SPIE* 10275, 1994, 1027502.
13. A. Shortland, N. Rogers and K. Eremin, *J. Archaeol. Sci.*, 2007, **34**, 781.
14. C. M. Jackson and S. Cottam, *J. Archaeol. Sci.*, 2015, **61**, 139.
15. T. Pradell, G. Molina, S. Murcia, R. Ibáñez, C. Liu, J. Molera and A. J. Shortland, *Int. J. Appl. Glass Sci.*, 2016, 7, 41.
16. H. Louw, *Constr. Hist.*, 1991, 7, 47.
17. C. R. Kurkjian and W. R. Prindle, *J. Am. Ceram. Soc.*, 2005, **81**, 795.
18. X. K. Shen and Y. F. Lu, *Appl. Opt.*, 2008, **47**, 1810.
19. C. Corkhill and N. Hyatt, *Nuclear Waste Management*, IOP Publishing Ltd., London, UK, 2018, Ch. 2, page 5.
20. H. D. Holland and D. Gottfried, *Acta Crystallogr.*, 1955, **8**, 291.
21. C. M. Jantzen and M. J. Plodinec, *J. Non. Cryst. Solids*, 1984, **67**, 207.
22. E. V. Armbrust, *Nature*, 2009, **459**, 185.
23. F. Li, Y. Xing, M. Huang, K. L. Li, T. T. Yu, Y. X. Zhang and D. Losic, *J. Mater. Chem. A*, 2015, **3**, 7855.
24. A. Arasuna and M. Okuno, *Geosci. Lett.*, 2018, **5**, 1.
25. M. Hildebrand, *Chem. Rev.*, 2008, **108**, 4855.
26. A. J. Milligan and F. M. M. Morel, *Science*, 2002, **297**, 1848.
27. M. Sumper and E. Brunner, *Adv. Funct. Mater.*, 2006, **16**, 17.
28. N. Kroger, R. Deutzmann, C. Bergsdorf and M. Sumper, *Proc. Natl. Acad. Sci. U. S. A.*, 2000, **97**, 14133.
29. R. R. Naik, P. W. Whitlock, F. Rodriguez, L. L. Brott, D. D. Glawe, S. J. Clarson and M. O. Stone, *Chem. Commun.*, 2003, **9**, 238.
30. M. Ozaki, S. Sakashita, Y. Hamada and K. Usui, *Protein Pept. Lett.*, 2018, **25**, 15.

31. H. Li, J. Wang, J. Cong, C. Wei, J. Li, H. Liu, S. Li and M. Yang, *Drug Delivery*, 2017, **24**, 1086.
32. T. Hatanaka, M. Ohashi and N. Ishida, *Biomater. Sci.*, 2018, **6**, 2316.
33. Z. Bao, M. R. Weatherspoon, S. Shian, Y. Cai, P. D. Graham, S. M. Allan, G. Ahmad, M. B. Dickerson, B. C. Church, Z. Kang, H. W. Abernathy, C. J. Summers, M. Liu and K. H. Sandhage, *Nature*, 2007, **446**, 172.
34. A. A. Ganesh, H. K. Raut, A. S. Nair and S. Ramakrishna, *J. Mater. Chem.*, 2011, **21**, 16304.
35. S. Nishimoto and B. Bhushan, *RSC Adv.*, 2013, **3**, 671.
36. K. Liu and L. Jiang, *Annu. Rev. Mater. Res.*, 2012, **42**, 231.
37. B. Dean and B. Bhushan, *Philos. Trans. R. Soc. A Math. Phys. Eng. Sci.*, 2010, **368**, 4775.
38. M. Liu, S. Wang, Z. Wei, Y. Song and L. Jiang, *Adv. Mater.*, 2009, **21**, 665.
39. I. Norifumi, *J. Jpn. Weld. Soc.*, 2009, **78**, 187.
40. X. Pu, G. Li and H. Huang, *Biol. Open*, 2016, **5**, 389.
41. L. L. Hench, *J. Mater. Sci.: Mater. Med.*, 2006, **17**, 967.
42. V. Miguez-Pacheco, L. L. Hench and A. R. Boccaccini, *Acta Biomater.*, 2015, **13**, 1.
43. H. Yu, J. Peng, Y. Xu, J. Chang and H. Li, *ACS Appl. Mater. Interfaces*, 2016, **8**, 703.
44. Q. Liang, Q. Hu, G. Miao, B. Yuan and X. Chen, *Mater. Lett.*, 2015, **148**, 45.
45. X. Li, Q. Liang, W. Zhang, Y. Li, J. Ye, F. Zhao, X. Chen and S. Wang, *J. Mater. Chem. B*, 2017, **5**, 6376.
46. K. Shizuma, T. Kajimoto, S. Endo, K. Matsugi, Y. Arimatsu and H. Nojima, *Nucl. Instrum. Methods Phys. Res. B*, 2017, **407**, 244.
47. L. Rutherford-Morrison, *Public Historian*, 2015, **37**, 76.
48. S. Tammas-Williams and I. Todd, *Scr. Mater.*, 2017, **135**, 105.
49. P. B. Vandiver, O. Soffer, B. Klima and J. Svoboda, *Science*, 1989, **246**, 1002.
50. F. Abreu, M. E. Cantão, M. F. Nicolás, F. G. Barcellos, V. Morillo, L. G. P. Almeida, F. F. Do Nascimento, C. T. Lefèvre, D. A. Bazylinski, A. T. R. De Vasconcelos and U. Lins, *ISME J.*, 2011, **5**, 1634.
51. A. Komeili, *FEMS Microbiol. Rev.*, 2012, **36**, 232.
52. T. Matsunaga, Y. Okamura, Y. Fukuda, A. T. Wahyudi, Y. Murase and H. Takeyama, *DNA Res.*, 2005, **12**, 157.
53. R. Uebe and D. Schüler, *Nat. Rev. Microbiol.*, 2016, **14**, 621.
54. G. Haszprunar, *Bolletino di Zool.*, 1992, **59**, 1.
55. K. S. Vecchio, X. Zhang, J. B. Massie, M. Wang and C. W. Kim, *Acta Biomater.*, 2007, **3**, 910.
56. L. Addadi, D. Joester, F. Nudelman and S. Weiner, *Chem.: Eur. J.*, 2006, **12**, 980.
57. Z. Tang, N. A. Kotov, S. Magonov and B. Ozturk, *Nat. Mater.*, 2003, **2**, 413.
58. B. A. Wustman, J. C. Weaver, D. E. Morse and J. S. Evans, *Connect. Tissue Res.*, 2003, **44**, 10.

59. S.-N. Wang, X.-Q. Zhu, X.-H. Yan, J.-F. Deng, R. Wang and X.-X. Wang, *CrystEngComm*, 2015, **17**, 2964.
60. X. Liu, J. Li, L. Xiang, J. Sun, G. Zheng, G. Zhang, H. Wang, L. Xie and R. Zhang, *Proc. R. Soc. B*, 2012, **279**, 1000.
61. J. Sun and B. Bhushan, *RSC Adv.*, 2012, **12**, 7617.
62. F. Barthelat, C. M. Li, C. Comi and H. D. Espinosa, *J. Mater. Res.*, 2006, **21**, 1977.
63. R. A. Metzler, D. Zhou, M. Abrecht, S. N. Coppersmith and P. U. P. A. Gilbert, *Phys. Rev. B*, 2008, 1.
64. K. S. Katti and D. R. Katti, *Mater. Sci. Eng., C*, 2006, **26**, 1317.
65. F. Marin, N. Le Roy and B. Marie, *Front. Biosci.*, 2012, **S4**, 1099.
66. K. S. Katti, B. Mohanty and D. R. Katti, *J. Mater. Res.*, 2006, **21**, 1237.
67. H. Gopalan and A. H. Chokshi, *J. Mech. Behav. Biomed. Mater.*, 2018, **78**, 96.
68. Q. L. Feng, F. Z. Cui, G. Pu, R. Z. Wang and H. D. Li, *Mater. Sci. Eng., C*, 2000, **11**, 19.
69. A. Sellinger, P. M. Weiss, A. Nguyen, Y. Lu, R. A. Assink, W. Gong and C. J. Brinker, *Nature*, 1998, **394**, 256.
70. P. Das, S. Schipmann, J. M. Malho, B. Zhu, U. Klemradt and A. Walther, *ACS Appl. Mater. Interfaces*, 2013, **5**, 3738.
71. F. Barthelat, *Bioinspir. Biomim.*, 2010, **5**, 035001.
72. L. B. Mao, H. L. Gao, H. Bin Yao, L. Liu, H. Cölfen, G. Liu, S. M. Chen, S. K. Li, Y. X. Yan, Y. Y. Liu and S. H. Yu, *Science*, 2016, **354**, 107.
73. J. Wang, J. Qiao, J. Wang, Y. Zhu and L. Jiang, *ACS Appl. Mater. Interfaces*, 2015, 7, 9281.
74. F. Bouville, E. Maire, S. Meille, B. Van De Moortèle, A. J. Stevenson and S. Deville, *Nat. Mater.*, 2014, **13**, 508.
75. F. C. Meldrum, *Int. Mater. Rev.*, 2003, **48**, 187.
76. M. H. Nielsen, S. Aloni and J. J. De Yoreo, *Science*, 2014, **345**, 1158.
77. W. J. E. M. Habraken, J. Tao, L. J. Brylka, H. Friedrich, L. Bertinetti, A. S. Schenk, A. Verch, V. Dmitrovic, P. H. H. Bomans, P. M. Frederik, J. Laven, P. Van Der Schoot, B. Aichmayer, G. De With, J. J. DeYoreo and N. A. J. M. Sommerdijk, *Nat. Commun.*, 2013, **4**, 1507.
78. J. Baumgartner, A. Dey, P. H. H. Bomans, C. Le Coadou, P. Fratzl, N. A. J. M. Sommerdijk and D. Faivre, *Nat. Mater.*, 2013, **12**, 310.
79. D. C. Green, J. Ihli, P. D. Thornton, M. A. Holden, B. Marzec, Y. Y. Kim, A. N. Kulak, M. A. Levenstein, C. Tang, C. Lynch, S. E. D. Webb, C. J. Tynan and F. C. Meldrum, *Nat. Commun.*, 2016, 7, 1.
80. B. Marzec, D. C. Green, M. A. Holden, A. S. Coté, J. Ihli, S. Khalid, A. Kulak, D. Walker, C. Tang, D. M. Duffy, Y. Y. Kim and F. C. Meldrum, *Angew. Chem. Int. Ed.*, 2018, **57**, 8623.
81. A. R. West, *Solid State Chemistry and its Applications*, 2nd edn, Wiley-Blackwell, 2014.
82. T. Ibn-Mohammed, I. M. Reaney, S. C. L. Koh, A. Acquaye, D. C. Sinclair, C. A. Randall, F. H. Abubakar, L. Smith, G. Schileo and L. Ozawa-Meida, *J. Eur. Ceram. Soc.*, 2018, **38**, 4922.

83. J. Xiao, M. Tomczyk, I. M. Reaney and P. M. Vilarinho, *Cryst. Growth Des.*, 2018, **18**, 4037.
84. S.-F. Wang and G. O. Dayton, *J. Am. Ceram. Soc.*, 1999, **82**, 2677.
85. S. Murakami, N. T. A. F. Ahmed, D. Wang, A. Feteira, D. C. Sinclair and I. M. Reaney, *J. Eur. Ceram. Soc.*, 2018, **38**, 4220.
86. H. Hilgenkamp and J. Mannhart, *Rev. Mod. Phys.*, 2002, **74**, 485.
87. N. Nitta, F. Wu, J. T. Lee and G. Yushin, *Mater. Today*, 2015, **18**, 252.
88. G. T. Grant, E. R. Morris, D. A. Rees, P. J. C. Smith and D. Thom, *FEBS Lett.*, 1973, **32**, 195.
89. S. Zilinskaite, A. J. R. Rennie, R. Boston and N. Reeves-McLaren, *J. Mater. Chem. A*, 2018, **6**, 5346.
90. I. Nowak and M. Jaroniec, *Top. Catal.*, 2008, **49**, 193.
91. A. Rumplecker, F. Kleitz, E. L. Salabas and F. Schüth, *Chem. Mater.*, 2007, **19**, 485.
92. W. C. Li, A. H. Lu, C. Weidenthaler and F. Schüth, *Chem. Mater.*, 2004, **16**, 5676.
93. W. Zhao, M. Lang, Y. Li, L. Li and J. Shi, *J. Mater. Chem.*, 2009, **19**, 2778.
94. Y. Wan and D. Zhao, *Chem. Rev.*, 2007, **107**, 2821.
95. A. E. Danks, S. R. Hall and Z. Schnepp, *Mater. Horiz.*, 2016, **3**, 91.
96. Z. Schnepp, *Angew. Chem. Int. Ed.*, 2013, **52**, 1096.
97. D. Walsh, L. Arcelli, T. Ikoma, J. Tanaka and S. Mann, *Nat. Mater.*, 2003, **2**, 386.
98. M. M. Rafi, K. S. Z. Ahmed, K. P. Nazeer, D. Siva Kumar and M. Thamilselvan, *Appl. Nanosci.*, 2015, **5**, 515.
99. Y. Y. Kim, C. Neudeck and D. Walsh, *Polym. Chem.*, 2010, **1**, 272.
100. D. Walsh, S. C. Wimbush and S. R. Hall, *Chem. Mater.*, 2007, **19**, 647.
101. R. Boston, A. Carrington, D. Walsh and S. R. Hall, *CrystEngComm*, 2013, **15**, 3763.
102. D. Walsh, S. C. Wimbush and S. R. Hall, *Supercond. Sci. Technol.*, 2009, **22**, 015026.
103. J. D. Jorgensen, M. A. Beno, D. G. Hinks, L. Soderholm, K. J. Volin, R. L. Hitterman, J. D. Grace, I. K. Schuller, C. U. Segre, K. Zhang and M. S. Kleefisch, *Phys. Rev. B*, 1987, **36**, 3608.
104. Z. Zhang, S. C. Wimbush, A. Kursumovic, H. Wang, J. H. Lee, H. Suo and J. L. MacManus-Driscoll, *CrystEngComm*, 2012, **14**, 5765.
105. S. R. Hall, S. C. Wimbush, Y. Shida and W. Ogasawara, *Chem. Phys. Lett.*, 2011, **507**, 144.
106. J. L. Konne, S. A. Davis, S. Glatzel and S. R. Hall, *Chem. Commun.*, 2013, **49**, 5477.
107. D. C. Green, R. Boston, S. Glatzel, M. R. Lees, S. C. Wimbush, J. Potticary, W. Ogasawara and S. R. Hall, *Adv. Funct. Mater.*, 2015, **25**, 4700.
108. S. R. Hall, *Adv. Mater.*, 2006, **18**, 487.
109. K. Cung, B. J. Han, T. D. Nguyen, S. Mao, Y. W. Yeh, S. Xu, R. R. Naik, G. Poirier, N. Yao, P. K. Purohit and M. C. McAlpine, *Nano Lett.*, 2013, **13**, 6197.

110. Z. Schnepp, J. Mitchells, S. Mann and S. R. Hall, *Chem. Commun.*, 2010, **46**, 4887.
111. Z. Schnepp, S. C. Wimbush, S. Mann and S. R. Hall, *CrystEngComm*, 2010, **12**, 1410.
112. Z. A. C. Schnepp, S. C. Wimbush, S. Mann and S. R. Hall, *Adv. Mater.*, 2008, **20**, 1782.
113. R. Boston, Z. Schnepp, Y. Nemoto, Y. Sakka and S. R. Hall, *Science*, 2014, **344**, 623.
114. S.-Z. Kang, T. Wu, X. Li and J. Mu, *Colloids Surf. A*, 2010, **369**, 268.
115. Z. Schnepp, S. C. Wimbush, M. Antonietti and C. Giordano, *Chem. Mater.*, 2010, **22**, 5340.
116. Z. Yang, Y. Zhang and Z. Schnepp, *J. Mater. Chem. A*, 2015, **3**, 14081.
117. Z. Schnepp, M. Thomas, S. Glatzel, K. Schlichte, R. Palkovits and C. Giordano, *J. Mater. Chem.*, 2011, **21**, 17760.
118. Z. Schnepp, M. J. Hollamby, M. Tanaka, Y. Matsushita, Y. Xu and Y. Sakka, *Chem. Commun.*, 2014, **50**, 5364.
119. W. Shenton, S. Mann, T. Douglas, M. Young and G. Stubbs, *Adv. Mater.*, 1999, **11**, 253.
120. K. M. Bromley, A. J. Patil, A. W. Perriman, G. Stubbs and S. Mann, *J. Mater. Chem.*, 2008, **18**, 4796.
121. R. Tsukamoto, M. Muraoka, M. Seki, H. Tabata and I. Yamashita, *Chem. Mater.*, 2007, **19**, 2389.
122. M. Knez, A. M. Bittner, F. Boes, C. Wege, H. Jeske, E. Maiß and K. Kern, *Nano Lett.*, 2003, **3**, 1079.
123. C. E. Fowler, W. Shenton, G. Stubb and S. Mann, *Adv. Mater.*, 2001, **13**, 1266.
124. S. Y. Lee, E. Royston, J. N. Culver and M. T. Harris, *Nanotechnology*, 2005, **16**, S435.
125. S.-W. Lee, C. Mao, C. E. Flynn and A. M. Belcher, *Science*, 2002, **296**, 892.
126. J. Richter, *Physica E*, 2003, **16**, 157.
127. J. He, Z.-W. Liu, W.-B. Fan, Z.-T. Liu, J. Lu and J. Wang, *J. Mater. Chem.*, 2010, **20**, 10107.
128. L. Shi, H. Yin, R. Zhang, X. Liu, J. Zi and D. Zhao, *J. Mater. Chem.*, 2010, **20**, 90.
129. L.-L. Lu, Y.-Y. Lu, Z. J. Xiao, T.-W. Zhang, F. Zhou, T. Ma, Y. Ni, H.-B. Yao, S.-H. Yu and Y. Cui, *Adv. Mater.*, 2018, **30**, 1706745.
130. T. Fan, X. Li, Z. Liu, J. Gu, D. Zhang and Q. Guo, *J. Am. Ceram. Soc.*, 2006, **89**, 3511.
131. X. Li, T. Fan, Z. Liu, J. Ding, Q. Guo and D. Zhang, *J. Eur. Ceram. Soc.*, 2006, **26**, 3657.
132. C. R. Rambo, J. Cao and H. Sieber, *Mater. Chem. Phys.*, 2004, **87**, 345.
133. T. E. Wilkes, M. L. Young, R. E. Sepulveda, D. C. Dunand and K. T. Faber, *Scr. Mater.*, 2006, **55**, 1083.
134. Z. Schnepp, W. Yang, M. Antonietti and C. Giordano, *Angew. Chem. Int. Ed.*, 2010, **49**, 6564.

135. W. Peng, S. Zhu, W. Wang, W. Zhang, J. Gu, X. Hu, D. Zhang and Z. Chen, *Adv. Funct. Mater.*, 2012, **22**, 2072.
136. M. R. Weatherspoon, Y. Cai, M. Crne, M. Srinivasarao and K. H. Sandhage, *Angew. Chem.*, 2008, **120**, 8039.
137. S. Kinoshita, S. Yoshioka, Y. Fujii and N. Okamoto, *Forma*, 2002, **17**, 103.
138. C. Mille, E. C. Tyrode and R. W. Corkery, *RSC Adv.*, 2013, **3**, 3109.
139. G. England, M. Kolle, P. Kim, M. Khan, P. Munoz, E. Mazur and J. Aizenberg, *Proc. Natl. Acad. Sci. U. S. A*, 2014, **111**, 15630.
140. Y. Tan, X. Zang, J. Gu, D. Liu, S. Zhu, H. Su, C. Feng, Q. Liu, W. M. Lau, W. J. Moon and D. Zhang, *Langmuir*, 2011, **27**, 11742.
141. W. Zhang, D. Zhang, T. Fan, J. Ding, Q. Guo and H. Ogawa, *Nanotechnology*, 2006, **17**, 840.
142. J. Huang, X. Wang and Z. L. Wang, *Nano Lett.*, 2006, **6**, 2325.
143. W. Zhang, J. Gu, Q. Liu, H. Su, T. Fan and D. Zhang, *Phys. Chem. Chem. Phys.*, 2014, **16**, 19767.
144. W. Ogasawara, W. Shenton, S. A. Davis and S. Mann, *Chem. Mater.*, 2000, **12**, 2835.
145. E. Culverwell, S. C. Wimbush and S. R. Hall, *Chem. Commun.*, 2008, **7345**, 1055.
146. O. C. Compton and S. T. Nguyen, *Small*, 2010, **6**, 711.
147. L. J. Cote, J. Kim, V. C. Tung, J. Luo, F. Kim and J. Huang, *Pure Appl. Chem.*, 2010, **83**, 95.
148. X. Zhou and Z. Liu, *Chem. Commun.*, 2010, **46**, 2611.
149. R. Boston, A. Bell, V. P. Ting, A. T. Rhead, T. Nakayama, C. F. J. Faul and S. R. Hall, *CrystEngComm*, 2015, **17**, 6094.
150. E. Smith, Z. Schnepp, S. C. Wimbush and S. R. Hall, *Physica C*, 2008, **468**, 2283.
151. R. Boston, K. Awaya, T. Nakayama, W. Ogasawara and S. R. Hall, *RSC Adv.*, 2014, **4**, 26824.
152. H. Dislich and P. Hinz, *J. Non-Cryst. Solids*, 1982, **48**, 11.
153. D. C. Green, S. Glatzel, A. M. Collins, A. J. Patil and S. R. Hall, *Adv. Mater.*, 2012, **24**, 5767.
154. O. G. Rojas and S. R. Hall, *Mater. Chem. Phys.*, 2017, **202**, 220.
155. L. Mottram, D. Z. C. Martin, N. Reeves-McLaren and R. Boston, *J. Am. Ceram. Soc.*, 2018, **101**, 4468.
156. R. Boston, P. Y. Foeller, D. C. Sinclair and I. M. Reaney, *Inorg. Chem.*, 2017, **56**, 542.

CHAPTER 2

Bioinspired Surfaces

A. M. COLLINS*[a] AND G. DEPIETRA[b]

[a] School of Physics, University of Bristol, Tyndall Avenue, BS8 1FD, United Kingdom; [b] Syngenta, Jealott's Hill International Research Centre Bracknell, Berkshire, RG42 6EY, United Kingdom
*Email: andy.collins@bristol.ac.uk

2.1 Introduction

Water is the most abundant molecule on the Earth's surface and life here has evolved to capitalise on this ubiquity. Fully 75% of the Earth's surface is comprised of water, either as ice or as a liquid. The total volume of the Earth's water is approximately 1 360 000 000 km^3, with 97.2% in oceans, 1.8% as ice, 0.9% in ground water, 0.02% in fresh water and 0.001% as water vapour.[1] The aqueous environment supports life, as in the oceans, but it is also the principal medium in which cellular and biological processes take place. The term "biomolecules" spans a wide range of compositions including small organic molecules, long-chain polysaccharides, proteins, DNA, lipids and more. Although compositionally diverse, these molecules have all evolved in some way to interact with water. As a prime constituent of nature, water provides a simple starting point to explore the forces and interactions that ultimately give rise to the hierarchical, functional and dynamic structures observed throughout biology. In this chapter, we will examine biological surfaces and the various methodologies employed to replicate them. With particular reference to biological matter at interfaces, we will move through the molecular to the nanoscale and describe the fundamental design rules that determine the macroscopic behaviour of surfaces. We will also provide some practical experiments that will enable the reader to enter

Inorganic Materials Series No. 4
Bioinspired Inorganic Materials: Structure and Function
Edited by Simon R. Hall

Published by the Royal Society of Chemistry, www.rsc.org

the world of bioinspired surfaces for themselves. These practical procedures are designed to provide basic starting points for surface replication and modification.

The chemical composition of water, two hydrogens covalently bound to one central oxygen, was suggested by Henry Cavendish in about 1781. Water is a polar molecule, having both positive and negative poles, and exists in a tetrahedral arrangement due to the presence of two unbonded lone pairs of electrons on the oxygen.[2,3] This polar structure is responsible for water's exceptionally strong self-attraction (cohesion) and numerous unusual properties that are responsible for its central role in life. An important feature of water is its ability to readily form hydrogen bonds (H-bonds), which occur when an electropositive hydrogen is weakly coordinated to an electronegative oxygen. The overall energy of the water molecule is determined by the local pressure, temperature and chemical potential of the adjacent molecules. Fluctuations in any of these parameters can prompt one hydrogen to dissociate from the water molecule, leaving behind an electron, and carrying a single free positive electrical charge into solution, which can then coordinate to another entity altogether. If a cationic hydrogen bonds to a negatively charged, anionic, OH^- then it reforms into a water molecule. If it coordinates through hydrogen bonding to a water molecule then that molecule will carry a cationic charge and is referred to as "protonated". An important distinction for the reader is that positively charged hydrogen ions are indeed single protons and this term is used interchangeably in chemistry. The concentration ratio of dissociated protons relative to negatively charged hydroxyl ions and whole water molecules is referred to as the pH. A greater number of protons relative to hydroxyl groups are referred to as acidic and the converse of a larger number of hydroxyl groups is referred to as basic. The pH of a solution has a strong influence over the biological processes that can take place in it. At a very simple level, the pH determines the level of protonation or acidic dissociation for a range of functional groups, which in turn strongly influences the behaviour of biomolecules possessing these moieties under aqueous conditions. In molecules, it is usual to consider the analogue of pH, known as the acidic dissociation constant, or pK_a. This is a function of the free energy change required for acidic dissociation to occur in a molecule and varies with the temperature. The level of dissociation for an organic molecule in aqueous conditions will be in a state of chemical equilibrium, which can be shifted left or right by the changing the pH of the reaction environment.

As a polar solvent, water accommodates acidic dissociation from itself, mineral acids (HCl, H_2PO_4, H_2SO_4 and others) and weak organic acids where the pH and pK_a values describe the relative concentrations.[4,5] The dissociation product K_a is given by

$$K_a = \frac{[H^+][HO^-]}{[H_2O]} \tag{2.1}$$

and

$$pH = -\log[H^+] \quad (2.2)$$

A further important addition to this is the notion of a dissociation constant for mineral salts, K_a, where one or more of the dissociated constituents is not a proton. Salt cations and anions present in a solution may have a kosmotropic (increasing order) or chaotropic (decreasing order) effect through coordination to polar or ionically charged aqueous molecules.[6] Dissociation constants and the associated solubility products of salts are important in discussing the process of biomineralisation onto surfaces.[7] For water, this can mean the ions interfere or disrupt hydrogen bonding by coordination to the free ions. The presence of ions in a solution raises the local osmotic pressure and lowers the amount of further solute that can be dissolved into it. The relative chaotropic and kosmotropic properties of a given ion place it along a spectrum known as the 'Hoffmeister series'. Salt in high concentrations is often used to precipitate proteins from solution. This technique is not a result of electrostatic screening and is attributed to the induced ordering of hydration layers at the solute surface as the water coordinates to the ionic species present on the protein surface. This ordering is driven by a reduction of entropy within the system. Anions have a greater influence than cations in terms of salting out behaviour, and anions generally retain their positions respective to one another in the Hoffmeister series when used to salt out different species of protein.[8]

The Hoffmeister series for cations from most to least precipitating is:

$$(CH_3)_4N^+ > Cs^+ > Rn^+ > NH_4{}^+ > K^+ > Na^+ > Li^+ > Mg^{2+} > Ca^{2+} \quad (2.3)$$

For anions, it is:

$$CO_3^{2-} > SO_4^{2-} > S_2O_3^{2-} > PO4^- > F^- > Cl^- > Br^- \sim ClO_4 > SCN^- \quad (2.4)$$

On the spectrum of molecular complexity requirements for life, along with water and salt, are the hydrocarbons that form biomolecules. Many of these may have little or no polar interaction with water but there are still intermolecular forces at work dictating how they interact, excluding direct chemical reaction, attractively or otherwise. Non-polar attractive interactions, known as van der Waals forces, are a collection of atomic, molecular, and surface interactions first proposed by the Dutch scientist Johannes Diderik van der Waals in the late nineteenth century. For his achievements, van der Waals received the 1910 Nobel Prize in Physics.[9] The van der Waals interactions are intermolecular forces, arising from electromagnetic fluctuations, that exist between all atoms and molecules, and which are responsible for many phenomena, such as the self-assembly of micelles, surface tension and capillary action. The van der Waals interactions can be subdivided into three groups: London dispersion forces, Debye interactions and Keesom forces.[10] The temporary formation of a dipole in non-polar molecules defines the London dispersion forces, whereas the Debye interaction describes the generation of an induced dipole in the presence of polar molecules; the forces arising by the interaction of two permanent dipoles are instead described by the Keesom forces. London dispersion forces arise from random fluctuations in electron density within collections of molecules and so, even in non-polar solvents free of any permanent dipoles there is still an attractive intermolecular force.

2.2 Thermodynamics of Molecular Scale Surfaces

In order for a chemical reaction to proceed there must be some form of thermodynamic driver for it to start. This is described simplistically as two simple molecules A and B, reacting to form a more energetically favourable product, C. The product C could be another molecule that is free in solution or it may be a solid precipitate, although as a single molecule its physical state is not so important. In both cases product C may change back into A and B with a steady state once the reaction has reached equilibrium. The kinetics of this reaction depends on the free energy, composition and geometry of the two molecules involved, but it is certain that they interact with one another. Expanding this model to include collections of A and B introduces the notion that there is a chemical gradient between A and B. Le Chatelier's principle may now be applied to the model, which states that an excess of A and B will drive the reaction to produce more of C.[11] Once A and B are depleted then the equilibrium may shift towards C dissociating to A and B. This may continue back and forth until there is balance between the reagents and products (eqn (2.5)).

$$A + B \leftrightarrow C \tag{2.5}$$

It is easy to picture mixing A and B in a beaker to form a homogeneous solution where the molecules will react quickly with one another. One might also imagine pouring an immiscible A on top of B in the beaker, like oil floating on top of water, and in this second scenario, we see that there is now a macroscopic interface, or boundary, between the two. In contrast to the homogeneous mixture, the heterogeneous system now has a spatial component that will influence the reaction dynamics along with the chemical thermodynamics and kinetics. At the most basic level it is possible to see that C will now be formed exclusively at the interface of A and B in the beaker and that if C does precipitate it will encounter very different equilibrium conditions as it sinks through B than it otherwise would in a homogeneous system.

Surfaces are often generally defined as the interfacial boundary of two or more differing physical systems. From a molecular perspective, they are boundaries that provide a significant environmental transition in chemical state and potential to the molecule encountering them. As discussed previously, molecular interactions encompass short-range attractive forces (van der Waals forces which are the summation of London dispersion and permanent dipole–dipole interactions) and longer range ionic (coulombic, permanent dipole and hydrogen bonding) interactions. In bulk, the forces between surrounding molecules on average balance one another, but at an interface there is a difference in energy between those at a surface and those remaining within the bulk continuum. Surface free energy is the work required to make more surface area for a solid or liquid from a bulk form. This is defined as force per wetted length or work per unit area in $mN\,m^{-2}$ or the equivalent $mJ\,m^{-2}$.[12] The surface energy, when referring to a solid exposed to gas, or the surface tension, when referring to a liquid exposed to gas, greatly influences the physical behaviour of two different media or materials when they are brought into contact with one another. The surface energy or tension is a summation of the polar and non-polar interactions of the surface molecules, and, for two different surfaces interfacing with one another, an interfacial surface tension or energy is formed.[13] Accordingly, interfacial surface tension is defined as the energy required to form a new unit area of interface between the two phases.

For two surfaces brought into contact with one another, the question is will they stay that way? In order for this to happen, some amount or work must be done to bring the two components together. For example, we could take a droplet of water falling onto a glass pane for a liquid–solid interface or one plastic and one metal sheet being pressed together as a solid–solid interface. A major determinant in order for the contact to continue is the magnitude of the difference in their surface energies and the magnitude of the resulting interfacial tension. When the two components of an interfacial layer have a similar free surface energy then the value of the resulting interfacial tension will be close to zero. For the plastic on metal example, a low interfacial tension value will mean there will be cohesive (attractive but not chemically bonded) interaction and the sheets will remain in contact. At the molecular level, this equates to the forces acting on the outward-facing side of a molecule in one of the sheets being matched by the forces acting on the inward-facing side. The molecule will be in equilibrium, with the thermodynamic drive to move towards the opposing surface in balance with the drive to move back within its own bulk. For a solid–liquid boundary, such as water on glass, the water will spread or "wet" widely over a surface if there is near-zero interfacial tension. Conversely, if there is a strong mismatch in surface energies at an interface then the difference will result in one of the phases being repelled, as there will be a thermodynamic impetus to minimise the interfacial surface area. A solid–solid interface will experience low friction with little or no cohesive interactions and a liquid–solid interface will result in the liquid balling up and forming a droplet. The contact angle of a droplet on a surface is a well-known method for determining the surface energy of a liquid or material.[14–17] By measuring the exterior angle formed at the point of contact of a droplet of a known surface tension liquid with a solid, ideally smooth, surface, it is possible to calculate that solid's surface energy. By further taking measurements with liquids that are polar and non-polar it is possible to calculate the contributions to that surface energy from hydrophobic and hydrophilic interactions. Surface molecules that have permanent dipoles can be thought of as acids or bases depending on the relative charge being correspondingly positive or negative. This gives rise to the polar contribution to surface energy and an area high in polar components will generally be hydrophilic or, rather, easily wettable by water or other polar liquids. Conversely, surfaces low in polar components, such as those that are organic aliphatic or aromatically based, will demonstrate hydrophobicity, where water will ball up upon the surface. Low or non-polar surfaces are referred to as having a larger dispersive contribution, as van der Waals interactions are dominant. For these surfaces, "oily" liquids will more readily wet over the surface. A standard method for measuring surface tension or energy involves measuring the contact angles of droplets of different liquids placed upon the surface having both high polarity (*e.g.* water) and zero polarity (*e.g.* diidomethane) and observing the relative wetting or beading of the liquid.

2.3 Contact Angle and Surface Free Energy Measurements

The contact angle is the angle formed at the intersection of the solid–liquid interface and is dependent on the interfacial energy (γ_{sl}) and the liquid–vapour interfacial energy (γ_{lv}). It is geometrically acquired by applying a tangent line from the contact point along the liquid–vapour interface in the droplet profile (Figure 2.1). The contact angle is the most commonly measured property in wettability studies, and indicates the degree of wetting, due to the interaction of a solid surface with a liquid.[18,19]

The relation between the surface free energy of the substrate, the surface tension of the liquid, the solid–liquid interface and the contact angle between substrate and liquid can be expressed by Young's equation:

$$\gamma_{sv} = \gamma_{sl} + \gamma_{lv} \cos\theta \tag{2.6}$$

where θ is the contact angle formed by liquid droplets on solid surfaces. In order to determine the relation between surface chemical composition and wettability, the dispersive components (van der Waals interactions) and the polar components (dipole–dipole and hydrogen-bonding interactions) need to be determined for both interfacial energies. Dipole–dipole interactions

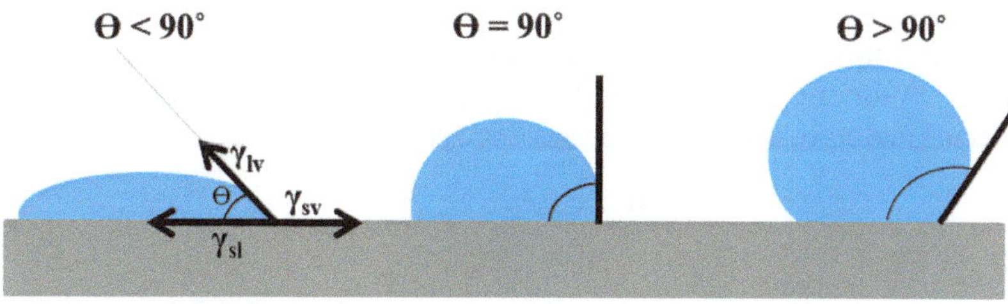

Figure 2.1 Liquid droplet contact angles on solid surface.

are electrostatic, arising between two polar molecules, analogous to the interaction between two magnets. The hydrogen-bonding interactions resemble a directional dipole–dipole interaction.[10] The relationship between surface chemical composition and interfacial energies is expressed by:

$$\gamma = \gamma^{D} + \gamma^{P} \tag{2.7}$$

where γ defines the total interfacial energy given by the dispersive and polar components: γ^D and γ^P, respectively. The dispersive and polar components of the surface can be determined by using the Owens–Wendt method,[20] by measuring the contact angle between the surface and test liquids with a well-known surface tension (Figure 2.1).

Contact angle values below 90° indicate high surface wettability, with the liquid spreading over the surface, whereas contact angle values above 90° are recorded on surfaces with low wettability, considering that a contact angle of 90° represents the upper limit for a hydrophilic surface. Contact angle values above 90° indicate that the liquid is minimising the contact with the surface by forming a compact liquid droplet. For instance, a contact angle of 0° defines complete wetting whereas contact angles above 150° are characteristic of superhydrophobic surfaces, where there is almost no contact between the liquid and the surface. An example of this would be the well-studied phenomenon known as the "lotus effect".[21–23]

At least two liquids with known dispersive and polar components are required to determine the surface free energy of surfaces, with one of the liquids characterised by a non-zero polar component. The two-component model allows calculation of the interfacial tension based on the polar and dispersive interactions between the surface and the test liquid. For example, the polar interactions contribute in a minimal way to reducing the interfacial tension in the case of water droplets on hydrophobic surfaces; this corresponds to high contact angle (*i.e.* >90°) and low surface wettability. Using a test liquid characterised only by dispersive components, for example diiodomethane, the dispersive components of the solid can be calculated. In this case, eqn (2.7) is simplified to:

$$\gamma_{l} = \gamma_{l}^{D} \tag{2.8}$$

From the contact angles formed by diiodomethane sessile liquid droplets on the solid surfaces, the dispersive components can be calculated using the eqn (2.9), where γ_l and γ_s define the liquid and solid interfacial energies, respectively. In this study, diiodomethane was chosen to determine the dispersive components, considering that this test liquid does not exhibit polar components due to its molecular symmetry.

$$\gamma_{s}^{D} = \frac{\gamma_{l}}{4}(\cos\theta + 1)^{2} \tag{2.9}$$

Conversely, to determine the surface polar components a test liquid exhibiting both polar and dispersive components, for instance water, is used.

By measuring the contact angle between water and the surface, the polar components can be determined by:

$$\sqrt{\gamma_s^D + \gamma_l^D} + \sqrt{\gamma_s^P + \gamma_l^P} = \frac{\gamma_l \cos\theta + 1}{2} \tag{2.10}$$

Furthermore, the polar and dispersive components can be used to determine the work of adhesion (W_a) thus:

$$W_a = 2\sqrt{(\gamma_l^D \gamma_s^D)} + 2\sqrt{(\gamma_l^P \gamma_s^P)} \tag{2.11}$$

The most widely used technique to measure contact angle is direct measurement of the tangent angle at the three-phase point on a sessile drop profile using a telescope goniometer.[18]

2.3.1 Volume Absorption Measurements by Sessile Drop Shape Analysis Methods

Absorption of a liquid in contact with a surface leads to reduction in the volume of the liquid, which can be monitored by drop volume measurements. An important factor to consider during volume absorption measurements is the volume of liquid that evaporates from the surface. Accordingly, the test needs to be performed in a saturated atmosphere to prevent loss of volume due to evaporation.[24] This is usually achieved by using a quartz cuvette, which is filled at the bottom with water in order to create a saturated environment and to prevent evaporation of the test droplets from the surface.

The volume is determined by the shape (width and height) of the droplet using the circle-fitting method in which the shape of the droplet is assumed to be in the form of a circular arc; this assumption is valid when small droplets volumes (*i.e.* between 1 and 10 μL) are used to perform the contact angle measurements. Droplets with a volume between 1 and 10 μL form spherical shapes on a solid surface, as a result of the interfacial tension, whereas larger volumes tend to form flatter shapes, due to the contribution of the effect of gravity. There is a length scale, defined by the so-called capillary length λ_c, above which gravity cannot be neglected compared to capillary forces (eqn (2.12)):

$$\lambda_c = \sqrt{\frac{\gamma}{\rho g}} \tag{2.12}$$

where g is the gravitational acceleration, ρ is the density of the fluid and γ is the surface tension of the liquid. Previous studies have demonstrated that, in the case of water droplets with a radius below 2.7 mm, the interaction with the surface is predominantly due to capillary forces and gravity effects can be neglected.[25] Accordingly, the smaller the droplet volume the more accurate the approximation with the circle-fitting method used for the volume calculation. Initial calibration is performed to identify the number of pixels per mm, by measuring the number of pixels in the known diameter of the syringe needle, which is subsequently used to identify the pixels defining the droplet shape. Three key points, namely left end-point, right end-point and apex of the droplet, are used to determine the width and height of the droplet placed on the surface. The width and the height are the values required in the circle-fitting method to calculate the volume of the droplet. This allows optical measurement of the volume reduction by recording reduction in the shape profile of the droplet. Automated image analysis software is used to determine the time evolution of the drop volume by determining the contact angle values. The droplet shape and contact angle are influenced by the changing characteristics of the surface due to the presence and adsorption of the liquid compared to the dry surface; the surfaces become more easily wetted when they are exposed to the liquid/liquid–vapour interface. The contact angle decreases with time, with the volume of the droplet reducing due to absorption by the surface if it is porous.

2.3.2 Application and Control of Surface Energy and Contact Angle

So far, this chapter has considered only the behaviour based on molecular surface interactions but beyond this the morphology and texture of the surface has a further large impact on the observed surface energy. This is the key to understanding biological and bioinspired surfaces. Perhaps the most well readily studied example of this is the "Lotus Effect" observed in leaves which demonstrate the property of superhydrophobicity.[13,21–23] A number of evolutionary adaptations give rise to this effect which helps the plant to clean itself by allowing water to roll off the surface of the leaf taking dust and contaminants with it.[22,26,27] The surface is waxy meaning it is low in polar contributions to the surface energy. The low wettability of the waxy surface is further enhanced by a surface patterning comprising an array structure of 10–20 μm sized conical papillae.[16] This loosely arranged morphology increases the surface area of the leaf such that it would require more work for a water droplet to spread across it and accordingly this correlates to an increase in the sum interfacial tension between the water and the leaf. There is an additional subtle difference in interaction depending on the water truly wetting the papillae as it moves over the surface (the Wenzel Model) or if the water is repelled to the point that air can occupy the space between the papillae and the water (Cassie Model).[13] The lotus leaf and the associated analogues found in nature are archetypal of a functional bioinspired surface. Biological superhydrophobicity and the replication of this effect has been covered comprehensively in the literature already and will not be covered extensively in this chapter.[13,16,21,22,28,29] Broadly, wetting and non-wetting mechanisms are utilised in nature for self-cleaning and examples of these, both hydrophobic and hydrophilic, are given in Figure 2.2.

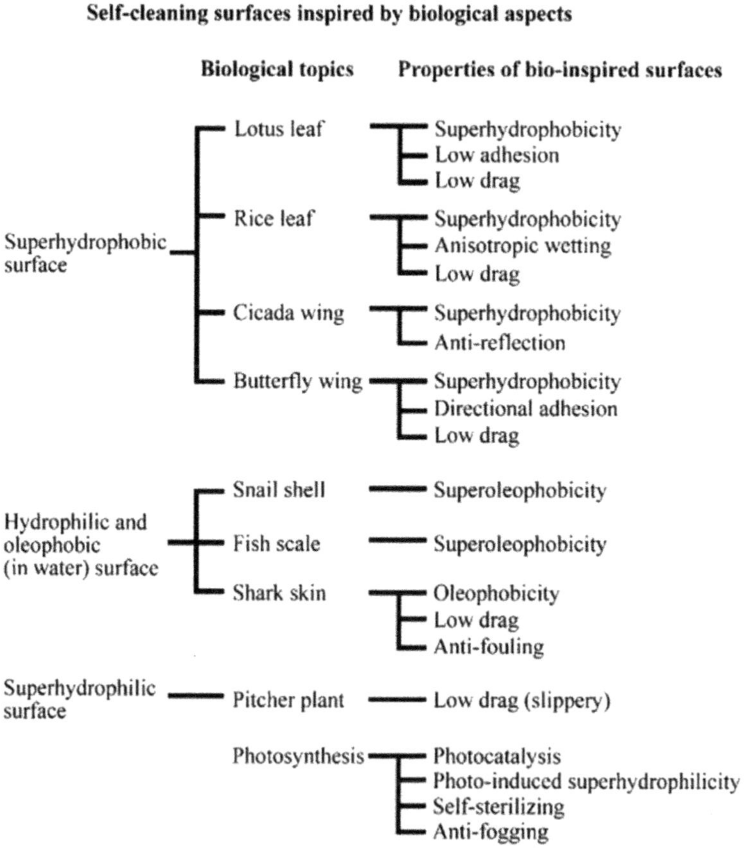

Figure 2.2 Overview of biomimetic self-cleaning surfaces inspired by biological objects. Reproduced from ref. 22 with permission from The Royal Society of Chemistry.

2.4 Bioinspired Non-wetting Surfaces

A non-wetting surface describes behaviour where the interface cannot be wetted by the contaminant medium and beads upon the surface. Often, this leads to liquid physically rolling off and carrying dirt and debris with it, trapped in the surface tension of the droplet. Naturally evolved self-cleaning mechanisms, like that of the lotus leaf depend on the interaction of liquids, both hydrophobic and hydrophilic, on the surface to remove or prevent attachment of a contaminant. This is important in plants, as leaves must remain clean to maintain optimal photosynthesis. A further benefit is the prevention of fungus and bacteria from taking hold within the plant. It is not only plants that benefit from water-repellent surfaces, and many water-dwelling spiders (arthropods) possess micro-setae that trap air and form a superhydrophobic interface on the water surface. It has been reported that the water strider has a buoyancy nearly 60 times its own body weight.[30]

The fisher spider, *Dolomedes triton*, breathes the air trapped within microsetae as a form of gill, called a plastron, while underwater and this adaptation also features within the spiders' tracheal network to prevent inhalation of water. Very generally, surfaces having greatly increased surface areas through micro- and nano-structures on the surface will be poorly wetting (Figure 2.3).

Sol–gel chemistry provides a facile route towards the replication of biological structures. The hydrolysis, alkoxylation and condensation reaction of an alkoxide to produce an inorganic monolith replica has been extensively demonstrated in the work of the Mann and Hall groups.[31–35] Biomimetic replication to form an inorganic copy can be achieved conveniently by coating

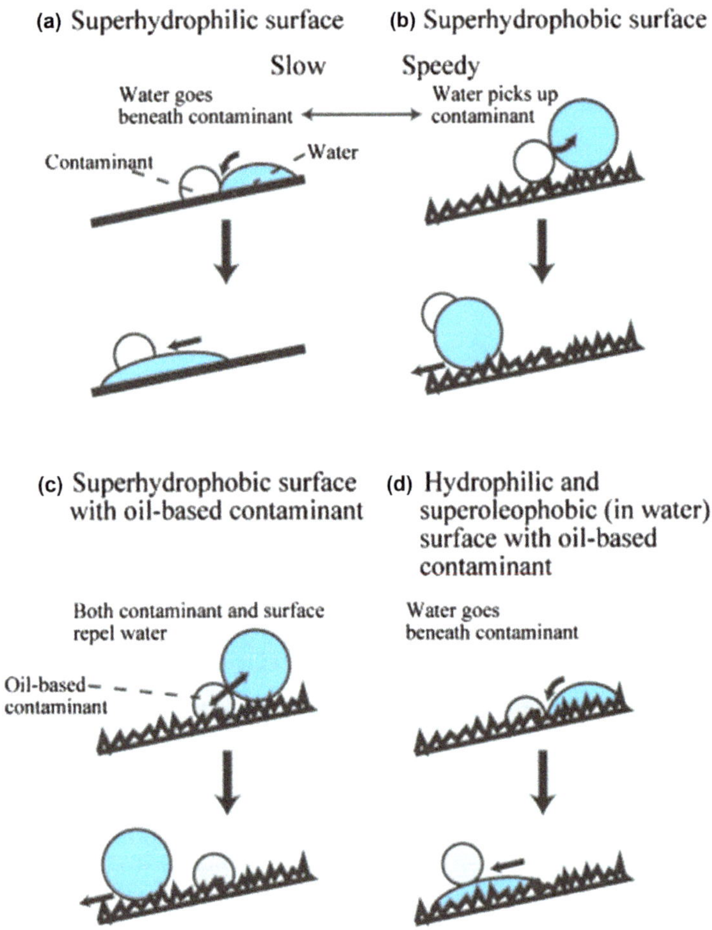

Figure 2.3 Schematic illustration of self-cleaning processes on (a) superhydrophilic surface, (b) superhydrophobic surface, (c) superhydrophobic surface with oil-based contaminant and (d) hydrophilic and superoleophobic (in water) surface with oil-based contaminant.
Reproduced from ref. 22 with permission from The Royal Society of Chemistry.

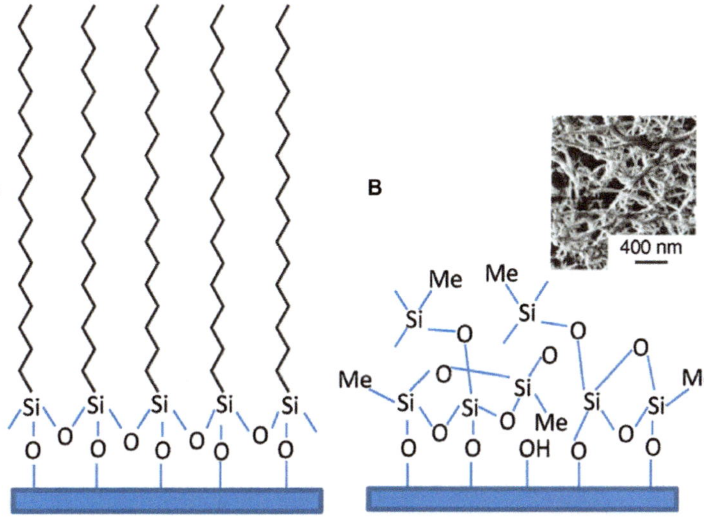

Figure 2.4 Schematic representation of a normal hydrophilic glass surface after treatment with (A) octadecyltrichlorosilane that results in the formation of a highly ordered self-assembled monolayer (hydrophobic) and (B) methyltrichlorosilane that results in the formation of a 3D polymethylsiloxane network (superhydrophobic). The inset shows a scanning electron miscrosope (SEM) image of the superhydrophobic surface. Reproduced from ref. 26 with permission from American Chemical Society, Copyright 2013.

a biological original with a silicon alkoxide, or variant thereof, then inducing the formation of a glass by hydrolysis. The underlying biological template can then be removed by chemical dissolution or thermal degradation. Pure silica replicas of biomaterials are generally hydrophilic at the chemical level, as the silicon oxide surface will have pendant oxygens or hydroxyl groups, which will readily coordinate to water molecules. At longer length scales, the texture contributes to the wettability and, as has been discussed above, roughness tends to make a surface less wettable. In the laboratory this can be achieved by preparing a glass slide by coating it with octadecyltrichlorosilane such that a non-polar monolayer is formed (Figure 2.4).[26] The chlorines on the silicon head group are displaced by surface oxygens in the glass to form Si–O bonds to the surface. A strong aqueous acid is used to initiate and catalyse the hydrolysis of the chlorosilane to form an intermediate silicic acid. The silicic acid form will further condense to form Si–O links to the glass surface and adjacent organosilane molecules. A further modification of this method is to produce superhydrophobicity by coating the surface with methyltrichlorosilane. Upon hydrolysis, this self-assembles into a three-dimensional network, which imparts a rough high energy surface within which air bubbles are trapped (Figure 2.4). These small bubbles are similar in action to the setae of the water spiders in effect. There are also non-polar contributions to the surface energy arising from surface-bound methyl moieties (Box 2.1).

Box 2.1 Preparation of a hydrophobic glass slide

This particular preparation has been optimised in order to retain the optical transparency of the slide. It is possible to get good hydrophobicity by directly soaking a clean glass slide directly in pure octadecyltriethoxysilane, which forms direct bonds by displacement of surface hydroxyls and reaction with surface-adsorbed water on the glass. After allowing the slide to sit for an hour, it can then be removed and rinsed using ethanol to remove the alkoxide. Both alkoxide and chlorosilanes are sensitive to moisture and are flammable. They should be stored under an inert gas and stock chemicals should be worked with under dry conditions in a fume cupboard. Do not touch any chemicals with bare hands and use appropriate personal protection equipment.

1. Wash a glass slide in deionised water and then rinse with ethanol. For extra cleanliness, you can sonicate the slide in ethanol. Allow the slide to air dry on a rack.
2. Once the slide is dry place it into a UV-ozone cleaner for 15 minutes.
3. Place the cleaned slide into a glass vial containing 10 mL of toluene. Make sure the toluene covers the slide.
4. To protect the reagent from moisture the octadecyltrichlorosilane will probably have a rubber septum or similar in the bottle neck. Use a graduated syringe to extract 0.35 mL of octadecyltrichlorosilane and add this to the toluene.
5. Add 0.25 mL of 1 M hydrochloric acid using a volumetric pipette.
6. Place a cap on the glass vial and allow the reaction to proceed for 1.5 hours.
7. Remove the slide and rinse three times each with toluene, ethanol, a 1 : 1 mixture of deionised water and ethanol and finally deionised water.
8. Dry the slide in the oven for 5 minutes at 100 °C.
9. The superhydrophobic variation is made by replacing the reagent in step 5 with 95 μL of methyltrichlorosilane. After this, follow step 6 and onwards as described. Contact angles of greater than 150 °C can be achieved, although this can be lowered if the reaction is maintained at 4 °C rather than at room temperature. This is due to the rate of inorganic nucleation being slightly lowered and results in a slightly smoother surface.

An excellent demonstration of self-cleaning mechanisms can be performed using these slides. Dust or small particles can be spread onto the methyltrichlorosilane-treated slide and a droplet of water added. When the slide is tipped even slightly the droplet will roll off the surface and should carry the particulates with it. If the surface is not strongly superhydrophobic then the water will slide off at higher tipping angles and will leave the contaminant behind on the slide.

2.5 Bioinspired Wetting Surfaces

Many biological interfaces are structured such that a protective layer of hydrophobic or hydrophilic liquid wets it completely and acts as a barrier to the opposite polar or non-polar medium containing a contaminant. Wetting is a key consideration in classical boundary lubrication, where a lubricant is required to maximally coat two surfaces moving against one another under force.[24,36] A surfactant, polymer or self-assembled monolayer applied at the boundary interface functions as a slip-plane and accommodates applied shear forces by allowing molecules to move over one another in a way that the underlying surfaces would not otherwise permit.[37] The efficacy of the lubricant is determined by its friction coefficient, μ, and the lower the value of the friction coefficient the less force is required to move one surface over the other. Lubrication is required biologically, and in the human body a variety of formulations adapted to suit particular functional purposes are observed.[38] Examples include synovial fluid, which coats cartilage in motile joints, the mucin layer between the eyelid and cornea and the glycoproteins and proteins in saliva, which ease the passage of food through the throat.[39–42] The majority of biolubricants are a mixture of phospholipids, which are surfactants, and glycoproteins, also known as mucins, which are a form of biopolymer.

2.5.1 Phospholipids

Phospholipid chemical structures are amphiphilic surfactant molecules where the polar head group is normally coordinated to a surface, while the non-polar section is presented to another medium. Predominantly a component of cell wall membranes, a discussion of the role these molecules play in cellular architecture is discussed in Section 2.5.3. Here we will look briefly at the role of the phospholipid in lubrication as a surfactant. Sodium lauryl sulfate (SDS) and cetyltrimethylammonium bromide (CTAB) are two surfactants commonly used in both industry and research in order to generate bioinspired surfaces. When dissolved in aqueous solution, SDS has an anionic (negatively charged) head group and CTAB has a cationic (positively charged) head group. A study by Wright and Dowson in 1976 found that aqueous solutions of these surfactants were as effective at lubricating cartilage as natural synovial fluid.[43] An initial mechanism proposed by Hills for biolubrication was the release of surface-active phospholipids, such as dipalmitoylphosphatidylcholine (DPPC) *in vivo* to sit interstitially between cartilage and other biosurfaces (Figure 2.5).[44] Polar head groups adhere to the biological surface and the long aliphatic chains slide over one another, tail against tail, on the opposite surface when load and force is placed upon them. It was found that hyaluronic acid aided the transport of poorly soluble lipids as a carrier solvent and that the presence of water improved lubrication greatly. The role of water in enhancing the lubrication is more complicated than a simple layer of interstitial hydration acting as an extra

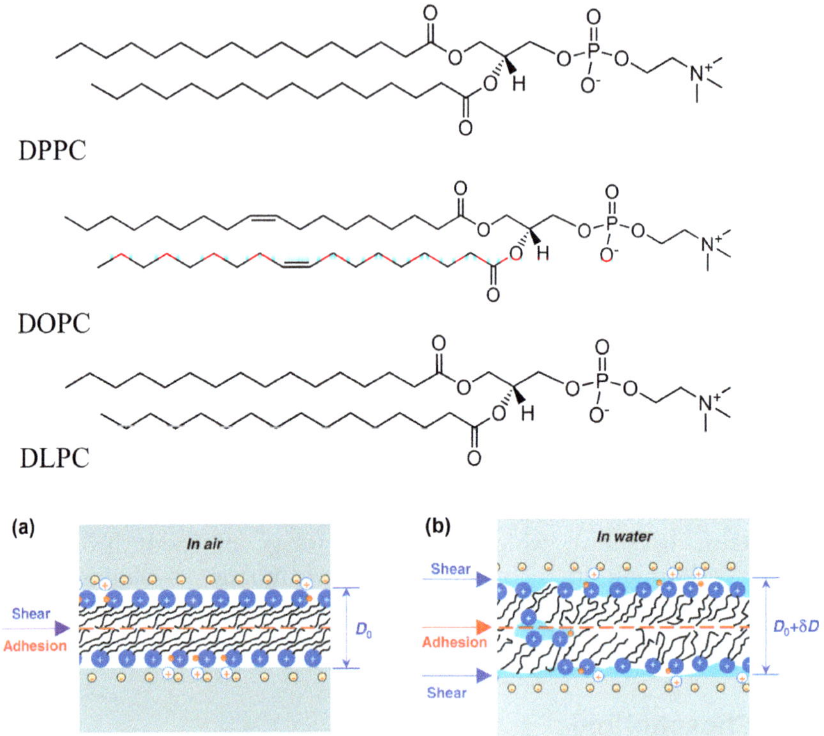

Figure 2.5 (Top) Chemical structures of DPPC, DOPC and DLPC. (Bottom) (a) In classic boundary lubrication, the plane of adhesion and shear lies at the interface between the tails. (b) Upon immersion in water, the surfactant head groups become hydrated, giving rise to a small swelling of $\delta D \sim 2.5$ Å and greatly enhancing surfactant lateral mobility. Some surfactant molecules could also undergo the flip-flop motion in which the molecules would turn over. In this case, the plane of adhesion is at the mid-plane, giving rise to adhesion comparable to that in air (a). However, shear sliding would take place at either of the interfaces between the head groups and the substrates decorated with molecular water puddles. Reproduced with permission from ref. 36 with permission from American Chemical Society, Copyright 2016.

slip-plane. The Briscoe group, investigating friction forces on mica lubricated with monolayers of a double-chained cationic surfactant *N*,*N*-dimethyl-*N*,*N*-diundecylammonium bromide (DDunAB), found the presence of water reduced the friction coefficient by two orders of magnitude, to 1% of the value in air.[24] The quaternary ammonium head groups in DDunAB become hydrated and this enhances the lateral molecular mobility, as the surfactant is not anchored to the substrate but can now 'flip-flop' around (Figure 2.5b). A requirement of boundary lubrication in biology is the capacity to remain effective under load for long periods of time. Hydration layers can get "squeezed out" under pressure. Double-chained surfactants generally form layers that are more robust to shear under pressure than single-chained molecules. This is reflected in nature as DPPC, 1,2-dioleoyl-sn-glycero-3-phosphocoline (DOPC) and 1,2-dilauroyl-sn-glycero-3-phosphocholine (DLPC) all possess a twin-tail architecture and have a high performance as biological lubricants.

2.5.2 Mucins

Mucin gels are found at all environmental boundaries in the body that are not already covered in skin, and are comprised predominantly of *O*-glycosated, long (of the order of micrometres) macromolecular peptide chains. Mucins possess a mucoadhesive gel structure, which coats and protects the mucosal surface (Figure 2.6). They range from 2 to 40 MDa in relative mass and can be as long as 10 micrometres (in saliva).[39,40,45,46] Mucins have a particular naming convention and are classed as human- (MUC) or animal-sourced (Muc) and are assigned a number based on the order of their discovery. Wet epithelia are lubricated through continued renewal of these polyelectrolytes by glycosidase and protease degradation and subsequent secretion of fresh mucins. Glycocalyx molecules on the apical surface of epithelial cells anchor the macromolecular chains in position and also act as a lubricant on the cell surface.[47] The gel layer imparts hydrophilicity to the corneal surface, which would be hydrophobic if uncoated. The gel properties are largely due to the presence of disulfide bonds in the cysteine-rich peptide domains. Gelling is also induced by the presence of cations, such as Ca^{2+}, in extracellular fluids and the interaction of hydrophobic regions. Biosynthesis of mucins in the cell takes place on time-scales of the order of a few hours, with secretion and hydration taking place over millisecond and second time-scales. Mucosal surfaces are a convenient route for drug delivery into the body through the application of topical medications. The eyes, nose, airways, oral and anal cavity are common routes for administering medication, through inhalation, suppositories and gels.[48] Medication administered in this way can be delivered rapidly to the bloodstream and so to the entire body. Chemicals and nanomaterials can be transported across the blood–brain barrier by translocation through the olfactory neurons or *via* transport along the retinal nerve.[49]

2.5.3 The Eye: A Biolubricant Wetting Example

The structure of the eye is a fine example of a biolubricated surface where multiple environmental and functional requirements have produced a multicomponent form of lubrication called the tear film lipid layer. Dry eye syndrome is a common problem for both contact lens wearers and the elderly population (15–25% of people over 65) such that many treatments for this condition replace or replicate some of the functions of these layers.[50] The tear film lipid layer covers the corneal epithelium and is a barrier to contamination from particulates, microorganisms and viruses. It is also required that it remains transparent in order for the eye to function properly.

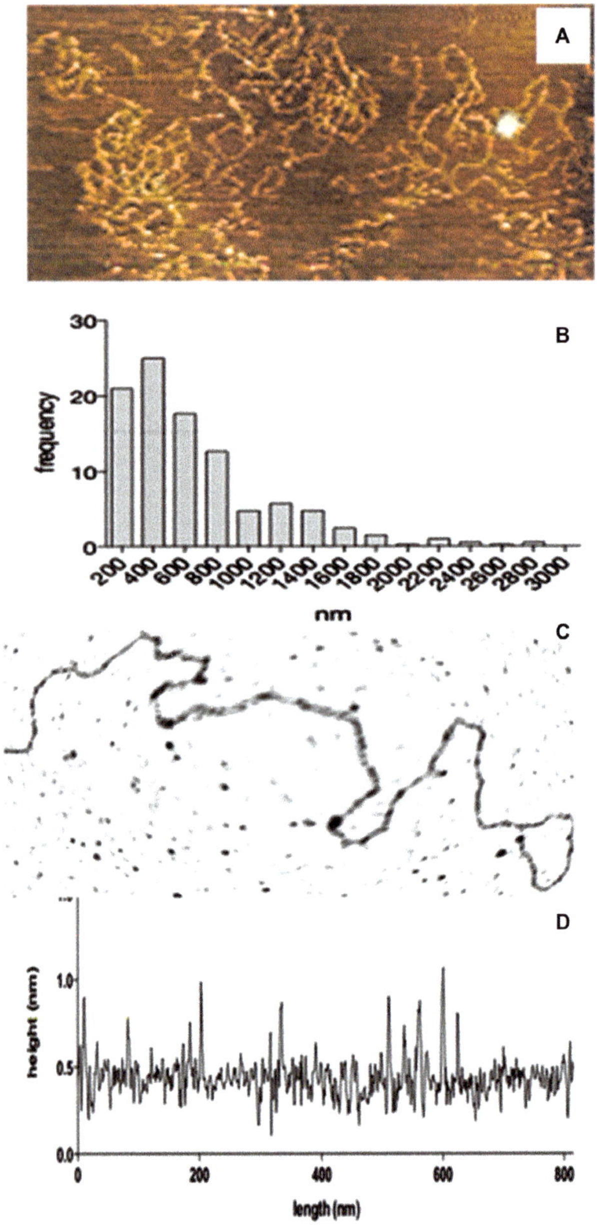

Figure 2.6 Mucin images, lengths and heights. (A) Hydrated ocular mucins on mica, height range 3 nm; (B) length distribution of the largest purified mucins; (C) topographic map of a single polymer emphasising the bead-like dark, high, domains of dense glycans and negative charges; (D) height profile: note the long domains (tens to hundreds of nanometres) of very low height, consistent with no, or sparse, glycosylation. The high peaks are likely to represent sequences with dense glycosylation. Glycans of ocular mucins are mostly very short.
Reproduced from ref. 40 with permission from The Royal Society of Chemistry.

Finally, it must wet the surface of the cornea efficiently while allowing the eyelid to move smoothly over the surface during blinking.[47,51,52]

The preocular tear film is composed of a multiple stratum of mucins, aqueous electrolytes and a lipid layer, each serving a specific function. These layers must all operate in concert as a dynamic system that is constantly undergoing loss of moisture to the environment and regular shear force. The mucin layer, 2.5–5 μm thick, is generated at, and anchored to, the surface of the cornea and promotes smooth wetting across the otherwise hydrophobic epithelial layer. On top of this is a ∼4 μm thick aqueous layer containing various electrolytes and proteins. The proteins are principally 2–2.5 mg mL^{-1} lysozyme and lipocalin. The lipocalin is an amphiphilic protein and is believed to aid the spreading of the tear film. At the exterior of the preocular tear film presented to the ambient environment is a relatively thin 15–160 nm tear film lipid layer (Figure 2.7). A lipid layer thickness below 60 nm is linked to dry eye complications.

The tear film lipid layer was long thought to provide a barrier to water loss through evaporation. Although it does function in this capacity to a limited degree, it has been shown that the lipid layer prevents the break-up of the underlying aqueous and mucin layers. In dry eye cases, where this film is inadequate, the film break-up can be observed macroscopically as dots or streaks on the corneal surface. Phospholipids comprise 5–13 mol% in whole tears, with 0.1 mol% traceably expressed from the meibomian glands. Meibomian lipids form the majority of the outermost tear film lipid layer, although analysis of the specific breakdown is difficult due to the small quantities that can be harvested from sampling. The most abundant lipid classes in the tear film are non-polar wax esters and cholesterol esters and polar (*O*-acyl)-ω-hydroxy fatty acids, phosphatidylcholines, phosphatidylethanolamines, lysophosphatidylcholines, sphingomyelins and ceramides.

Artificial tears are a huge industry worth $540 million annually.[50] The majority of formulations have active ingredients that to some extent mimic mucins and are carbohydrate based. Carboxymethylcellulose and hydroxypropylmethylcellulose form the base ingredient for many artificial tear and personal lubricant formulations. This is followed by polyols, such as polyethylene glycol, which have gel-forming properties but demonstrate slip behaviour in aqueous conditions. Hyaluronic acid is also widely used, as this supports the transport and dispersion of phospholipids having poor aqueous solubility to the ocular film layer. Hydroxypropylmethylcellulose-based lubricants lose efficacy if they are allowed to dry out but readily regain their properties on addition of further moisture.[38]

Interfacial energy has been introduced here as the most basic interaction between two surfaces in proximity, before more dynamic processes are considered. In addition to the thermodynamically driven balance of interfacial tension, there will also be direct chemical reactions occurring across a boundary. In Section 2.6, we will examine the molecular interactions that give rise to adhesive behaviour between two surfaces in contact, and the implications for bioinspired surfaces.

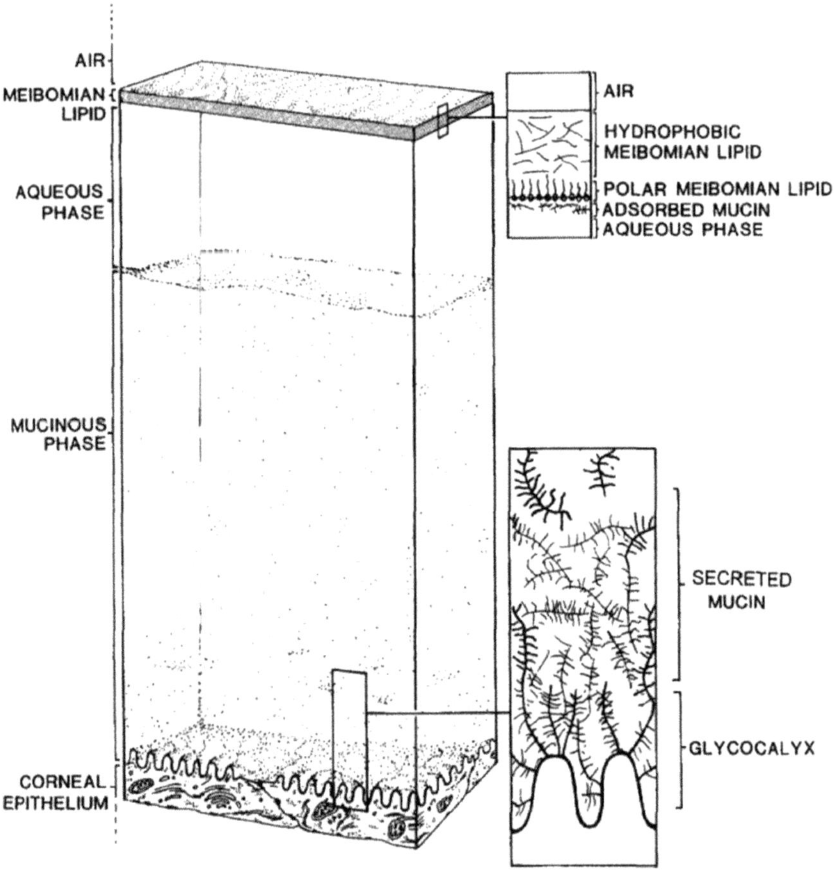

Figure 2.7 Diagram showing the composition of the preocular tear film. Insets show interactions occurring at the glycocalyx, mucinous phase and meibomian lipid/aqueous phase.
Reproduced from ref. 45 with permission from Elsevier, Copyright 1997.

2.6 Adhesion at the Molecular Level: Synthetic and Natural

We have already seen that wetting can allow a liquid to spread over a surface, but this is not the same as a directly attractive bond. The "stickiness" of a surface can be thought of as either adhesive or cohesive or a mixture of the two. Reactive adhesion is the presence of a direct covalent chemical bond being formed between the two items being stuck together. Non-reactive adhesion, also known as cohesion, is through physical adsorption such as by van der Waals dipole interactions or electrostatic contributions such as ionic charge.

2.6.1 Biological Adhesives

Adhesion is observed throughout nature and falls, very broadly, into three main categories.

1. Attachment of body parts or cellular structures to one another. This is essential for the development of hierarchical living organisms.
2. Attachment of one organism to another for mating, parasitism or capturing prey.
3. Attachment of an organism to a non-living surface, such as a mollusc or plant sticking to a rock.

Throughout these three categories nature largely employs a protein-to-protein or a protein and polysaccharide combination to instigate an adhesive bond.[53–55] This is, of course, an over simplification and the search for bioderived adhesives is an active and varied field. Research is active in determining promising candidates for biomimetic adhesives from plants, animals, fungi and algae.[55–59] Aquatic organisms, such as types of kelp, echinoderms, crustaceans and molluscs, also show great promise in yielding stronger and more adaptable adhesives.[60,61] Sea-dwelling organisms have evolved to adhere under water and salt concentrations that would defy many synthetic commercial adhesives currently available. In particular, the ability to bond two surfaces permanently in a wet, high salt concentration environment would have a great impact in the medical field for wound closure and implant surgery. One promising area of study is the development of bioinspired adhesives based on mollusc foot proteins, normally shortened to "mfp". The mussel byssus is anchored in place by a number of plaque-terminated fibrils on which a range of protein-based glues are excreted. The mfp family, isolated from various mussel species, is a well-researched example of a bioinspired adhesive and several different variants of this protein, mfp 1–6, have been identified. Once the protein has been expressed then a number of post-translational reactions occur that increase the binding strength. Notably, the addition of particular amino acids has been found to have a radical effect on the bond strength. Mfp-3 and mfp-5 are found in high concentrations on the plaque surface and have been found to contain from 20 to 30 mol% of 3,4-dihydroxyphenylanaline (DOPA). The amino acid DOPA features a catechol moiety and under weakly basic conditions, as is found in seawater, this can oxidise to form a quinone. Both the catechol and quinone forms of DOPA can form a wide range of covalent and non-covalent bonds with a range of organic, inorganic and metallic substrates.[54] The surface of a rock, for example, is rich in oxides and polar mineral surfaces, which lend themselves well to hydrogen bonding from protein hydroxyl groups. The presence of aromatic groups in the protein also allows non-polar attractive interactions. The ratio of catechol to quinone forms is kept in equilibrium by acidic mediation of thiol groups in the mfp-6 protein in order to maximise the bond strength (mfp-6 mediated). The mollusc bioadhesive has further adapted a mechanism to remove the weak boundary layers of hydrated ions that would otherwise prevent direct, intimate contact. The catechol moiety is flanked by lysine residues that are cationic in nature. Hydrated ions at the substrate surface will coordinate to the residues and allow the protein to come into contact with the surface (Figure 2.8).[62]

The efficiency of bioadhesives is such that they present a challenge to isolation, as they need only be secreted in small amounts by the organism. For example, in order to isolate enough protein cement for study from the crustacean *Balanus crenatus*, over 150 individuals were needed to extract less than 1 mg of purified adhesive.[63] Although intensive, it is a common approach to isolate a compound of interest after observing an organism in the environment. The difficulty of direct extraction for physico-chemical characterisation has led to the development of advanced proteomic and transcriptomic techniques for rapidly surveying and identifying promising adhesive proteins.[53] As a result, there are an ever-increasing number of sequences for these proteins available in public databases. Knowing the protein sequence is only part of the full picture and metal ions, pH and polysaccharide content in the secreted adhesive all contribute to adhesive performance and strength.

Many proteins, as a building component inherent in living systems, operate an adhesive function to bind cells within organic matrices. Cell proliferation, differentiation and response take place within a structural framework, the extracellular medium, in which the cells are bound.[56,64] Physically, this binding occurs through glycoproteins such as fibronectin, collagen, vitronectin and laminin. These initiate binding and cytoskeletal reorganisation on contact with a matrix promoting cell differentiation and survival. At larger length scales, proteins such as collagen are the predominant constituent of connective tissue in the musculoskeletal system. There are many different types of collagen, which named by roman numerals assigned to their structural function. The word collagen itself is derived from the Greek word for glue and this reflects the early historical use of rendered animal hide to produce adhesives. This is a practice that is still common today, particularly in the repair of antique woodwork, where isinglass, a fish bladder collagen extract, is commonly employed.[57] The five most common types of collagen in the human body are skin/tendon, cartilage, reticulate, basal lamina and cell surfaces, numbered types I to V, respectively. Type I, skin and tendon, makes up around 90% of the total collagen. The basic structure is a triplet of polyproline peptides that are covalently bound together to form a triple-stranded helical tecton. Through hydrogen bonding, these tectons self-assemble into larger fibrils termed α-helices and larger assemblies called β-pleats. Much research continues in the field of synthetic biology looking at the selection rules for self-assembly of these molecular tectons into larger structures.[65–69] By careful selection of a particular peptide sequence to form a trimeric base unit, it is possible to assemble cage-like assemblies around 100 nm in diameter that are morphologically reminiscent of naturally occurring clathrate structures. Clathrates can move across lipid

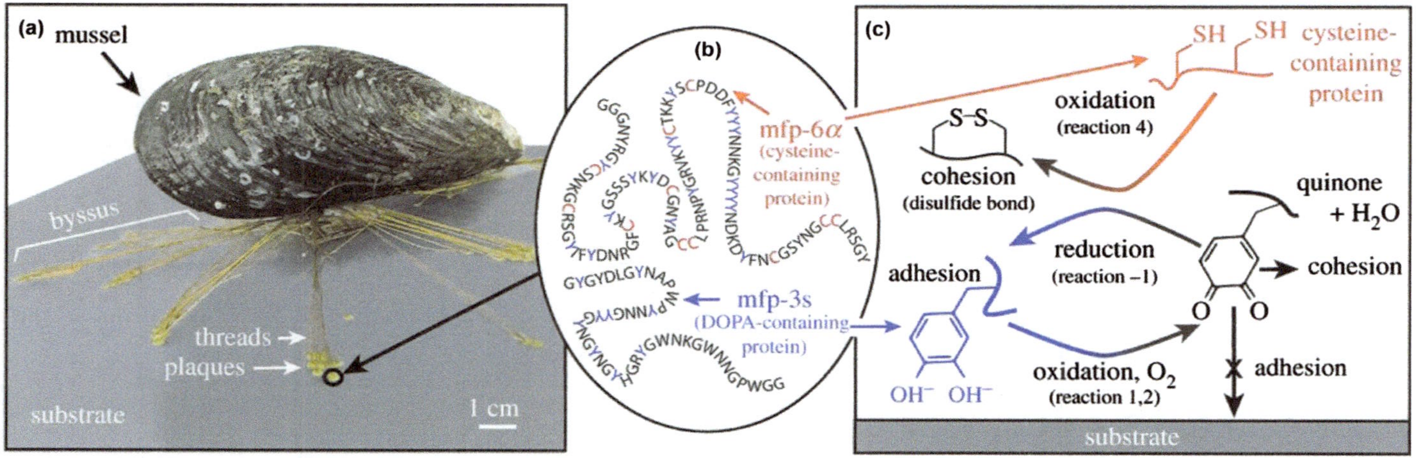

Figure 2.8 Adaptive aspects of *Mytilus californianus* adhesion in oxidising seawater conditions. (a) Thread–plaque pairs forming the adhesive byssus that attaches *M. californianus* to the substrate. (b) Schematic of representative DOPA-containing adhesive proteins (mfp-3s) and cysteine-containing antioxidant proteins (mfp-6a) in the plaques. (c) DOPA (blue) oxidising to dopaquinone (black) in a DOPA-containing protein (*e.g.* mfp-3s), reducing O_2 to H_2O in the process. Reduced DOPA adhering readily to a substrate where its oxidised form, dopaquinone, is primarily cohesive. Cysteine thiols (red) in an antioxidant protein (*e.g.* mfp-6a) reducing dopaquinone back to DOPA.
Reproduced from ref. 62 with permission from the authors and the Royal Society.

membrane boundaries and are similarly assembled from fibrous trimers of short amino acids. *De novo* design of biological structures is at the heart of synthetic biology and the formulation of protocells capable of new biological actions and functions extends the toolbox for biotechnological exploration beyond the already diverse metabolites, enzymes and proteins found in nature.[70]

To form into basic glues, collagen fibrils can be irreversibly denatured into random coil structures by hydrolytic breakdown in hot water (55–63 °C) to form gelatine. Most mammalian collagens denature between 40 and 41 °C, with fish-sourced collagens denaturing at temperatures between 15 and 29 °C for deep cold water- to shallow warm water-dwelling fish, respectively. Traditional glue formulations are made by extracting gelatine from animal hides or bones by prolonged soaking in water at the correct temperature. Manipulation of the pH by the addition of acids or bases can further disrupt H-bonding in the collagen and cleave larger chains. Typically, higher molecular weight gelatines will be formed from lower temperatures and less extreme pH values. The extracted gelatine is then isolated and dried to form a waxy substance that can be melted and applied to surfaces to form a bonding glue. A simple demonstration of proteinaceous glue manufacture can be made using milk.[71] Casein forms the majority of protein content in milk and in pH-neutral milk this largely hydrophobic molecule remains soluble as a complex with Ca^{2+} cations. By adding ethanoic acid, or rather vinegar, to milk, the Ca^{2+} coordinates to the carboxylic acid group of the vinegar to form calcium ethanoate. The casein precipitates from the milk to form curds, which can be filtered, neutralised, partially dried and used as a glue. The casein polymerises upon drying to form a plastic-like solid, which can be surprisingly strong. This is analogous to the manufacture of cheese where the enzyme rennin is used in place of vinegar to precipitate solids (Box 2.2).

The use of calf rennin, predominantly an aspartic acid protease called chymosin, in cheese manufacture is a good example of a biological process operating at the molecular level that is used on an industrial scale.[72] Chymosin is produced in the fourth stomach of ruminants such as cows and acts to curdle milk so that it remains in the cow stomach and is digested over a longer period. More recently, it has been possible to use recombinant bioengineering techniques to produce non-animal-derived chymosin, although it does change the flavour and texture of the cheese produced. The process of curdling is a result of the enzyme cleaving specific points in milk casein, which then allows the hydrophobic domains to cluster and precipitate. The coagulation is driven largely by hydrophobic interactions rather than by the direct formation of covalent bonds. Another enzyme used with increasing ubiquity, which does form direct cross-links between proteins, is transglutaminase.[73] Protein–glutamine γ- glutamyltransferase, also known as 'meat glue', catalyses the formation of an isopeptide bond between the group of γ-carboxamides of glutamine residues (donor) and the first-order ε-amine groups of different compounds (Figure 2.9). Transglutaminases are found in plants, invertebrates, mammals and microbes, and are involved

Box 2.2 How to make milk into glue

1. Heat 100 ml of semi-skimmed milk in a beaker but do not boil. Semi-skimmed milk tends to yield a superior quality of glue compared to full fat or skimmed, as there is a high level of protein but a lower quantity of fats. Fat molecules that are present when the casein dries can impair the formation of protein-to-protein bonding.
2. Add 20 mL of vinegar and stir until lumps form. Take off the heat, but keep stirring until no more lumps precipitate.
3. Let the solids settle to the bottom of the beaker and decant the top solution to leave the remaining solids. Squeeze out as much liquid as possible from the solids.
4. Add 15 mL of water and stir until the mixture is smooth.
5. Neutralise the solution using a sodium hydroxide solution and check the pH using pH paper or a pH meter. pH paper may be better, as a probe will need to be cleaned thoroughly. A quicker method is to add magnesium or calcium carbonate salts that will neutralise the vinegar. When the mixture stops evolving CO_2 bubbles then it will be neutral.
6. The mixture can now be used as a glue by applying it to surfaces and allowing it to dry. A word of caution– the glue will probably not keep well and should be used for demonstration purposes only!

metabolically in immune reactions, coagulation and photosynthesis. It can be harvested from pig and cow tissues, mostly blood, but this is an expensive process and industrial quantities are produced microbially. Depending on the source, the enzyme can differ in molecular weight, but most require calcium to be present as a cofactor in order to function. Transglutaminase was initially isolated from the *Streptoverticillium* sp. microbe but it has since been found that expression in *Streptoverticillium mobaraense* yields an enzyme that does not require calcium to link proteins. This is useful in the food industry and molecular gastronomy where the presence of salts may impair the final flavour. Macroscopically, the enzyme is applied to the surface of two meats, which are then placed together, exactly as if they were being glued to one another. In a matter of hours, isopeptide bonds form and bind the two meats together. More broadly, the enzyme is used to modify viscosity, elasticity and water binding in foods.

2.6.2 Synthetic Reactive Molecular Adhesives

In everyday experience, the difference between true adhesion and cohesion is illustrated by the range of reactive and non-reactive glues, respectively. Taking their cues from the physical behaviour of the natural adhesives discussed above, glues that form direct and strong chemical bonds to

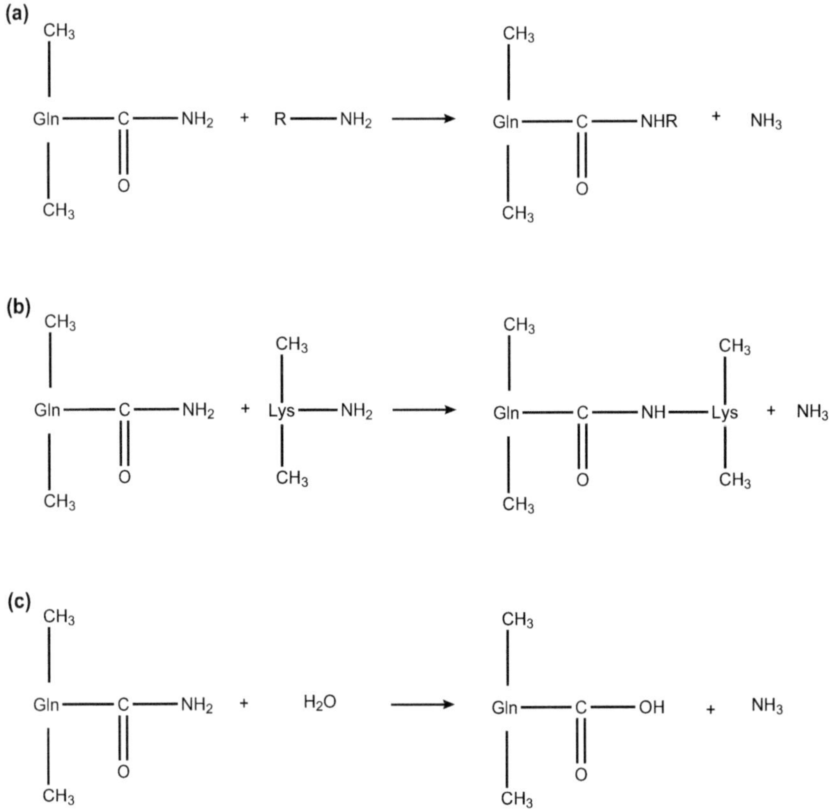

Figure 2.9 The reactions catalysed by transglutaminase include (a) acyl transfer reaction, (b) cross-linking reaction between Gln and Lys residues of proteins or peptides and (c) deamidation.
Adapted from ref. 73 with permission from Springer Nature, © 2018 Springer Nature Switzerland AG. Part of Springer Nature.

surfaces include epoxy resins, acrylics and cyanoacrylates (super glue), silicones and polyurethanes. Each of these reactions has particular advantages depending on the target use and these are given in the examples below.

2.6.2.1 *Epoxy Resins*

Epoxy resins are familiar as being a two-component glue that needs to be mixed together before application.[74,75] One component is a polyepoxide and the other is a cross-linking agent comprising an organic molecule with two or more functional groups, such as amines, alcohols or organic acids. A common example of this is the reaction of epichlorohydrin with bisphenol A to produce a diglycidylether in the presence of a basic catalyst. Once mixed together nucleophilic attack from the lone pair on the hydroxyl group will initiate opening of the epoxy ring to form a C–O–C bond (Figure 2.10).

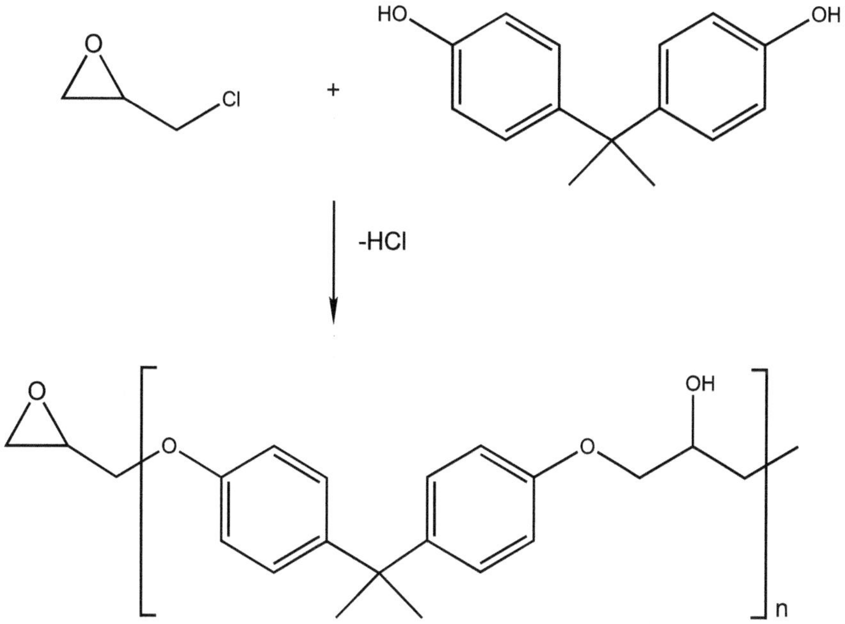

Figure 2.10 Schematic diagram of the epoxy ring-opening mechanism to form a C–O–C bond.

The epoxy group will undergo a similar reaction to bond covalently with a wide range of nucleophilic functional moieties, often those found in the surface the resin is being applied to, which is why resins of this type are excellent adhesives. This class of resin can bind to metals, glasses and wood, with a bond strength normally higher than other types of adhesive. The resulting properties of an epoxy resin can be controlled by careful selection of the functional group, the density of those groups per monomeric unit and the larger chemical structure of the resulting polymer. Resins can be produced with high thermal stability, mechanical strength and chemical resistance, and epoxies are used industrially in aerospace and automotive composite material, consumer electronics and anti-corrosion coatings in high volumes.

2.6.2.2 *Acrylic*

Acrylic adhesives comprise a large class of resins and plastics based upon the polymerisation of monomeric propenoates. The basic structure of the monomer unit includes an unsaturated carbon–carbon bond, which is terminated with a carboxylic acid or ester. In the presence of a free radical initiator, the diene bond is opened and forms a direct bond to the next monomeric unit, generating a further radical and propagating the living polymerisation mechanism. It is also possible to instigate polymerisation in acrylates through exposure of an acrylate solution to UV light. The speed of the light-induced polymerisation can be increased by the addition of photoinitiators in the adhesive formulation. Poly(methyl methacrylate) (PMMA) is a well-understood system and since its development in the 1930s has been used as a lightweight, transparent replacement for glass. PMMA, and many other polymers of this type, demonstrate excellent transparency, strength and scratch resistance and, importantly, can be moulded and extruded if the polymer is taken above its glass transition temperature (110–120 °C).[76,77] Despite concerns about the toxicity of methacrylic acid it remains a useful adhesive in biological and surgical applications.[78] Various formulations are routinely used as bone cements and for dental bonding procedures, as it is non-toxic once in the polymer form.[63] Methacrylic acid and its derivatives are particularly attractive for use in bonding to bone or dentine as their acidic character makes them self-etching to the surface. The application of an organic acid to a hydroxylapatite (bone) or fluoroapatite (dentine) surface will weakly dissolve and interpenetrate into the inorganic matrix. This makes for a strong bond once the monomer is polymerised *in situ* as there will be a concentration gradient moving from the pure inorganic material to the pure polymer phase. Bonds of this type between materials are not only strong but also help to integrate materials that have different mechanical moduli. If two differing materials are bonded directly face to face then this is likely to be a failure point of the bond under load. By affording a graduated change in material properties across the joined faces, stress loads are better distributed and resisted.

2.6.2.3 *Cyanoacrylate*

Cyanoacrylate, also known more commonly in various formulations as superglue, was originally developed as a clear resin for gun sights in 1947.[79] It was not used in this capacity and was initially discarded, as it proved too sticky for use, but it was rediscovered for bonding in aerospace applications in the late 1950s. Shortly after its introduction to the market in the early 1960s, it was found to be excellent at closing wounds and has been employed in surgery and field medicine since then, in varying formulations. Cyanoacrylates are formed by the condensation of a cyanoacetate with formaldehyde in the presence of a salt or organic base catalyst. Exposure of a cyanoacrylate to water will initiate a rapid polymerisation reaction as illustrated in Figure 2.11. The nitrile group on the monomer is polarised and will readily accept a nucleophilic attack from lone pair electrons in a weak base such as amine or water. Hydroxyl and amine groups are particularly rich on

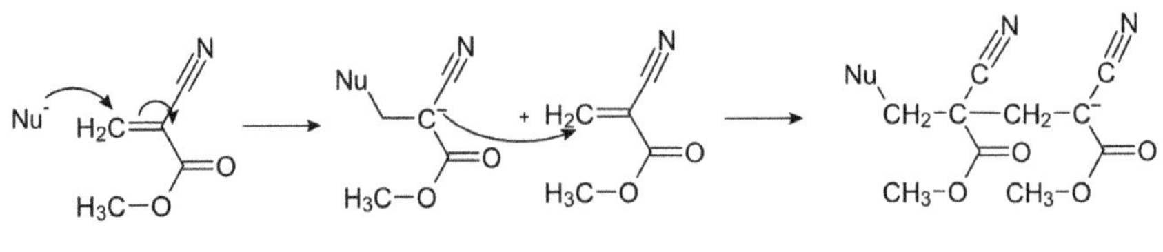

Figure 2.11 Schematic of the polymerisation reaction of cyanoacrylates with water.

proteinaceous surfaces, such as skin, so that direct covalent bonding occurs upon contact with the liquid monomer. The polymerisation reaction is further driven by moisture in the air and does not require a catalyst, pressure or heat. The reaction is exothermic and proceeds faster the shorter the alkyl chain in the monomer is, to form strong and dense but relatively brittle polymeric networks. Early formulations based on methyl-2- and ethyl-2-cyanoacrylates were phased out, as the heat of reaction damaged the tissue that they bonded to. It was also found that formaldehyde and cyanoacetate were formed as a product of hydrolytic breakdown of the polymer, which caused inflammatory responses at wound sites. Medical grade formulations were approved for use in Europe in the 1980s and are predominantly based on 2-octyl- or *n*-hexyl-2-cyanoacrylates. Poly(cyanoacrylates) formed from these larger monomers degrade more slowly biologically and show greatly reduced inflammatory responses. The larger alkyl chains or groups in these formulations result in a prolonged set time with a lower exothermic output to form more flexible polymer networks with a higher ultimate breaking strength compared to the shorter chain formulations.

2.6.2.4 Polyurethanes

Polyurethanes were discovered in the late 1930s by direct reaction of diisocyanates with polyester diols.[58] The isocyanate group is highly electrophilic and will react with a wide range of nucleophilic electron donors such as hydroxyl- or amine-containing molecules. For a polymer to form it is necessary to use a diisocyante and a monomer with two or more nucleophilic groups. The range of polymers that can be formed from this combination is extremely diverse and depends on the alkyl chain length, the degree of aromaticity or conjugation and the density of cross-linking. In general, polymers formed with a high degree of cross-linking or with shorter alkyl chain lengths will be stiffer. Conversely, longer chain lengths and lower cross-linking densities will produce polymers that are more flexible. As with the cyanoacrylate and epoxide examples, the reactive adhesive properties are a result of the readiness of the monomeric reagents to couple directly with nucleophilic moieties on a substrate surface. Many formulations are thermoplastic and lend themselves to extrusion moulding or blowing.[75] When in a melt form or with a low enough viscosity, polyurethanes can also be utilised as non-reactive pressure adhesives.

2.6.2.5 Synthetic Non-reactive Molecular Adhesives

Continuing the analogy of everyday adhesives, non-reactive adhesives are pressure-sensitive putties like the brand Blu Tack®, thermosetting polymers such as ethylene-vinyl acetate, polysiloxanes and emulsion-delivered polymers like the child-safe glue polyvinyl acetate.[59] Office favourite "Post-it® notes" stick because of a monolayer of elastomeric acrylate copolymer microspheres. The similarity between all these non-reactive adhesives is that they are highly viscoelastic during application of either pressure or heat, and on being forced into proximity with a surface. A bioinspired variant of non-reactive molecular adhesion has been exploited in the development of a "water glue" hydrogel. The solid phase of the hydrogel is composed of chitosan, alginate and hyaluronan, which flows over and adheres to a covalently bonded intermediate layer. Upon drying down, the polysaccharides form into a tough and strong interface layer.[80] A pressure of ~1–10 Pa applied for between 1 and 5 seconds is all it takes for a sample to reversibly adhere to a surface. The relative propensity for a pressure-sensitive adhesive to stick to a given substance is known as the "tackiness", or simply "tack". The polymer molecules are placed close enough to those of a surface to experience attractive intermolecular forces and they remain stuck until pulled free. The binding capacity of a non-reactive adhesive to a given surface depends on the affinity of the surface energy with the viscoelastic medium being forcibly "wetted" upon it. For this reason, it is impossible to get a putty like Blu Tack to stick to a low surface energy material such as Teflon, a polymeric fluorocarbon. This flow model is known as the rheological theory of pressure-sensitive adhesion. In this model a substance may be predicted to act as a pressure-sensitive adhesive if it exhibits an elastic modulus (G') between 0.01 and 0.1 MPa and has a glass transition temperature between −100 and 0 °C.[58] It is important to note that meeting these criteria does not necessarily mean a polymer will behave as an adhesive, more that all polymers demonstrating adhesive behaviour fall largely into this range. This is known as Dahlquist's criterion. At the molecular level, pressure-sensitive adhesives must strike a balance between the seemingly contradictory properties of flow under force and the mechanical retention of shape once that force is released. This is accomplished by having minimal cross-linking between polymer chains such that flow is permitted during stress. Inter-chain linkages can be covalent, like the sulfur–carbon bridges that form during the vulcanisation process in rubber. This can result in a stiffer material with poor flow properties. Most synthetically designed polymers take advantage of hydrogen bonding between polymer chains. These bonds can be broken under force-induced flow, only to reform in a different position once the stress is removed. This is exemplified in a number of copolymer blends crafted to demonstrate pressure-sensitive adhesion and a few examples are presented below.

2.6.2.6 PVP–PEG Blends

Neither poly(*N*-vinylpyrrolidone) (PVP) or poly(ethylene glycol) (PEG) exhibit adhesive properties in and of themselves. However, when blended together in specific ratios or particular molecular weights the resulting mixture demonstrates excellent adhesion.[81] PVP has a relatively large molecular weight of $\sim 1.5 \times 10^9$ g mol^{-1} and PEG a smaller one of ~400 g mol^{-1}. The terminal protons at the ends of the PEG oligomers coordinate through hydrogen bonding to the electronegative oxygen on the pyrrolidone group. In this way, the PEG bridges across PVP chains as a cross-linker and introduces a large intermolecular free volume, which facilitates fluidity. The presence of water within the free volume disrupts the hydrogen bonding between polymeric chains and results in what is termed in the literature as "detackification". This is useful as it provides a means for controllable deactivation of an adhesive. Stimuli-responsive adhesives are useful for engineering application-specific functionality. One area in which this is desirable is in medicine, where triggering a bandage or wound dressing to come off without having to pull it free and cause distress to a patient is highly desirable. A thermoswitchable variant was developed by replacing the PVP with poly(*N*-vinylcaprolactam) (PVCL) to form a biocompatible blend. A key feature of this was a significant increase in the density of the PEG cross-linking, which gave rise to improved adhesion. The replacement of PVP with PVCL also significantly changed the thermal behaviour of the polymer blend. The lower critical solution temperature (LCST) is the temperature below which a polymeric component in a blend is miscible in water. PVP has a calculated LCST of 170 °C and so no phase separation is observed in these blends around mild variations of room (~20–25 °C) to physiological temperatures (37 °C). That is, over this range no thermoresponsive change in adhesion is observed for a PVP–PEG blend. The LCST for PVCL is around 32–37 °C and above this temperature a water–PVCL solution will begin to cloud. In a PVCL–PEG blend, the gel becomes thermoreversibly adhesive over the range 40–60 °C and this means that the blend can be triggered to de-bond by the use of water at the correct temperature. Above the LCST in the blend the hydrogen bonding in the PVCL is disrupted and there is phase separation resulting in de-bonding of the copolymer blend. Upon drying and lowering of the temperature, the hydrogen bonds are re-established with the PEG and the blend is adhesive once again (Figure 2.12).

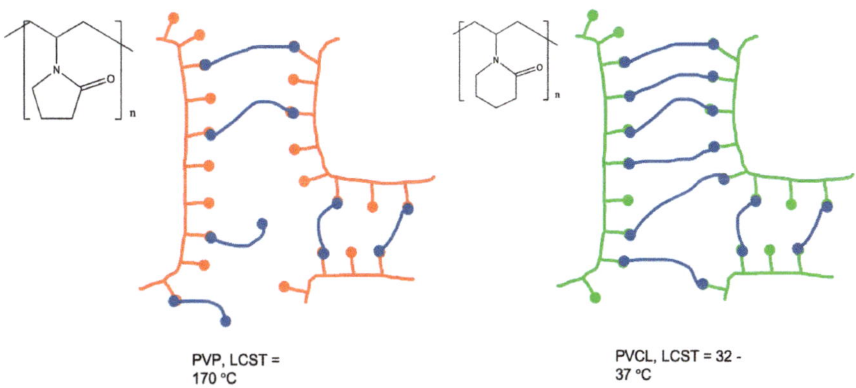

Figure 2.12 Chemical structures of poly(*N*-vinylpyrrolidone), poly(*N*-vinylcaprolactam) and schematic presentation of the corresponding networks formed with poly(ethylene glycol) oligomer, PEG-400.
Reproduced from ref. 81 with permission from American Chemical Society, Copyright 2014.

2.6.3 Silicone or Polysiloxane

Poly(siloxanes), also known as silicones, are long-chain polymers where silicon–oxygen bonds are the main backbone, along with further organic side groups. Silicone is used in various forms as a surgical biomaterial as it is bio-inert and, in most formulations, is non-toxic and does not trigger an immune response in the body.[82,83] A further benefit for using silicones is the high degree to which they retain these properties across a range of mechanically different formulations. Silicones can be prepared as viscous liquids through to solid plastic or elastomeric rubber-like formulations and still be used for human implantation. Siloxanes used for implants do not lend themselves to blending with other polymers. This can be accomplished by mixing but the guest polymers of molecules within the siloxane tend to leech out under physiological conditions. For this reason, most chemical modifications for biocompatibility or mechanical properties are performed at the synthesis stage. Siloxanes are formed by the hydrolysis of a chlorosilane precursor to form a silicic acid oligomer. The hydroxyl groups on the oligomers will then undergo a condensation reaction to produce the siloxane and hydrochloric acid as a by-product. The production of polydimethylsiloxane (PDMS), which is used extensively in cosmetics, food additives, lubricants and soft lithographic techniques, is shown in eqn (2.13). Soft lithographic techniques are discussed later in this section.[84] Siloxanes used in everyday pressure-sensitive adhesives, such as building caulk, use an acetic acid-based silane precursor. On setting, the elastomer generates relatively benign acetic acid rather than highly acidic HCl gas.

$$n\mathrm{Me_2SiCl_2} + n\mathrm{H_2O} \rightarrow (n\mathrm{Me_2SiO})_\mathrm{n} + 2n\mathrm{HCl} \tag{2.13}$$

Dimethyl silicone oil, or derivatives of it, form the basis for a wide range of elastomers. The dimethyl formulation is the simplest and substitution of the methyl moieties by more complex functional groups in the silicon chloride precursor will tailor the mechanical properties (Box 2.3).

The mechanical properties of a siloxane elastomer depend greatly on the number of oxygens coordinated to the central silicon within the oligomeric unit. Silicic acid, $Si(OH)_4$, has four hydroxyl groups, which can each undergo a condensation reaction to form Si–O–Si bonds with other silicic acid oligomers. If an aqueous solution of silicic acid is allowed to dry then the oligomers are driven to undergo a polymerisation reaction to form glass, which is a long-range SiO_2 network comprising four silicon–oxygen bonds per repeating unit. Glasses are characterised by being hard and brittle. However, by substitution of one or more of the OH groups on the oligomeric Si with a Si–C bonded functional organic group, the glass that forms takes on a greater degree of plasticity as the organic character increases. The repeating unit for PDMS has two oxygens and two methyl groups and so, instead of a glass network, the Si–O units form into rotationally flexible backbone chains. Further control of the mechanical properties is imparted by using different side organic groups that can be cross-linked or

Box 2.3 How to make dimethylsilicone

1. You will need to do this synthesis in a fume cupboard, as chlorine gas will be evolved during the hydrolysis step. Have an ice bath on hand ready to cool the reaction, as it will be exothermic.
2. Place the ice bath on top of a magnetic stirrer and set up a reflux condenser with a 250 ml three-necked round-bottomed flask. Put a magnetic stirrer bar in the round-bottomed flask.
3. Fit the condenser in the middle neck and place rubber septum caps in the other two. Ensure water goes in at the bottom of the condenser and out of the top and that the rubber tubing is firmly attached. Make sure everything is clamped properly and that clips are fitted onto the glassware to hold it in place. The reflux condenser will be open at the top to the fume cupboard.
4. Flush the round-bottomed flask with nitrogen to remove moisture through one of the necks in the round-bottomed flask. This can be done with a gas line fitted to a needle pushed through the rubber septum. The organosilicon chloride precursor is water sensitive and you will want to control the rate of hydrolysis.
5. Using a needle and syringe in the fume cupboard, transfer 20 mL of dichlorodimethylsilane into the round-bottomed flask through one of the rubber septa.
6. Again using a needle and syringe, transfer 40 mL of diethyl ether to the round-bottomed flask under magnetic stirring.
7. Through a needle in the rubber septum, slowly begin to add water dropwise, using a syringe. As the hydrolysis begins, HCl will be generated. The reaction is exothermic and will be vigorous if the water is added too quickly.
8. After the addition of around 10 mL, less HCl will be visibly produced and the rate of dropwise addition can be increased. After all the water has been added, stir the solution for a further 5 minutes.
9. Turn off the magnetic stirrer and reflux condenser. Remove the condenser from the flask and remove the stir bar. Decant the contents of the flask into a 250 mL separating funnel.
10. The product will be in the ether layer, which will be floating on top, with the water layer on the bottom.
11. Separate the aqueous layer and discard (after making sure you have the right layer – water is on the bottom!).
12. With the product-containing ether layer still in the funnel, add 100 mL of a 1 M aqueous sodium bicarbonate solution and mix. This will remove any excess acid in the product. Take care to let any CO_2 evolved in the rinsing step out of the separating funnel. Repeat this washing step two more times.
13. Add a final wash of pure water and separate.

14. Transfer the product-containing ether layer to a 250 mL Erlenmeyer flask containing anhydrous magnesium sulfate. This will remove any remaining water from the product. Make sure you put a stopper in the flask to stop the ether evaporating and add more diethyl ether if needed.
15. Once the solution is dry, decant the product-containing ether through a filter paper and into a 250 mL beaker or round-bottomed flask. In the beaker the ether can be evaporated using a water bath to leave your isolated oily transparent dimethylsilicone oil product. In a round-bottomed flask, you can use a rotary evaporator to isolate the product.

functionally active in some way. A wonderful example of mechanical control within a siloxane, and of the importance of hydrogen bonding, can be demonstrated by making "Silly Putty". Silly putty was discovered concurrently during the Second World War by James Wright at General Electric and Earl Warrick at Dow Corning. At the time, finding a substitute for natural rubber, which needed to be farmed or imported prior to vulcanisation, was extremely important. Although the viscous formulation was not suitable as a replacement for rubber, it captured the imagination of children and adults alike. Silly putty is formed trivially by mixing dimethylsilicone oil with boric acid and picks up its name from its non-Newtonian fluid behaviour. It flows like a viscous liquid if left but can bounce if thrown on the floor. These properties are a result of hydrogen bonding between the silicon-bound oxygen and the relatively electron-deficient (compared to the silicon) boron. The transient nature of hydrogen bonding affords fluidity to the borosilicate solution as the molecules flow slowly around one another. Hydrogen bonds are constantly broken and reformed during flow and this gives rise to the apparent viscosity of silly putty. If force is applied faster than the natural flow rate then the hydrogen bonding maintains the rigidity of the putty and it behaves as an elastomer. Commercial formulations are made with fillers such as iron oxide, which act as thickening agents and improve the consistency from slimy to rubbery. One of the "play" components of silly putty is using it to remove the ink from newspaper. This is a good example of the borosilicate formulation operating as a pressure-sensitive adhesive. If the experimentalist is quick and has firm enough putty then the removed ink can be restamped onto another surface. Although a simple demonstration, this is the same operating principle for how elastomeric stamps are utilised for the manufacture and patterning of self-assembled monolayers.

For the purposes of the creation of bioinspired surfaces *via* direct replication, the class of siloxane materials offers flexibility, both physically in in terms of material elasticity and chemically in terms of ease of modification. It is significantly more cost-effective to use a soft lithographic approach where possible in comparison to e-beam or photolithographic methods.

Soft lithography is a top-down technique pioneered by the Whitesides group.[85,86] A soft material, normally a silicone elastomer formed from a polyvinyl siloxane base and a cross-linking agent, is used to produce a negative copy of the master structure. This is accomplished by pouring a fluid elastomer pre-polymer onto the master surface and allowing it to set. After curing, the flexible but solidified elastomer is peeled from the original surface and can be used repeatedly as a mould for replication. The practice is ubiquitous to the point that replication kits are readily available and predominantly Sylgard 184 is the formulation used. This particular product is popular as the elastomer produced, PDMS, is optically transparent down to 300 nm. This makes it suitable as a component in lab-on-a-chip devices where microscopes often need to observe what is happening on devices. Other formulations are available that offer increased toughness or resistance to solvents at the cost of opacity. Sylgard 184 and similar kits are sold as a three-component system comprising a silicon oil base (predominantly dimethylvinyl-terminated dimethylsiloxane), a cross-linker (dimethylhydrogen siloxane) and a metal-based (normally a sparing amount of a platinum complex) catalyst.[87] The latter is often provided premixed into the cross-linker and to form the elastomer the base and curing agent are mixed in a 10 : 1 ratio, respectively. Thorough mixing is required to promote homogeneity as the elastomer sets. This can lead to air bubbles and when casting a mould, it is important to degas the elastomer pre-mix by exposing it to a vacuum.

Soft lithography is well suited to the repeatable replication of large (centimetre or more) area patterns and surfaces, with resolution to around 5 nm being possible. This method is advantageous over a technique such as photolithography in that it can accommodate curved, deformed and unusual surfaces, although features having aspect ratios above 2 may be distorted depending on the stiffness of the elastomer used. The procedure is also mechanically gentle meaning that malleable structures, such as leaf papillae or insect wing hairs, will tolerate the copying process with little damage, thus making soft lithography well suited to the replication of biological features. Because siloxanes are largely chemically inert, a wide variety of materials can be placed within the mould to form positive replicas of the original structure. Predominantly, this may be another homogeneous polymer that is either a pre-polymer mix poured into the mould or a thermosetting melt. Commonly, polyurethanes are used for forming positives due to their durability and toughness, although this should by no means rule out using anything else that can be poured into the mould. Polar solvents such as water, glycol, ethylene glycol, dimethylsulfoxide (DMSO) and nitromethane will not cause swelling or distortion in siloxane elastomers. An increase in organic character results in increased swelling, and alcohols and acetone induce up to a 5% volume increase. Benzene and chloroform result in a volume increase of up to 30%, which becomes greater if the elastomer is exposed to heptane, pentane or tetrahydrofuran (THF) (Box 2.4).

Siloxanes tend to be strongly hydrophobic post-synthesis and in order to make them biocompatible the surface must be altered to accept the

Box 2.4 Making a basic PDMS replica

The method described here for making an elastomeric stamp is based on using a Sylgard 184 kit. This method may not be suitable for the replication of delicate biological surfaces. For example, the use of exposure to vacuum and heat in this process can distort and melt surface waxes. An adaptation more suitable for delicate replicas will be provided in Section 2.7.

1. Get a disposable cup and disposable spoon or knife. Silicone oils are highly viscous and cannot be easily washed off anything.
2. Weigh 10 g of silicone elastomer base into the disposable cup. It is a good idea to put some paper towels on top of the scales you are using in case any of the oil spills or drips.
3. Weigh in 1 g of the cross-linking agent provided with the kit.
4. Mix thoroughly using the disposable knife for a couple of minutes.
5. Place the sample to be replicated in a polystyrene or glass petri dish with the surface you want to copy facing up. Where possible fix the sample in place using glue – the degassing step can dislodge items when the mixture is exposed to vacuum.
6. Pour or spoon the siloxane pre-polymer mixture over the top of the sample and fill the petri dish to about three-quarters of its total depth.
7. Place the sample in a bell jar lined with tissue paper and connect to a vacuum pump. The bell jar should have a tap valve fitted so that you can control the pressure.
8. Turn on the pump and evacuate the bell jar. Keep an eye on the siloxane pre-polymer as it begins to degas. It will begin to bubble and possibly foam over the sides. If this appears to be too vigorous, then open the tap valve and the foaming will slow. Eventually, as trapped air leaves the sample, the bubbling will stop. The degassing step is very important where high resolution is required, as air trapped during the curing process will impair the quality of the cast.
9. Turn off the pump and return the bell jar to normal pressure. Remove the petri dish containing the mould.
10. Place in an oven at 70 °C for 2 hours during which time the PDMS will cure.
11. Remove from the oven and allow to cool. There will be ~1% shrinkage in the solid elastomer compared to the liquid form. Remove the elastomer from the petri dish. You may need to use a scalpel to cut the mould section out.
12. Peel the sample from the elastomer carefully – it can be difficult to remove softer samples if they tear and are left behind in the mould

but, generally, the elastomers are very resistant to mechanical manipulation or gentle scraping.

13. You now have a negative mould in which you can cast various positive copies. A material like dental wax can be melted and poured in. Once cooled the wax is removed from the mould although this will be delicate and will not work well for features with large aspect ratios. Positives that are more robust can be made by pouring an epoxy resin into the mould and allowing it to set.

Having replicas of biological surfaces made from known materials makes it easier to compare how behaviour such as contact angle is influenced by morphology, topology or chemical effects.

adherence of biomolecules. There are both "wet" chemical methods and "dry" plasma- or laser-based methods for the introduction of biocompatible groups. Treatment of a polysiloxane surface with hot or cold O_2 gas plasma generates surface oxides in the form of CO or SiO groups. On exposure to water these oxides will form pendant carboxylic acid or hydroxyl groups, which greatly improve contact angle and wettability.[83] The presence of polar groups on the underlying hydrophobic substrate allows the surface to accommodate hydrogen-bonding and electrostatic interactions. This is very important for contact lenses, which need to allow liquid shear over their surface and for the transport of oxygen to the eye. Plasma-activated PDMS-based elastomer surfaces are also highly adhesive to themselves and to substrates with similar surface energies, such as glass. This property is useful in the manufacture of PDMS-based microfluidic devices, as the components can be tightly bonded together. Functionalities that are more exotic can be introduced by using different gas plasmas such as CF_4/O_2 mixtures, and oligomeric surface groups can be bonded directly through plasma-initiated polymerisation. Wet chemical methods can be used for surface oxidation in a similar manner to O_2 plasma exposure. It is also possible to introduce covalently bonded macromolecules through free radical-initiated grafting to the surface. Many silicon elastomer-based microfluidic devices that feature biochemistry or cells require some sort of internal chemical modification either to prevent fouling or to present the correct biochemical environment.[88,89]

2.7 Soft Lithography of Soft Surfaces

Soft lithography is useful for replicating biological structures, although a modification of PDMS stamping is required to capture three-dimensional and "hairy" morphologies seen in nature.[23,90–92] Early replication attempts using elastomeric stamps were not as good for capturing the hierarchical structures present, often in plant surfaces with wax-coated cell papillae. Crystals of wax self-assemble upon leaf surfaces in a range of morphologies, including tubules, rods and platelets. The morphology of the crystal formed is largely dependent on the chemical composition. A good example is the nucleation of octacosan-1-ol, the main component of waxes found on wheat, to form platelets on highly ordered pyrolytic graphite layers identical to those found in the native plant. Vacuum exposure in the degassing step causes plant cells to dehydrate and deform. Curing at elevated temperatures also distorts the soft waxes at the surface. Various methodologies have since been explored to counter these issues with successful replication of leaf and other wax structures by templating and gentle rinsing of the residual plant waxes from the mould using chloroform. Many of the degassing and temperature issues are solved by using a polyvinylsiloxane formulation that is fluid enough to wet the master sample completely and which sets rapidly at room temperature (Figure 2.13). The method presented in Box 2.5 allows copying of biological samples with overhangs and high aspect ratio bifurcated structures, such as hairs.

2.8 Biomolecular Surfaces

Biomolecules may be loosely categorised into four main groups, namely, nucleic acids, proteins, lipids and polysaccharides, and all of these are influenced by or demonstrate hydrogen-bonding or polar interactions of some sort. Further to these interactions with water, the presence of hydrocarbons and other p-block elements introduces further competing molecular forces and a greater order of possible complexity. The aversion of hydrocarbons for water is known as the hydrophobic effect and in concert with polar or charged hydrophilic interactions comprises the major driving force for all biological structure at the molecular level. A simple example in nature is the spontaneous self-assembly of micelles, liposomes and bilayers which form when phospholipids are mixed with water.[36,94,95] Phospholipids are amphiphilic molecules possessing an ionic head group and one or more long-chain aliphatic hydrocarbon tail groups. The head groups are solvated into the water while the tail groups are repelled by it and cluster together. Depending on the concentration of the phospholipid, it may form into a range of structures driven by the competing strength of its hydrophobic or hydrophilic interactions. Curvature in these assemblies is also influenced by the length and number of tail groups associated with each head group. The lipid bilayer and micelle structures that form in nature comprise the walls and boundaries of cells and cellular compartments and are usually around 5 nm thick. Naturally occurring phospholipids range between 14 and 24 carbon atoms in length with a mean of 16. The thinnest membrane layer that will form is 18.3 nm, as the smaller chain lengths (around 10 carbons long) required typically form surfaces with a high curvature and so will form micelles.[96] The manifest self-organisation of compartments capable of regulating complex chemical and metabolic processes within themselves is known as autopoiesis. The mechanistic adaptation of metabolic processes for continued cellular function takes place *via* the screening and transition

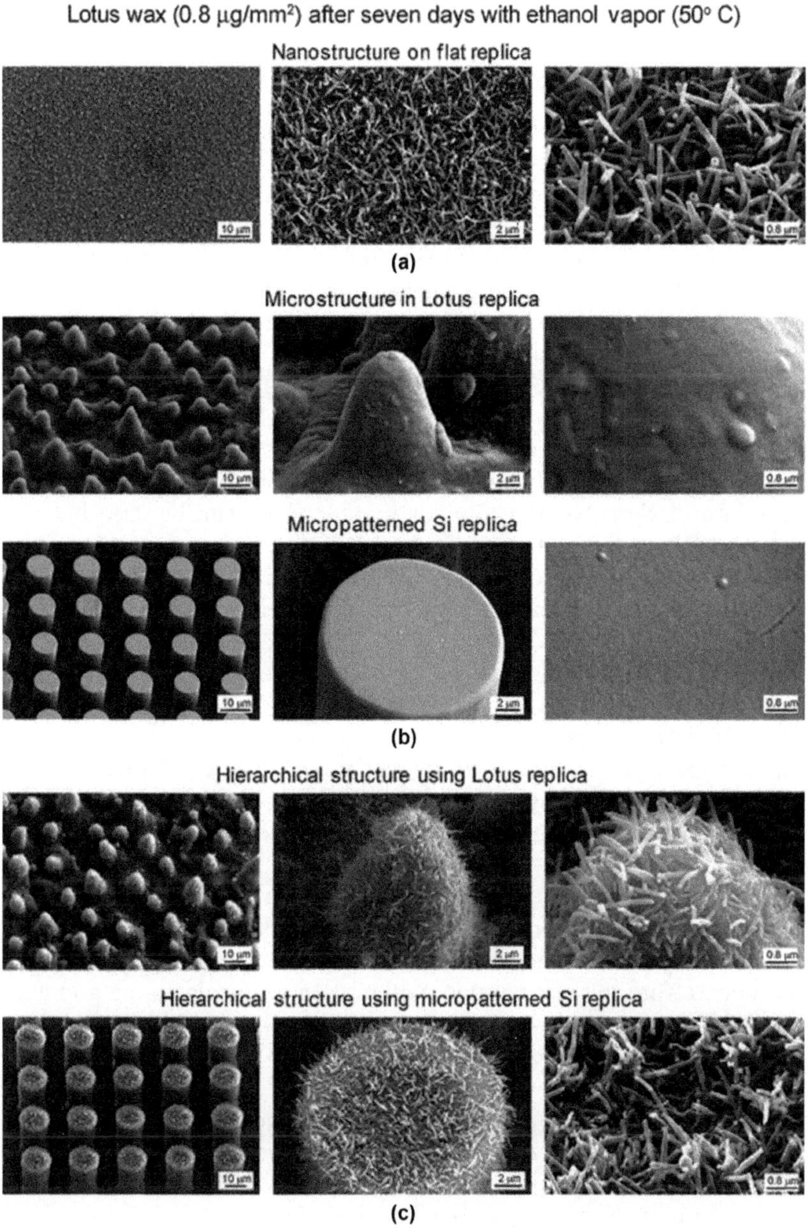

Figure 2.13 SEM micrographs taken at 45° tilt angle (shown using three magnifications) of (a) nanostructure on flat replica, (b) microstructures in lotus replica and micropatterned Si replica and (c) hierarchical structure using lotus and micropatterned Si replicas. Nano and hierarchical structures were fabricated with mass of 0.8 mg mm^{-2} of lotus wax after storage for seven days at 50 °C with ethanol vapour.
Adapted from ref. 23 with permission from The Royal Society of Chemistry.

Box 2.5 Making a high-fidelity replica of high aspect ratio biological features.

This method is similar to the PDMS elastomer technique in Box 2.4 except a softer polyvinylsiloxane (PVS) base is used for the negative mould. This is a more suitable technique for leaf and "hairy" surface replication. It is important to note that the original biological masters can only be used once as the casting stage and rinsing means that surface waxes will be stripped from the original.[93] Samples with very high aspect ratio features, such as hairs, may also experience damage during the peeling stage, although often-folded cuticles and hair papilla cells can be templated multiple times as long as they do not dry out. You will need a PRESIDENT light body® ISO 4823 kit from Coltène/Whaledent. This is a soft dental impression kit and will come in a self-mixing dispenser. You will also need an Epoxydharz®, Nr. 236349 epoxy resin kit from Conrad Electronic. Other formulations, such as Spurr's methacrylate, can be used but it is important that the negative mould can be cured at room temperature and that both the siloxane and epoxy are of low viscosity. This is to ensure that air bubbles are not trapped in either the negative master or the positive replica.

1. Place the leaf to be replicated into a petri dish with the side to be copied facing up.
2. Use the PRESIDENT kit to dispense the PVS pre-polymer onto the surface of the biological master.
3. Press the pre-polymer gently into the surface using a glass slide to apply pressure and to spread it out.
4. The elastomer will set in around 2 minutes at room temperature and should be left for at least 10 minutes to ensure the mould sets.
5. Carefully peel the negative mould from the sample.
6. Waxes from the plant surface will remain in the elastomer and need to be removed. In a fume cupboard, wash the mould in chloroform for 3–5 minutes and allow to dry for 10–20 minutes.
7. Mix the epoxy resin with the correct ratio of hardener as outlined in the kit instructions. Pour this into the mould. For plants that are hairy or have high aspect features, it is recommended that you expose the cast epoxy pre-mix in the mould to 1 mbar pressure for 1 minute to degas any trapped air bubbles.
8. Allow the epoxy to cure for 24 hours then peel the cast positive replica from the elastomer mould.

The negative moulds are rather tough and will reliably produce high-fidelity copies for at least ten casting cycles.

of molecules and material at the cell boundary.[70] A lipid bilayer is a dynamic surface and, although it maintains separation of aqueous components across the interior and exterior of its thickness, it is possible for the molecules within to move about the surface. This is possible as there is no direct chemical bond between the constituent molecules and the lipid bilayers behave as a fluid-like nanocomposite. They are held together by the aggregate force of the clustered amphiphilic polar and non-polar attractions or repulsions. This property affords a lipid bilayer the capacity to bend, flex, deform and accommodate embedded proteins across its thickness. Proteins are the machinery through which chemical communication between bilayer interior and exterior phases takes place.[97]

The folding of a protein that can fit into a lipid membrane is dependent on the harmonious interaction of polar and non-polar forces with themselves and with the local environment.[95] A protein is composed of a chain of amino acids biochemically synthesised in a particular sequence called a primary structure. The chemical motif of an amino acid is an amine group directly bonded to a carboxylic acid, and the relative polarity within the amino acid can be controlled through pH. Viewed alone, a carboxylic acid may dissociate its proton to give a weakly delocalised electronegative group and an amine can accept a proton to become positively charged. Alternatively, a proton-dissociated anionic carboxylic acid moiety can accept a potassium or sodium cation to form a salt. In order to synthesise a protein, two or more amino acids must combine to form the sequence of the primary structure. When the amine group on an amino acid is in close proximity to the carboxylic acid group of an adjacent amino acid a condensation reaction occurs to form a peptide bond (Figure 2.14).[98] That is to say, the carbon of the carboxylic acid accepts a nucleophilic attack from the lone pair electrons on the amine group. This forms a direct covalent bond between the two amino acids, which subsequently eliminate a hydroxyl group and hydrogen to form a water molecule and the peptide residue.

There are 22 different amino acids that are incorporated into proteins by natural synthesis or by naturally occurring biological mechanisms.[99] These are known as the proteinogenic amino acids and each of them features a different functional group bound to the amine, which significantly influences the pK_a values of the amino acid head group. So far, ionic and polar contributions have been discussed, but the presence of hydrocarbons introduces a number of non-polar interactions. The organic functional groups terminating the individual amino acids in the peptide have distinctly different pH-dependent, polar and non-polar interactions. The various repulsive and attractive forces arising within long chains of amino acids in aqueous conditions will then fold this sequence upon itself to give what is called a secondary structure, often a simple coil such as an α-helix or β-pleat, although many more architectures exist.[65,66] At the larger scale these proteins can be broadly categorised as globular, providing biomechanical functionality or processing of some sort, or fibrillar, providing structure and

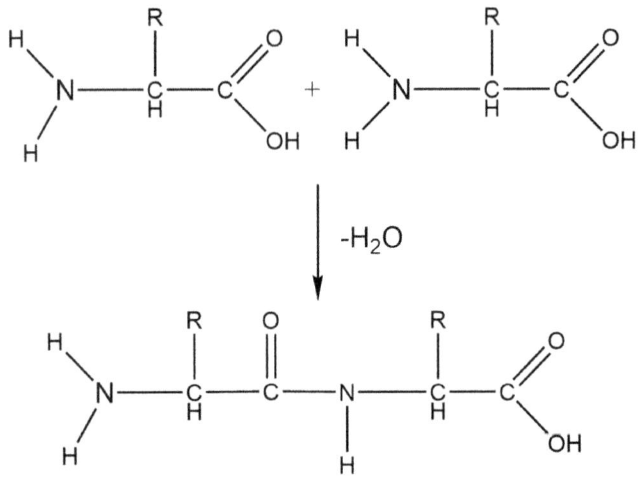

Figure 2.14 Schematic of the reaction of two amino acids to form a peptide bond.

operating in some physically mechanical way. Globular proteins are on average around 100–200 residues in length and comprise roughly half-and-half hydrophilic and hydrophobic domains. The optimal domain size for proteins of this type is ~4.5 nm and this is of the same order as the lipid bilayer membranes. Evolutionary pressure has led to the nanoscale being the optimal domain for biological function.[100] Most transmembrane proteins will be strongly hydrophobic on their exterior in order to situate themselves within the lipid bilayer. The interior may then be rich with groups that allow for the recognition or passage of small molecules. An area that is dense with transmembrane proteins will demonstrate ordering, and an example of this is bacteriorhodopsin, an integral membrane protein found in the characteristically purple bacterium *Halobacterium salinarum.* Bacteriorhodopsin forms closely packed trimeric arrays within lipid patches covering ~50% of the bacterial surface and the light-sensitive *cis*-retinal present within the protein is what gives the membrane its characteristic purple colour.[89] In nature this protein acts as proton pump and, under illumination, an H^+ will be pumped from the cytoplasmic interior of the bacterium to the extracellular side. This allows the bacteria to continue metabolic functions driven by sunlight in anaerobic conditions (Figure 2.15).

2.8.1 Proteins at Surfaces

Proteins are "sticky" and will form a coating upon most surfaces upon immersion or on exposure to a biofluid containing them. They are large macromolecules that are strongly amphiphilic in character. For a single, large protein there are a number of possible spatial arrangements and combinations of steric, polar and non-polar regions that may be orientated relative to a given surface. When larger numbers of different proteins are

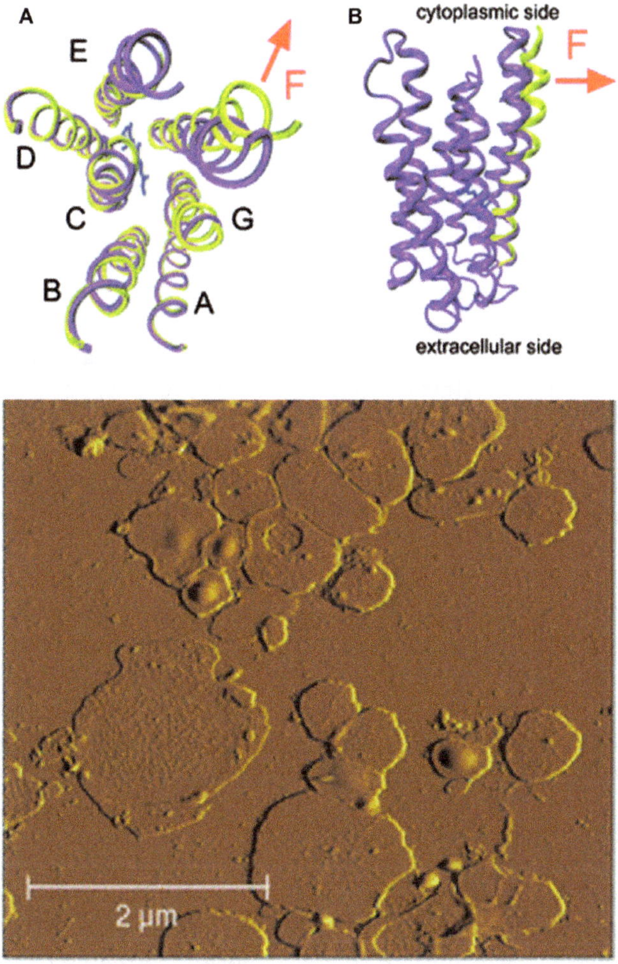

Figure 2.15 A schematic showing the structure of the bacteriorhodopsin protein assembled from peptide chain coils to form a channel through which protons are pumped under illumination. The outward tilt of helix F in the direction of the red arrow is linked to the opening of the cytoplasmic half-channel which alters the initial bacteriorhodopsin morphology (purple conformation) to become wedge-shaped (yellow conformation). (A) View from the cytoplasmic side, looking down the proton path through the membrane. The retinylidene residue is shown in blue. Only the transmembrane helical domains are shown without the interlinking amino acid loops. (B) Side view looking from within the plane of the purple membrane highlighting the large-scale conformational change resulting in the wedge shape of bacteriorhodopsin in the M2-state. (Bottom) A tapping-mode amplitude AFM image of BR-D85T membranes at neutral pH (7.0) on mica.
Adapted from ref. 101 with permission from The Royal Society of Chemistry.

also considered then the likelihood of a particular combination having a strong affinity for a surface becomes a near certainty. In contrast to analogous interactions for basic surfactant molecules, a protein can undergo conformational transformations once adsorbed to a surface, which can radically alter the desorption behaviour. Often, this means the protein is irreversibly bound and to study the kinetics of asymmetric adsorption behaviour requires experimental approaches beyond waiting for a protein/surface system to equilibrate.[102] Adsorption to a surface is the first step for many biological processes, including the blood coagulation cascade and many types of transmembrane signalling. "Sticky" transmembrane proteins also serve as anchors for cell-to-cell interactions; one example of this is the adhesion receptor protein integrin. At the microscopic scale, cells grow and function within a network composed of fibrous proteins, such as collagens, carbohydrates and polysaccharides, proteins and mixtures of proteins and carbohydrates known as glycoproteins. This network is generated by the cells growing within it. This is commonly referred to as the extracellular matrix, which organises and guides cell proliferation and mediates intracellular signalling. Integrin is a heterodimer formed from α-helix and β-pleat units coordinated together rather than covalently bonded, which sit across a phospholipid membrane.[64] The extracellular-facing components of the integrin are activated by contact with proteins on the surface of an adjacent cell or surface, usually within an extracellular matrix. The binding of a protein causes integrins to cluster on the surface, increasing the affinity of the area for further binding. This response triggers the cell to form another protein, actin, which forms into stress-bearing fibrils comprising what are termed focal adhesions – anchor points of contact between the two cells. What works within the extracellular matrix will also work for any surface coated with proteins.

That surfaces can be so readily adhered to by proteins is at once highly useful or highly problematic in the field of biomedicine, depending on the application. Simple immersion of a material for implantation into a biological medium will initiate the adherence of proteins to the surface. This is vital for something like a tissue scaffold, where a tightly bound proteinaceous interface is necessary for revascularisation. Conversely, the prolonged build-up and aggregation of protein fouling on implants exposed to the blood stream could lead to thrombosis.[6,103] One everyday example of protein fouling is in contact lenses, which must be replaced routinely. Lenses very quickly accumulate a biofouling layer that cannot be removed by everyday lens care. Making a surface resistant to protein adsorption represents a major challenge as the diversity of protein composition and morphology means that at least a small percentage will have a strongly matching affinity for adherence to the surface. A number of strategies exist to treat surfaces so that they reject the adherence of proteins. This includes the formation or self-assembled monolayers and polymer grafting. PEG derivatives of various molecular weights are commonly used for this purpose.[104] PEG has long been established as a protein-inert macromolecule and is employed as a stabilising agent in a number of pharmaceutical formulations.[105] PEG-coated micelles and nanoparticles have a greatly increased physiological retention time; remaining in the vascular system rather than adsorbing proteins and triggering the bodies clearance systems to sequester them in the kidneys for excretion in urine.[106,107] In order to understand how anti-biofouling surfaces work it is important to take a look at the processes taking place in proteinaceous build-up on a surface.

In addition to the nature of the surface itself, the external factors that influence adsorption are the temperature, buffer composition, ionic strength and pH. A protein in an aqueous environment will have water, cations and anions coordinated to it. By adhering to a surface there will be a gain in entropy for the protein as the water molecules and ions are displaced. At higher temperatures, this becomes more favourable and proteins will diffuse randomly to surfaces in higher numbers, increasing the rate of deposition. The functional groups comprising the peptide chains will have a positive, negative or neutral charge, as determined by their respective pK_a values and the pH of the solution. The sum of these charges will be the overall electrostatic charge of the protein although the charge presented to the surface will be dependent on the orientation of the biomolecule. For a given protein there will be a pH value known as the isoelectric point, pI, where the number of positive (cationic) functional groups is balanced by an equal number of negative (anionic) ones. At low pH, that is, a pH value below 7 and tending towards pH = 1, the solution will be highly acidic and proteins will generally have a net positive charge. At high pH, above pH = 7 and tending towards pH = 14, proteins will have a net negative charge. Depending on the protein, it is therefore possible to instigate deposition on a charged substrate using either pH or ionic strength.

The ionic strength of a solution is defined as the total molar concentration of all ionic species present in a solution. The distance over which the electrostatic charge of the protein can interact with other charges is called the Debye length, and this distance is determined by the charge on the surface of the protein along with any counter ion species associated to those charges. For a colloidal particle, like a protein, this outer sphere of influence is called an electrical double layer. As the ionic strength of a solution is increased, the Debye length is correspondingly decreased such that electrostatic attraction of oppositely charged entities is reduced. Correspondingly, for two entities of a similar charge the repulsion may be screened enough to allow non-polar attractive interactions to dominate. The Debye–Hückel model of precipitation through charge screening only holds up to a concentration of ~ 0.1 M, and above this the effect of ionic strength on solubility is ion specific. Both the charge and the species of ion present can accelerate the rate of protein aggregation or deposition in a solution in a process known generally as "salting out" which is commonly performed at multi-molar concentrations of salts.[108] For example, a well-used method for the isolation of DNA from human nucleated cells is performed by digesting cells using a protease then salting out the bulk of proteins using 6 M NaCl. Salting out is defined as the solubility of a non-electrolyte substance decreasing with an increase in salt concentration.

It is important to note that the protein itself will behave dynamically and undergo conformation changes due to alterations in the surrounding environment.[6,107,109] This multivariate influence and response is what makes the study of protein adsorption to surfaces challenging compared to smaller molecules such as surfactants. For every environmental variable that is changed the protein itself can demonstrate a completely different behaviour. Proteins are of the order of nanometres in length and seldom present themselves against a surface in a rigid symmetric or spherical geometry. Deformation on contact or an initial anisotropic morphology will dynamically influence how the protein aggregates to the surface. The most common shapes found in nature are rod-like or elliptical, although a multitude of morphological variety can be found. Immunoglobulin (Ig) is Y shaped, where the trunk of the protein is an anchor to a cell surface and the spars form the receptor site. If the orientation of the protein were reversed, with the anchor pointing to solution and the receptor end bound to the surface, then the protein will not fulfil its function. Although nature has evolved to orient biomolecules correctly, it is harder to control this synthetically. Van der Waals, electrostatic and charge interactions, hydrogen bonding and entropy gain of losing solvation ions or molecules compete as forces on the biomolecule. A simplified model of this process is to assign a general positive or negative charge to areas, or patches, corresponding with particular types of amino acid sequence present at the protein surface. In Figure 2.16 an example of β-lactoglobulin is presented adhering to a negatively charged surface.[6] The charged patches on the surface represent the general charge for collections of charged amino acids, such as glutamic acid, histidine, arginine, lysine and aspartic acid. In low concentrations there is a tendency for the electrostatic charge most complementary to the charge of the solid surface to dominate. At higher concentrations the protein-to-protein interactions will begin to influence the orientation as they push or attract one another in balance with the surface. Conformation changes upon surface adherence take place over minutes to hours as the native state gains entropy. The gain is facilitated through shifting of the secondary structure and further loss of solvation molecules to reach a point where the biomolecule is irreversibly bound.

It has recently become possible to isolate proteins and enzymes in a pure liquid state that retains full chemical functionality.[110–113] Previously, the consensus was that hydration of some sort was required to retain function and proteins were formerly isolated as powders that needed reconstitution or as aqueous dispersions. For a collection of proteins in isolation and free from water the intermolecular interactions are strong at short range (hydrogen bonding, electrostatic interactions and van der Waals), but weak beyond the size of the molecule. When exposed to heat a solid protein phase will generally denature before forming a melt under ambient pressure. Aqueous protein solutions have limits on the possible range of

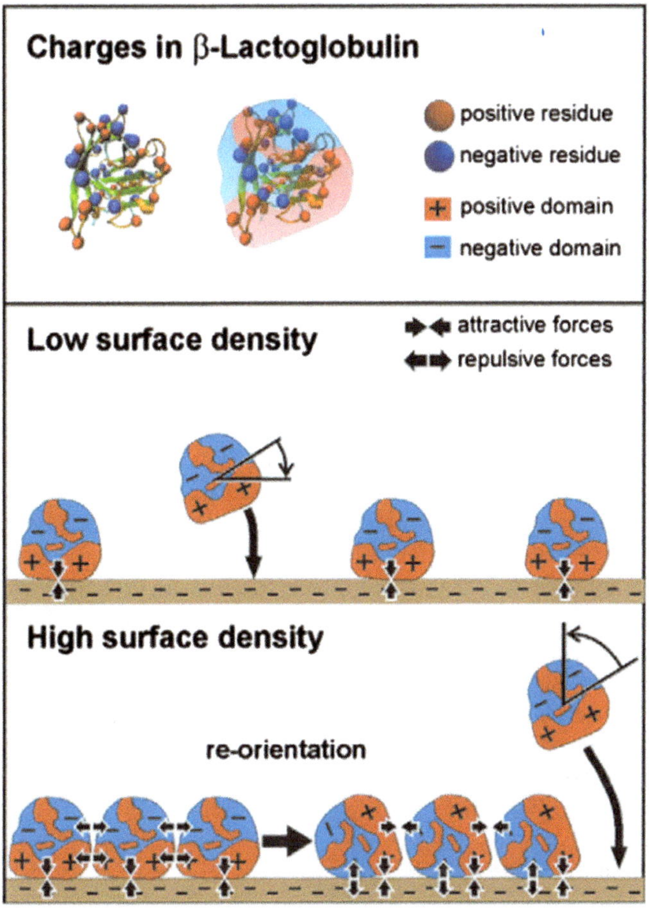

Figure 2.16 Schematic representation of orientational changes of surface-adsorbed proteins. (Top) A cartoon representation of β-lactoglobulin depicts the distribution of positively (red spheres) and negatively (blue spheres) charged amino acids. After strong simplification, the protein is represented as a globular entity consisting of positive and negative domains. (Middle) At low surface densities the protein orientation is solely determined by surface–protein interactions. (Bottom) At high surface densities increasing protein–protein interactions can trigger orientational changes leading to a decrease of protein–surface interactions. Adapted from ref. 6 with permission from Elsevier, Copyright 2011.

concentration, pH, ionic and thermal stability, depending on the surface charges, conformation and size of the protein. A pure liquid melt phase of globular proteins was first demonstrated using proteins such as myoglobin and ferritin.[110] The process to form melts of this type is done in three steps. First, the protein surface is given a net positive charge using *N*,*N*′-dimethyl-1,3-propanediamine covalently bound to the acidic amino acid chains on the macromolecule exterior. Under aqueous conditions, an anionic carboxylate or sulfonate surfactant with an 11-unit long PEG tail is added. The PEG surfactant coordinates to the surface of the supercharged protein to form a net charge-neutral construct. In the final step, water is removed from the system by lyophilisation to leave behind a pure form of the construct manifest as either a melt or a waxy solid. Myoglobin protein melts demonstrated retention of function for oxygen binding but with greatly improved resistance to thermal decomposition (408 °C compared to 315 °C in the native protein) as well as unexpected optical polarisation properties and unique rheological behaviour. The capability to produce a near-pure protein melt affords new lines of development in biomaterials. Biocatalytically active, self-supporting, protein–surfactant polymer films have been prepared by slow addition of the anionic polymer surfactant 4-nonylphenyl-3-sulfopropyl ether to aqueous solutions of supercharged cationic proteins. Ferritin, apoferritin, myoglobin, alkaline phosphatase, glucose oxidase and green fluorescent protein can all be treated in this way and will form globular clusters ~100–200 nm in diameter upon vacuum drying. By exposing these clusters to glutaraldehyde vapours during the drying process the clusters cross-link to one another and form a free-standing film.[111]

It might be expected that the kinetics of protein adsorption could be plotted as a monotonically plateauing function of number proteins on the surface against time. For irreversible binding, a straight-line gradient would be expected during the initial stages of deposition, followed by a slow decrease in the deposition rate as unoccupied binding sites become sparse.[6] This can be mathematically modelled at a simple level using a Langmuir adsorption model as the basic formula and then adjusting this to account for the more complex behaviour arising in proteins:

$$\frac{d\theta}{dt} = k^{on} \cdot C_s \cdot \left(1 - \frac{\theta}{\theta_{max}}\right) - k^{off \cdot \theta} \tag{2.14}$$

where θ is the protein coverage, θ_{max} is the maximum coverage for which no more binding sites are available, k^{on} and k^{off} are the on-rate and off-rate constants and C_s. The plot would be similar if proteins reach an equilibrium state where the rate of adsorption is matched by that of desorption. What actually occurs is that there is an initial oversaturation on the surface followed by desorption and eventually a steady state is achieved of either static monolayer coverage or adsorption and desorption. This type of adsorption kinetics profile has been noted for surface deposition of polymer and colloid solutions. This behaviour is termed the time delay model and is described by a surface becoming temporarily oversaturated by polymer at the surface, which is deposited in random orientations. The surface is assumed to be entirely covered in polymers having slightly different affinities and corresponding surface energies. After an initial time delay, during which the overshoot phase of deposition occurs, further conformational rearrangements of the polymer molecules upon the surface take place. These align the molecules into more energetically favourable states and displace more weakly bound molecules back into solution. Proteins, as long-chain amino acid-based biopolymers, can be modelled in a similar way, although the kinetic models are more complex; the dynamic quaternary structure of a protein is harder to model than a structurally more simplistic linear chain molecule. One commonly used model for protein adsorption is the "Vroman effect", named after Leo Vroman, who conducted investigations into the deposition of blood plasma proteins at solid interfaces. These experiments showed that fibrinogen rapidly deposited at surfaces to a maximal coverage point before some of the protein desorbed back into solution. This overshoot behaviour was observed consistently over variations of protein concentration in the blood plasma. It was determined that some of the surface-bound fibrinogen was being displaced by the higher molecular weight kininogen, which had a higher surface affinity. The Vroman effect holds for solutions containing different types of proteins, but a similar effect is also observed in monomeric protein solutions. The reason for this is that for a solution containing only one type of protein the protein unit itself will have higher and lower energy facets presented, depending on orientation. Just as with polymers, various conformations and orientations exhibit different affinities that will compete for the surface and displace anything more weakly bound. This behaviour was researched extensively in studies by Daly *et al.* and Wietz *et al.* who monitored the adsorption kinetics of lysozyme at various concentrations and on hydrophobic and hydrophilic surfaces and over varying pH.[114] Lysozyme may be loosely pictured as being shaped like a rugby ball and end-on interactions will be weaker than side-on ones in surface contact. Broadly, a lysozyme protein on its side takes up 1.5 times more area than one presented end-on to the surface, meaning that as the protein transitions into a more strongly bound orientation it will displace other proteins. In both the Daly and Wietz models it was found that adsorption overshoots take place when the adsorption rate is higher than the rate of conformational change from end-on to side-on. This description only accounts for the electrostatic interactions and a third affinity "site-on" must be taken into account when specific chemical moieties present on the protein exterior react directly with a substrate.

The affinity of a protein layer is further influenced by the strength of protein–protein interactions in the adsorbed layer. A study on β-lactoglobulin, a major constituent in milk, looked at how this protein adsorbed to a hydrophilic glass surface and found that during the initial deposition stage β-lactoglobulin is irreversibly bound and that a buffer rinse failed to remove the proteinaceous layer. If the β-lactoglobulin was allowed to deposit beyond the overshoot point then spontaneous reversal of the behaviour from irreversible to reversible was seen. It must be stated that this was not simply the removal of the second layer of proteins from a permanently stuck underlayer on the glass; all deposited material was removable. The explanation for this is that the cumulative lateral interactions across the protein layer eventually result in conformational changes at the interface that lessen the strength of the surface binding.

2.8.2 Biofouling and Bioinspired Antimicrobial Surfaces

On the macroscopic scale, protein build-up and biofouling manifests as a difficulty of eluting the protein, and secondary structure changes can be seen by circular dichroism or infrared spectroscopy techniques. Irreversible binding can often confer an enhanced resistance to denaturation. For these reasons protein films can be difficult to remove from surfaces and this effect has a macroscopic physical and sometimes economic impact. A good example of this is in the dairy industry, where every item of equipment used in the processing and handling of milk must be regularly cleaned and sterilised. Around 20% of milk is whey and this consists of a class of protein called caseins and in cow's milk the major whey proteins are β-lactoglobulin and α-lactalbumin. At temperatures between 75 and 110 °C a spongy white deposition begins to form on metal surfaces consisting of 50–70% protein (mostly β-lactoglobulin), 30–40% minerals and 4–8% fat. This solidified matter is the anchor point for more serious fouling in the form of bacterial colony formation and biofilm formation. Bacterial biofilms are more difficult to remove once formed, as the bacteria are encased in an extracellular matrix of less soluble polysaccharide and glycoprotein matter. For most dairy equipment, cleaning is performed on a daily basis and produces 1.3 litres of wastewater for every litre of milk product produced. Much of the water is generated by what is called "cleaning in position" (CIP), where pipes and spray balls are fitted directly into tanks or milk tankers to facilitate regular rinsing. The technique was developed in the 1950s as a cost-effective method of preventing biofouling in lines and liquid food storage equipment. Cleaning in position removes the need to spend time dismantling and then reassembling the closed systems needed for dairy production. Even so, 4–6 hours per day can be spent in non-production to allow for the CIP process.[115] This involves spraying surfaces with a sequential rinse of alkali and then acid solution, normally a sodium hydroxide solution, followed by a nitric or phosphoric acid. A 1 wt% NaOH solution denatures protein bound on the surface by inducing a net negative charge in the amino acid sequence, which causes the protein to unfold and solubilise within solution. The alkali is then rinsed away with water and the acid solution is pumped through. Acids can also denature proteins by inducing a net positive charge in amino acid chains at low pH, but in this case the presence of the acid is to remove encrusted salts. In the milk industry the removal of proteins is followed by sterilisation by either Cl_2 or O_3 treatment.

Acid and alkali rinses seem like a rather extreme treatment to prevent surface fouling. Harsh chemical treatments or "hard scrubbing" are often not desirable in industrial or laboratory-scale processes. Heterogenous catalysts, for example, must remain clean in order to function but could be irreversibly hydrolysed then oxidised by exposure to strong bases.[116] Similarly, many catalysts are modified at the nanoscale to maximise the surface area and correspondingly the turnover rate at the reactant–substrate interface.[27] A "hard scrub" of this surface would destroy the gains made through precision texturing to the near-molecular level. Returning briefly to the contact lens example given earlier, it would be bad indeed to use strong acids or alkalis for cleaning reusable lenses as the chances of eye damage if incorrectly rinsed would be very high. Some commercially available contact lenses still require cleaning in a hydrogen peroxide solution that oxidises organic matter on them while leaving the siloxane lens material alone.[117] In order to make the lenses safe to return to the eye the peroxide solution must by catalytically degraded to water in a lens case containing a metal disc, normally iron. A safer and more widely adopted solution uses boric acid, which is the only type of acid that does not damage the eye in comparison with similar concentrations of other inorganic acids.[118]

Cleaning of a contact lens is a relatively simple problem when compared with the more serious issue of preventing infection in medical tubing and catheters. Many intravenous and urinary tract catheters are manufactured from polyurethane resins that are susceptible to biofilm formation.[119] Although necessary for a great many medical procedures, this provides a route for common, clinically relevant pathogens, such as *Pseudomonas aeruginosa* and *Escherichia coli*, to translate directly from the external hospital environment to the internal physiological environment. Infections contracted within hospitals are known as nosocomial and, as the reality of antimicrobial resistance continues to escalate, the risk of acquiring a nosocomial infection increases. Eighty per cent of nosocomial infections are related to formation of biofilms on polymer surfaces.[120] As antimicrobial treatments lose their efficacy it has become a focus of research to derive physiologically safe but clinically effective physical mechanisms to inhibit the accumulation of bacterial plaques on these surfaces.

A more recent approach to the prevention of fouling is to impregnate surfaces with lubricants. Recently, it has been demonstrated that catheter materials can be impregnated with silicone oils that prevent bacterial adhesion to surfaces. Polyurethane tubing is coated in polydimethylsiloxane using a standard Sylgard 184 kit and the viscosity of the pre-polymer is modified using a hexamethyldisiloxane solvent to ensure the correct mechanical and swelling properties.[121] The PDMS coats the surface and acts as a reservoir for a 5 cSt silicone oil in which the catheter is soaked. Polydimethylsiloxane swells readily in silicone oil such that a catheter swollen in this way can replenish oil to the surface if it is worn away. The presence of the oil prevents the formation of bacterial plaques and this has been shown to be efficacious against *Escherichia coli* and *Staphylococcus epidermidis*. The presence of the oil prevents the normal mechanisms of attachment by which bacteria form a biofilm and colonise the surface. This type of treatment has also been used to counteract macrofouling by mussels on the exterior of boat hulls. This is a huge problem in the maritime industry and inlet pipes and filters can become clogged with mussels.[61]

Ultimately, lubricants and chemical treatments on a surface will be depleted, dissolved or worn away. It is not desirable to use chemical antibiotics, as bacterial colonies quickly adapt and develop resistance. In nature there are many surfaces that have a bactericidal action through physical effects, called the *contact killing mechanism,* to which microbes cannot easily develop resistance.[122] This is because the killing action is a physical one rather than chemical one, where the bacteria coming into contact with a surface are impaled upon arrays of nanostructured pillars. The action of chemical antibiotics is to interfere with some aspect of bacterial metabolism. The most common class are β-lactamases, which interfere with and inhibit the generation of peptidoglycan in cell wall manufacture during cell division.[123] Unfortunately, bacteria evolve rapidly to compensate for this and other classes of antibiotics. Antimicrobial resistance is one of the most pressing global issues today with many frontline antibiotics becoming ineffectual within years of their release. Antimicrobial infections claim ∼700 000 lives a year and, at current rates of resistance development, this is predicted to rise to 10 million lives a year by 2050. The search for effective new antibiotics continues and there are a number of methodologies to kill pathogens based on contact killing mechanisms. This includes the manufacture of nanoscale sterilising agents that are safe for surface treatment *in vivo* through to biomimetic artificial surfaces.[124,125] Soaking a surface in antibiotics is not desirable as, over time, this will encourage colonisation by resistant strains.

Engineered surfaces covered in rods 50–250 nm in diameter, 80–250 nm in height, with a pitch of between 100 and 250 nm will generally be resistant to colonisation by a range of pathogens. Both Gram-negative and Gram-positive bacteria are susceptible to this mechanism, which induces a large curvature in the bacterial membrane. However, this mechanism is broadly more efficacious against Gram-negative bacteria, which have a peptidoglycan layer 5–8 nm thick, which is around 4–5 times thinner than Gram-positive bacteria. The terms Gram-negative or -positive describe how a particular crystal violet staining technique in microscopy (developed by Hans Christian Gram) stains the bacteria, and this in turn is related to the structure of the bacterial cell wall. Gram-negative bacterial walls comprise lipid and lipoprotein membranes covering the cell interior and exterior of a single layer of peptidoglycan, ∼40 nm thick, which sits on the exterior of the cytoplasmic membrane. Gram-positive cells have a far thicker peptidoglycan layer. The induced cell membrane curvature in contact with the pillar is larger than the breaking stress of the lipid and peptidoglycan layers and so the membrane is ruptured. Cicada wings, gecko skin, dragonfly wings and lotus leaves all have antibacterial properties, which are often complementary to self-cleaning mechanisms arising from the high surface area of the nanostructured surface (Figure 2.17). The effect has been replicated synthetically in silica, titania, gold, polystyrene, PMMA and PDMS using a variety of techniques, ranging from reactive ion etching, plasma deposition, hydrothermal etching and soft lithography. Research continues to fine-tune the efficacy of particular structures towards specific bacteria, but a consistent observation is that the density of pillars must be enough that the bacteria must be forced to sit on the top of them. It is also a requirement that the pitch of the pillars induces the largest possible deformation curvature.[122]

2.9 Creating and Characterising Biological and Bioinspired Surfaces Using Atomic Force Microscopy

Atomic force microscopy (AFM) is a high-resolution imaging technique belonging to the family of scanning probe microscopy (SPM) techniques.[19] The development of AFM started in 1981 when Binning and Rohrer at IBM developed a scanning tunnelling microscope (STM) to study conducting and semiconductor surfaces at atomic resolution.[127] The researchers recorded a quantum mechanical tunnelling current between an ultrafine probe mounted on a piezoelectric scanner and the sample, when the probe was positioned within a nanometre of the surface and a voltage was applied between them.[128] The first AFM instrument was invented in 1986 by Binning, Quate and Gerber, with the practical demonstration of an atomic force microscope, by using a vibrating cantilever, reported for the first time by Wickramasinghe in 1987.[129] AFM represents a widely applied technique due to its instrumental properties, which exploit a flexible cantilever with a sharp tip or probe at the end. The probe has a radius of curvature (ROC) in the range of nanometres (typically 5–50 nm) and is used to scan the specimen surface. When the tip is brought close to the sample surface, the cantilever is deflected due to forces occurring between the tip and the sample.

In the biological field, the success of STM and AFM was rapid. This represents a significant contrast with the electron microscope, which required several years before becoming an established method for biological structural characterisation.[130] The first STM images of DNA were reported relatively quickly after the invention of the technique.[131–134] The publication of these studies caused significant interest in the technique; however, the initial inability to understand the presence of artefacts in the non-conductive structures presented doubts on the validity of the STM images.[135,136] Conversely, AFM became the first choice for characterisation of biological samples, because it does not require the sample to be conductive thus overcoming the difficulties encountered with STM in biological imaging. Bustamante and collaborators published the first DNA images obtained by AFM in air, initiating an upsurge in applications of AFM to biological molecules.[137–141]

The different studies reported AFM images of DNA obtained at a similar resolution with different preparatory methods, indicating that specimen preparation is a key component in the success of AFM characterisation of biological samples. The use of AFM, as a tool for biological imaging, increased significantly in the decade after the first DNA images were published, with rapid development of different techniques. AFM imaging found wide application in addressing numerous biological questions and was successfully applied to characterising DNA and RNA strands, nucleoprotein complexes and nucleic acid aggregates both in air and in liquid.

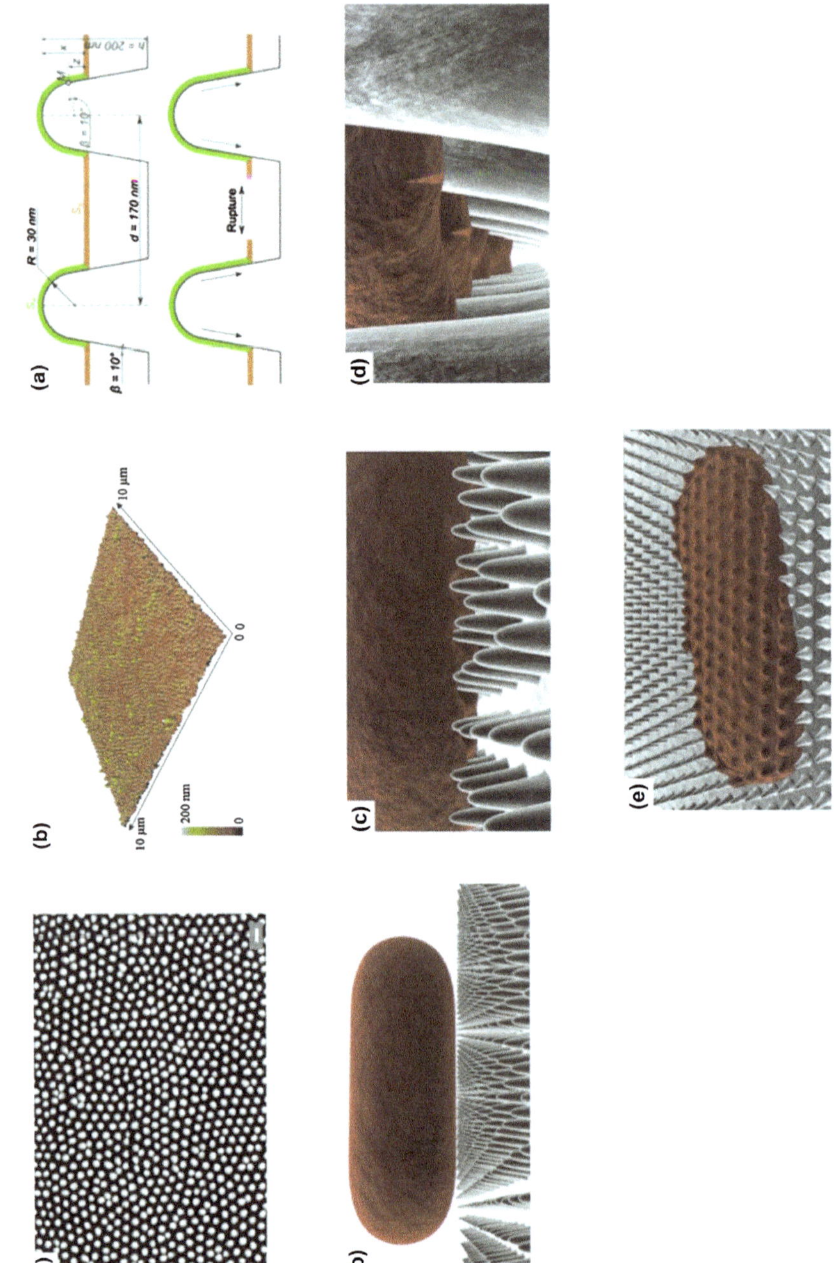
(a)
(b)
(c)
(d)
(e)
10 μm
10 μm
200 nm
0
0 0
(a)
R = 30 nm
d = 170 nm
β = 10°
h = 200 nm
Rupture
(b)

AFM imaging has stimulated new applications in biological systems.[142–144] One example is the chemical modification of the AFM tip using various strategies, such as surface modification and cross-linking, which has then been used to investigate the covalent linkage of biomolecules. Another example is the use of AFM in single-molecule force-pulling experiments to probe the molecular energy landscape of biomolecular events.[145–147] Other applications of AFM in imaging include cells, pharmaceutical molecules and structural and mechanical properties in materials science.[148,149] The rapid success of AFM is predominantly due to the possibility of working under physiological conditions, with a spatial resolution significantly higher than that obtained with light microscopy, the primary tool for biological studies in the past.[105,150,151] AFM provides additional capabilities and advantages over other microscopic methods, such as SEM, in the study of micro- and nano-structures by providing direct measurement of the three-dimensional (3D) shape, with both high lateral and vertical resolutions. Due to the facility of working in fluid, AFM has been used to follow real-time processes. One example is the observation of epicuticular wax crystals, as well as the observation of the dynamic crystallisation processes of organic molecules under real environmental conditions, as demonstrated by Koch *et al.*, who were able to use AFM to image wax regeneration on the leaf surface after removal of the original wax layer.[152]

AFM operates by exploiting the interatomic forces between the atoms of the surface and the atoms of a tip, integrated at the end of a spring-like cantilever, which also functions as the probe of the AFM *via* an integrated sharp tip. When the tip is brought close to the sample surface, the cantilever is deflected due to forces occurring between the tip and the sample. To allow for changes in topography of the sample surface, a feedback mechanism is generally employed to adjust the distance between the tip

Figure 2.17 (Top) Cicada (*Psaltoda claripennis*) wing surface topography. (a) Scanning electron micrograph of the surface of a cicada wing as viewed from above (scale bar, 200 nm). (b) Three-dimensional representation of the surface architecture of a cicada wing constructed from AFM scan data and coloured according to height. (Bottom) Biophysical model of the interactions between cicada (*P. claripennis*) wing nanopillars and bacterial cells. (a) Schematic of a bacterial outer layer adsorbing onto cicada wing nanopillars. The adsorbed layer can be divided into two regions: region A (in contact with the pillars) and region B (suspended between the pillars). Because region A adsorbs and the surface area of the region (SA) increases, region B is stretched and eventually ruptures. (b)–(e) Three-dimensional representation of the modelled interactions between a rod-shaped cell and the wing surface. As the cell comes into contact (b) and adsorbs onto the nanopillars (c), the outer layer begins to rupture in the regions between the pillars (d) and collapses onto the surface (e).
Adapted from ref. 126 with permission from Elsevier, Copyright 2013 Biophysical Society.

and the surface. Traditionally, the sample is mounted on a piezoelectric tube that allows the force to be controlled in the z direction, while scanning in the x and y directions (Figure 2.18).

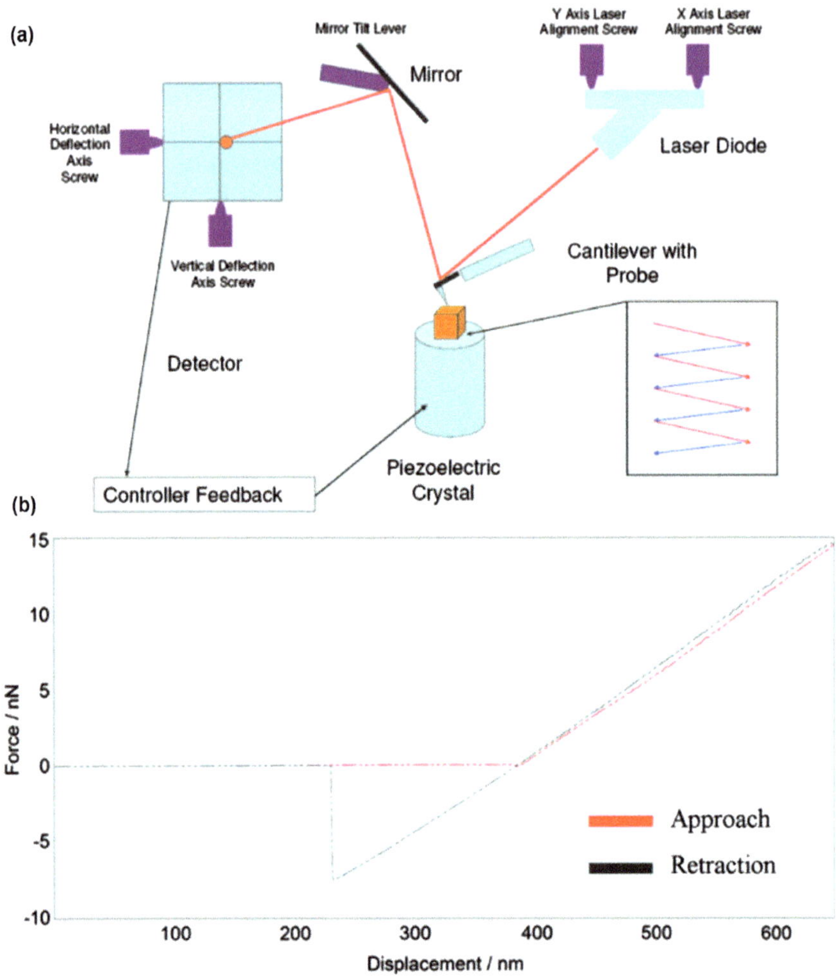

Figure 2.18 Schematic set-up of an atomic force microscope. (a) A schematic showing the basic set-up for a contact-mode AFM experiment. A sample is moved back and forth on a piezoelectric stage underneath a tip maintained at a constant height above the surface. The deflection in the cantilever is measured by bouncing a laser off the back of the cantilever and aiming it into a photodetector with a mirror. (b) An example plot of force against movement in the z direction for a tip approaching a mica surface. The red line is the force experienced by the tip as it moves towards the sample surface and then pushes into it. The black line is the force experienced by the tip as it is pulled back out. Adapted with permission from ref. 87 with permission from Elsevier, Copyright 2012.

2.9.1 AFM Hardware and Set-up

The profile of the AFM tip is critical in obtaining good images. In particular, the tip should be as sharp as possible. Tips are commonly fabricated from silicon or silicon nitride and can sometimes be coated with a special material to suit a particular application. The tip is mounted on a thin (2.7 μm) cantilever (length between 100 μm and 500 μm) whose deflections provide a measure of the interaction. Commercial AFM tips, with a physical size relative to the cantilever of 17 μm are produced in three geometries, conical, tetrahedral and pyramidal. Conical tips (with a ROC down to 5 nm) can be made sharp with a high aspect ratio (the ratio of tip length to tip diameter), making them especially useful for imaging features that are deep and narrow. The pyramidal and tetrahedral tips have lower aspect ratios with the tip ROC typically ranging from 5 to 50 nm.[153] Diamond tips mounted at the end of cantilevers were used in the early stages of AFM experiments, whereas nowadays commercially available microfabricated tips are from silicon nitride or silicon.[154–156] Tips can also be replaced by colloidal particles with a well-defined spherical shape in the "colloidal probe technique", which was applied for the first time in 1991 by Ducker *et al.* and Butt.[157,158] The colloidal probe technique operates with exactly the same procedure as a "standard" AFM tip. Similarly to the latter, a voltage, applied by a piezoelectric translator, controls the movement of the probe across the sample, with the surface topography characterised by recording the cantilever deflection with a detection system.[158] Spherical colloidal probes, available in the range between 500 nm and 50 μm, represent a powerful tool in the quantitative characterisation of surface forces that can also be analysed in a selective approach due to the possibility of modifying the chemical composition of the probe through the functionalisation process.[159] The colloidal probe technique is a well-established approach to study surface forces, providing advantages in the study of interfacial forces. The main advantage of colloidal probes is represented by a contact radius much larger than the probe–sample separation distance, which allows measurement of forces ranging from pico- to nano-Newtons. Contamination and degradation of the colloid surface represent the most common problems that limit the lifetime of a colloid probe and, therefore, to avoid the presence of artefacts in the images it is good practice to collect data using new probes.[160]

The deflection of the cantilever is normally measured using the optical lever technique, which employs a beam produced by a laser diode. The reflection produced during the characterisation is enhanced by using a thin metal layer, typically gold, on the reflective side of the cantilever. The laser is focused onto the end of the cantilever and the reflection of the beam is monitored by a sensitive detector consisting of a four-quadrant photodiode, as the cantilever characterises the surface topography. The topographic features of the surface can be mapped out by measuring the pixel-by-pixel deflection of the cantilever as the sample and tip are scanned relative to each other.[149] A piezoelectric scanner is used to move the cantilever tip in three dimensions with atomic resolution. Either the tip or the sample is scanned using this approach. The piezoelectric effect is the production of a voltage upon the deformation of a piezoelectric material, for example, a quartz crystal; the inverse piezoelectric effect is the deformation of such a crystal upon application of a voltage. Typically, the sensitivity of a piezo transducer in AFM is of the order of 5 Å per volt. The piezoelectric materials used in AFM instruments are usually lead zirconate titanate (PZT) which, unlike quartz, exhibits a permanent electric dipole. Piezoelectric materials are polycrystalline and must be poled in an electric field at high temperature in order to exhibit a piezoelectric effect.[161] Two parameters control the feedback loop in an AFM instrument, namely, the proportional gain and the integral gain. These work by feeding back a portion of the output signal from the cantilever position detector to the actuator controlling the position of the tip or, in some AFM models, the sample. This is done with the aim of reducing the difference between the set point value, initially set by the operator, and the actual measured value; the difference between the set point value and actual measured value is known as the error signal. The proportional gain value is the high-frequency feedback control, which controls the amount of fed-back signal proportional to the error signal, while the integral gain value is the low-frequency feedback control, which governs the fed-back signal calculated from the error signal over a specific time interval.[162] The feedback system uses the information from the detection system to control the height of the probe and to adjust the scanner vertical position, with the topography mapped out by measuring the pixel-by-pixel deflection of the cantilever.

2.9.2 Surface Forces and Tip–Sample Interaction

The probe has a nanometre, sharp ROC and is used to scan the specimen surface. When the tip is brought close to the sample surface, the cantilever is deflected due to forces occurring between the tip and the sample. The forces can be of multiple origins and can include electrostatic repulsion/attraction, magnetic interactions, van der Waals interactions, adhesion and capillary forces. Depending on the magnitude and polarity of surface charges, electrostatic forces can be either attractive or repulsive. In the measurement of surface forces, the total force between the two surfaces is determined. The origin of all surface forces is caused by several components, with the components considered as being additive. These components are the van der Waals forces, the electrostatic double-layer force, the hydration repulsion (between hydrophilic surfaces) and the hydrophobic attraction (between hydrophobic surfaces). To analyse the electrostatic double-layer force, the latter needs to be separated from all other components. This is usually done by assuming a certain distance dependency for the van der Waals forces and the electrostatic force. In addition, the electrostatic force can be identified by means of the influence of the salt concentration present in the medium.[159] Accordingly, the sample–tip interaction may be controlled by the environmental conditions in which the experiment is performed, for example pH, ionic concentration and temperature.[163] The study of electrostatic forces represents a specific area in the SPM family and in particular electrostatic force microscopy was developed for characterisation of the potential across the surface by applying a voltage between the tip and the surface; this technique found widespread application in the study of semiconductor properties of materials.[164] In ambient conditions, while scanning in air, hydrophilic surfaces adsorb a thin film of water, which causes the formation of a meniscus, as the AFM tip approaches the surface. The meniscus is caused by attractive (capillary) forces that show distance dependence, which in AFM is determined by the tip–sample separation. In particular, the presence of a meniscus represents a limiting factor in the force that needs to be applied on the surface by the AFM tip. In fact, the imaging of fragile samples requires the force between the tip and the sample to be minimised, in order to reduce the possibility of generating damage on the surface, but the reduction of that force is limited by the existence of a meniscus force. Moreover, the mechanical properties of several materials are sensitive to the presence of water vapour molecules in the atmosphere; examples of materials affected by water vapour in the local environment are granular materials such as powders and sand.[165] In an AFM experiment where the tip may move towards, into contact with and then away from the surface, unloading of the AFM probe from the sample surface is governed by the adhesive interaction between the tip and the surface, with the adhesion force generated by a combination of van der Waals', electrostatic and capillary forces. As described, the presence of water films on the surface generates the formation of a meniscus, which may dominate the adhesion force, depending on the chemical groups present on the surface of the tip and on the surface of the sample. At ambient conditions, a water neck forms between the AFM tip and the substrate due to capillary condensation and adsorption of thin water films at surfaces. This attractive interaction depends on the relative humidity and the hydrophilicity of the tip and the sample (Box 2.6).

2.10 Function and Form in Bioinspired Surfaces on the Macroscale

Bioinspired functional surfaces are characterised by intricate architectures across the nano-, micro- and meso-scales, with tailored functional material properties in two dimensions.[54] These surfaces are presented on biological structures which in and of themselves further enhance some form of evolutionary function. While intensive research has provided us with complex multifunctional materials, preparation of self-shaping and stimuli-responsive materials are still in the embryonic phase, with initial studies of shape-changing materials focused mainly on alloys and piezoelectric materials.[168,169]

Box 2.6 Imaging of DNA and other biomolecules

AFM is highly versatile for determining the physical and spatial characteristics of proteins and other biomolecules on surfaces.[138,140,166] The ability to image in water allows imaging of biomolecules in their native state and the researcher can directly observe chemical interactions. By anchoring a protein to a tip and then further anchoring the other end of a protein to the analytical substrate, the structure can be physically unwound by raising the tip. The cantilever deflection correlates directly to the binding forces within the peptide sequences. There are a great many more experiments that can be implemented using AFM and associated techniques. Here is a simple technique for preparing DNA for AFM analysis in liquid using an AFM in tapping mode. Freshly cleaved muscovite mica is used as an atomically flat sample surface for imaging. Siloxy groups present on the surface possess a negative charge, which under aqueous conditions will repel electronegative molecules like DNA. To overcome this, a divalent cationic salt can be used that coordinates to the mica surface and further provides an electrostatic anchor for the phosphate backbone of the DNA. This bonding is strong enough to keep the DNA in place even as a tip in tapping mode is oscillated over it. Mg^{2+} present in the form of a chloride salt is ideal for this, as magnesium chloride is present in most biological buffer solutions. Ni^{2+}, Co^{2+} and Zn^{2+} will also effectively bind DNA to mica under aqueous conditions. Conversely, proteins and biomolecules having a net positive charge will be attracted to the mica surface without the need for salt. Ferritin is rich in lysine residues, which provide an overall positive charge at neutral pH values. An example of what you should observe is given in Figure 2.19.

1. Prepare a stock buffer containing 20 mM trisaminomethane (known as Tris), 50 mM KCl and 5 mM $MgCl_2$. Adjust the pH to between 7.4 and 8.0.
2. Dissolve DNA in an aliquot of buffer to a concentration of $\sim$2 μg μL^{-1}.
3. Freshly cleave a mica surface. This can be done by taking a square of muscovite mica and gently bending one corner back and forth using tweezers. The mica will begin to delaminate and, with care, it is possible to peel two layers apart. Take care not to touch the inner surfaces, one of which will be used as the substrate.
4. Mount one of the cleaved surfaces on your AFM puck or holder.
5. Add a droplet of the DNA in buffer solution to the surface of the mica. Leave it to stand for a few minutes but do not let the droplet dry out.
6. Gently wash the mica surface by applying droplets of stock buffer to the surface of the mica. Tip the mica and allow the excess liquid to

drain off. This is important as unbound DNA may stick to the tip during imaging.

7. Before the surface dries out add a droplet of buffer or transfer into the aqueous imaging AFM tip holder. Your sample is ready for imaging in tapping (non-contact) mode and the precise set-up will depend on the make and model of the AFM you will be using.

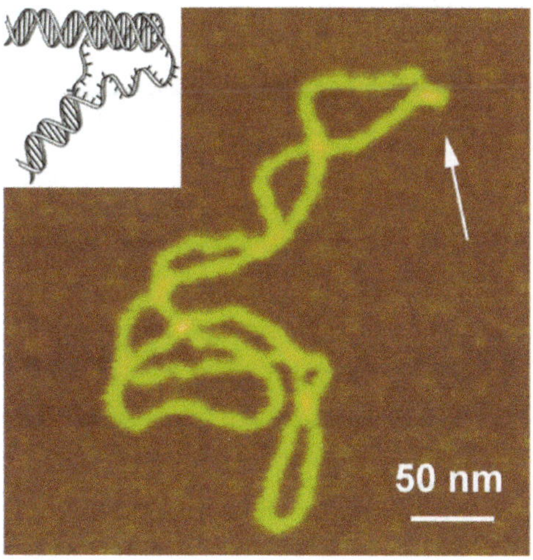

Figure 2.19 AFM images of plasmid pCW2966 containing a 46 base pair mirror repeat forming H-DNA at acidic pH. H-DNA regions are indicated with arrows. The DNA sample was incubated in 50 mM sodium acetate buffer at pH 5.0 and then deposited onto APS-mica in the same buffer. The images for plasmid-containing H-DNA were acquired in air with a NanoScope III microscope operated in tapping mode. A sharp kink at the base of a thick protrusion is indicated with an arrow. The inset shows the model for H-DNA.
Adapted from ref. 167 with permission from Elsevier, Copyright 2010.

In nature, many responsive materials actuate their responses as a consequence of changes in the environment, for instance, temperature or relative humidity. Examples of the changes caused by variation in humidity are the curling of a wet leaf as it dries or the process of self-burial of seeds.[170] A more "charming" example is given by the opening and closing of the scales of a pine-cone caused by hygroscopic actuation for seed dispersal (Figure 2.20). The movement of the pine-cone scales is caused by changes in the structural orientation of the cellulose microfibrils, which have a greater resistance to extension and accommodate deformation, behaviour that can be compared to that of a thermally responsive bimetallic material. Particularly important is the transmission and amplification of strain across the

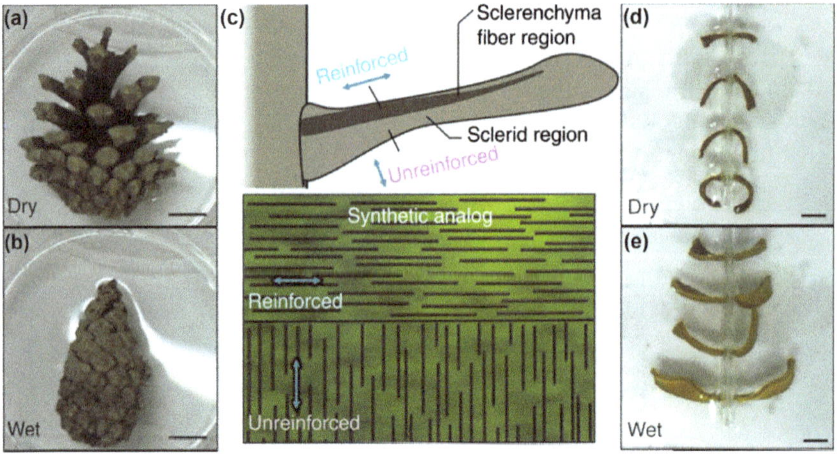

Figure 2.20 Pine-cone biomimicry by Studart and Erb, using ultra-high magnetic response plates. (a), (b) The natural principles of expansion and contraction in response to humidity. (c) (top) Diagram of the positioning of fibrous stiff cells (sclerenchyma) and short amorphous cells (sclereids) which give rise to the anisotropic movement of the pine-cone and (bottom) a schematic showing the synthetic reproduction of these properties with magnetic orientation of reinforcement. (d), (e) The results: responsive synthetic pine-cone scales produced with selective orientation within a gelatine matrix, in a similar bilayer structure, in both dry and wet states.
Reproduced from ref. 171 with permission from The Royal Society of Chemistry.

length scales, which can translate the response to external stimuli into large-scale movements such as twisting, bending and opening.

The orientation of cellulose caused by changes in the surrounding environment is also characteristic for different plants, such as the opening of eudicot seedpods.[172] In nature, behaviours that are more complex are produced with more elaborate cellulose arrangements, in which each constituent cell is encircled by cellulose in a helical arrangement characterised by a nanostructure that results in a macroscopic coiling effect. This cellulose arrangement can be observed, for instance, with the seed delivery mechanism of *Erodium circodia,* which uses a thin support to attach the seeds to the main body of the plant and curls into a spring as it desiccates, storing elastic energy; at a critical point, this material cracks and the elastic energy is released. In particular, as the material dehydrates, the coil becomes tighter, storing more and more tension until the material eventually fractures. Similar to the mechanism employed by *E. circodia*, self-actuating synthetic systems, with inherent active and passive regions transforming through origami folding, have been reported for different stimuli methods employing polymer shrinking; in the literature, examples of shape-memory materials and magnetic materials are also reported.

Cellulose, which represents the main component of paper, has been studied as a self-actuating material and, in particular, the hygroscopic properties of bilayer natural cellulose composites have been studied with the aim of developing self-actuating components triggered by different stimuli such as swelling across neighbouring layers. The cellulose "bilayer" design provides useful advantages that can be used to accommodate the strain in the material and this approach has been reported for paper–plastic and wood veneer–glass epoxy composites.[173–175] While the properties of cellulose can provide advantages in the design of "new" materials, the underlying principles of the deployment of cellulose-based architectures is still not well understood.

A better understanding of the science behind this deployment will permit the development of more affordable approaches to manufacturing complex architectures by exploiting the programmed morphing mechanism for active shape change. The development of self-actuating functional materials represents an effective space-saving solution for structures that can be deployed across a variety of length scales, from microelectromechanical systems (MEMS) to heart stents to space-mast/boom.[176,177] Researchers at the University of Bristol have characterised the deployment mechanisms for different paper architectures, which can be employed to obtain specific folding/unfolding designs to use in synthetic composite material systems. In particular, the studies conducted in Bristol have shown the influence of porosity, moisture, surfactant concentration, temperature and hornification in the deployment of sample architectures. Understanding the role of the substrate, its architecture and the environmental stimulus are essential to better design internal actuators for morphing structures that can be used in the next generation of active four-dimensional (4D) fibrous materials. When folds are created in paper, extension and localised compressive deformation occur simultaneously. Accordingly, understanding the nature and magnitude of the internal damage is key information in the design of multi-deployable architectures. Studies have shown that in the folded region observed on the paper the fibres are not torn when the folds are created but deform and reposition themselves to minimise the local strain field. Conversely, some other types of paper developed cracks in the brittle filler materials since they are not able to cope with the localised strain of folding. In the case of contact with a liquid medium, the cracks serve as a low resistance flow path for penetrating liquids. It was hypothesised that the cellulose fibre network can rearrange the local architecture to absorb the global energy of folding with the microfibrils on a nanoscale providing flexibility to the individual fibres. The changes happening at a nano, micro and macro level avoid failure of the fibres upon the application of mechanical stress, with the individual layers held by hydrogen bonds that offer reversible adhesion by hydration and dehydration. The ability of the fibre network to rearrange itself locally in order to minimise the bending stress permits the creation of stored energy as the trigger for subsequent deployment of the folded region, similarly to the "strategy" deployed by the pine-cone previously described.

Cellulose represents the main component of paper fibre with the latter forming an intricate network which gives rise to the hierarchical structure of paper.[178] Cellulose molecules have a crystalline and an amorphous region, where hydrogen bonding within the crystalline part is very stable compared to that in the amorphous region.[179] The exposure of cellulose to water or moisture has a plasticising effect on the amorphous region, which causes a weakening of the inter-fibre hydrogen bonding. Additionally, intercalating water molecules in the fibrous structure of cellulose form hydrogen bonds between fibrils, which, along with the water retained in inter- and intra-fibre pores, causes swelling of the fibres, which are ultimately responsible for the "bilayer" effect, previously described, which propagates through the thickness direction of the paper as the water front moves in the cellulose fibres. The volume fraction of the fibres in the cellulose network plays an important role in recovering the initial folded configuration of the paper; the configuration is also influenced by the fibre orientation, fibre-to-fibre connectivity and filler content. With regards to the latter, studies have demonstrated that the filler materials provide a mechanism for substrate stabilisation once hydrated, which minimises the wrinkles upon drying; this contributes towards effective shape recovery in the paper. Accordingly, in order to design functioning self-actuating materials, a profound knowledge of fibre orientation and fibre-to-fibre connectivity is essential.

The study conducted by Mulakkal *et al.* investigated the deployment of a "model" folded paper architecture, using the medium of water as the stimulus for self-actuation (Figure 2.21).[180] The work was inspired by the passive nastic movements exhibited by biological systems and, in particular, by the pine-cone scales described at the beginning of this section, which achieve complex deployment with cellulose fibres. The aim of the study was to capture the "design rules" which can be translated into a synthetic fibrous material based on the active morphing of a natural fibrous system. The work performed demonstrated that controlling the cellulose architecture, which is ultimately responsible for controlling the fluid ingress in the paper structure, permits the design of fibrous structures for programmable deployment.

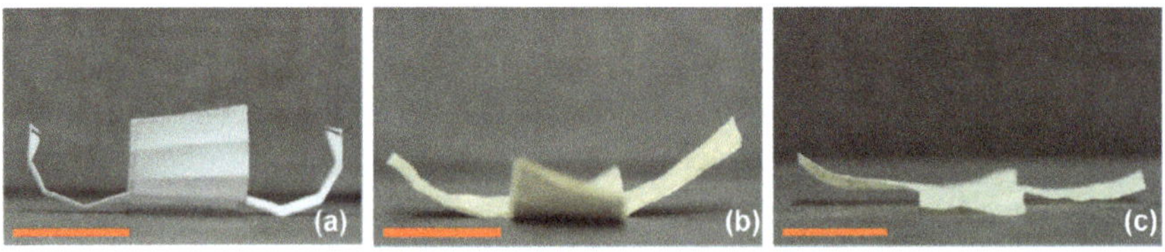

Figure 2.21 Photographs showing the recovery of sample architectures after deployment: (a) printer paper 90 gsm, (b) Lokta brand paper 3×30 gsm and (c) Lokta brand paper 30 gsm. Scale bar, 15 mm. Depending on the weight of the paper the addition of water allows the paper architectures to reversibly unfurl to varying degrees.
Reproduced from ref. 180. DOI: 10.1088/0964-1726/25/9/095052.

An important factor to consider is the scalability, which is always a critical parameter for successfully adapting a system across different length scales, with the layering of individual lamina sheets into a typical composite structure explored. The authors of the work performed in Bristol are currently exploring an experimental procedure to determine the "blocking force" for different paper architectures; the latter governs the response of the fibrous architecture in response to the hydration and dehydration environment.

References

1. P. Gleick, *Water in Crisis: A Guide to the World's Fresh Water Resources*, Oxford University Press, New York, 1993.
2. O. Björneholm, M. H. Hansen, A. Hodgson, L. M. Liu, D. T. Limmer, A. Michaelides, P. Pedevilla, J. Rossmeisl, H. Shen, G. Tocci, E. Tyrode, M. M. Walz, J. Werner and H. Bluhm, *Chem. Rev.*, 2016, **116**, 7698–7726.
3. L. G. M. Pettersson, R. H. Henchman and A. Nilsson, *Chem. Rev.*, 2016, **116**, 7459–7462.
4. J. Clayden, N. Greeves, S. Warren and P. Wolfers, *Organic Chemistry*, Oxford University Press, Oxford, 2017.
5. P. Atkins, J. de Paula and J. Keeler, *Atkins' Physical Chemistry*, Oxford University Press, Oxford, 11th edn, 2017.
6. M. Rabe, D. Verdes and S. Seeger, *Adv. Colloid Interface Sci.*, 2011, **162**, 87–106.
7. S. Mann, Molecular recognition in biomineralization, *Nature*, 1988, **332**, 119–124, vol. 332.
8. F. A. Long and W. F. McDevit, *Chem. Rev.*, 1952, **51**, 119–169.
9. W. Stillwell, in *An Introduction to Biological Membranes*, Elsevier, 2013, pp. 29–41.
10. J. Yu, W. Wei, E. Danner, R. K. Ashley and J. N. Israelachvili, *Nat. Chem. Biol.*, 2012, 7, 588–590.
11. M. Hillert, *J. Phase Equilib.*, 1995, **16**, 403–410.
12. G. A. Pilkington and W. H. Briscoe, *Adv. Colloid Interface Sci.*, 2012, **179–182**, 68–84.
13. M. Yamamoto, N. Nishikawa, H. Mayama, Y. Nonomura, S. Yokojima, S. Nakamura and K. Uchida, *Langmuir*, 2015, **31**, 7355–7363.
14. L. Yin, L. Zhu, Q. Wang, J. Ding and Q. Chen, *ACS Appl. Mater. Interfaces*, 2011, 1254–1260.
15. V. M. Swinerd, A. M. Collins, N. J. V. Skaer, T. Gheysens and S. Mann, *Soft Matter*, 2007, **3**, 1377.
16. G. Wang, Z. Guo and W. Liu, *J. Bionic Eng.*, 2014, **11**, 325–345.
17. *So You Want to Measure Surface Energy?*, KRÜSS Technical Note, 1999, https://www.kruss-scientific.com/fileadmin/user_upload/website/literature/kruss-tn306-en.pdf.
18. Y. Yuan and T. R. Lee, Contact Angle and Wetting Properties, in *Surface Science Techniques*, ed. G. Bracco and B. Holst, Springer, Berlin, Heidelberg, 2013, pp. 3–34.

19. G. Depietra, The plant seed surface: a functional nanomaterial, PhD Thesis, University of Bristol, 2016.
20. D. K. Owens and R. C. Wendt, *J. Appl. Polym. Sci.*, 1969, **13**, 1741–1747.
21. Y. Y. Yan, N. Gao and W. Barthlott, *Adv. Colloid Interface Sci.*, 2011, **169**, 80–105.
22. S. Nishimoto and B. Bhushan, *RSC Adv.*, 2013, **3**, 671–690.
23. K. Koch, B. Bhushan, Y. C. Jung and W. Barthlott, *Soft Matter*, 2009, **5**, 1386 1393.
24. W. H. Briscoe, S. Titmuss, F. Tiberg, R. K. Thomas, D. J. Mcgillivray and J. Klein, *Nature*, 2006, **444**, 3–6.
25. S. C. Russev, N. Alexandrov, K. G. Marinova, K. D. Danov, N. D. Denkov, L. Lyutov, V. Vulchev and C. Bilke-Krause, *Rev. Sci. Instrum.*, 2008, **79**, 104102.
26. J. X. H. Wong and H. Z. Yu, *J. Chem. Educ.*, 2013, **90**, 1203–1206.
27. K. R. Phillips, G. T. England, S. Sunny, E. Shirman, T. Shirman, N. Vogel and J. Aizenberg, *Chem. Soc. Rev.*, 2016, **45**, 281–322.
28. M. Nosonovsky and B. Bhushan, *Ultramicroscopy*, 2007, **107**, 969–979.
29. K. Koch and W. Barthlott, *Philos. Trans. R. Soc. A Math. Phys. Eng. Sci.*, 2009, **367**, 1487–1509.
30. R. Hensel, C. Neinhuis and C. Werner, *Chem. Soc. Rev.*, 2016, **45**, 323–341.
31. A. Collins, D. Carriazo, S. A. Davis and S. Mann, *Chem. Commun.*, 2004, **4**, 568–569.
32. A. M. Collins, C. Spickermann and S. Mann, *J. Mater. Chem.*, 2003, **13**, 1112–1114.
33. J. M. Bulpett, A. M. Collins, N. H. M. Kaus, P. T. Cresswell, O. Bikondoa, D. Walsh, S. Mann, S. A. Davis and W. H. Briscoe, *J. Mater. Chem.*, 2012, **22**, 15635–15643.
34. S. R. Hall, V. M. Swinerd, F. N. Newby, A. M. Collins and S. Mann, *Chem. Mater.*, 2006, **18**, 598–600.
35. D. C. Green, S. Glatzel, A. M. Collins, A. J. Patil and S. R. Hall, *Adv. Mater.*, 2012, **24**, 5767–5772.
36. W. H. Briscoe, *Curr. Opin. Colloid Interface Sci.*, 2017, **27**, 1–8.
37. H. J. Kim, K. J. Seo, K. H. Kang and D. E. Kim, *Int. J. Precis. Eng. Man.*, 2016, **17**, 829–841.
38. C. S. Dezzutti, E. R. Brown, B. Moncla, J. Russo, M. Cost, L. Wang, K. Uranker, R. P. Kunjara Na Ayudhya, K. Pryke, J. Pickett, M.-A. LeBlanc and L. C. Rohan, *PLoS One*, 2012, 7, e48328.
39. M. Berry, A. Harris and A. P. Corfield, *Investig. Ophthalmol. Vis. Sci.*, 2003, **44**, 567–572.
40. M. Berry, A. P. Cor and T. J. Mcmaster, *Soft Matter*, 2013, **44**, 1740–1743.
41. P. Fratzl, O. Kolednik, F. D. Fischer and M. N. Dean, *Chem. Soc. Rev.*, 2016, **45**, 252–267.
42. C. J. Reeves, A. Siddaiah and P. L. Menezes, *J. Bio- Tribo-Corrosion*, 2017, **3**, 1–27.

43. V. Wright and D. Dowson, Lubrication and cartilage, *J. Anat.*, 1976, **121**, 107–118.
44. B. A. Hills, *Proc. Inst. Mech. Eng. Part H J. Eng. Med.*, 2000, **214**, 83–94.
45. A. P. Corfield, S. D. Carrington, S. J. Hicks, M. Berry and R. Ellingham, *Prog. Retin. Eye Res.*, 1997, **16**, 627–656.
46. B. T. Käsdorf, F. Weber, G. Petrou, V. Srivastava, T. Crouzier and O. Lieleg, *Biomacromolecules*, 2017, **18**, 2454–2462.
47. L. E. Downie and J. P. Craig, *Clin. Exp. Optom.*, 2017, **100**, 438–458.
48. M. Laksitorini, V. D. Prasasty, P. K. Kiptoo and T. J. Siahaan, *Ther. Delivery*, 2014, **5**, 1143–1163.
49. A. F. Hubbs, R. R. Mercer, S. A. Benkovic, J. Harkema, K. Sriram, D. Schwegler-Berry, M. P. Goravanahally, T. R. Nurkiewicz, V. Castranova and L. M. Sargent, *Toxicol. Pathol.*, 2011, **39**, 301–324.
50. M. Moshirfar, K. Pierson, K. Hanamaikai, L. Santiago-Caban, V. Muthappan and S. F. Passi, *Clin. Ophthalmol.*, 2014, **8**, 1419–1433.
51. T. J. Millar and B. S. Schuett, *Exp. Eye Res.*, 2015, **137**, 125–138.
52. L. Cwiklik, *Biochim. Biophys. Acta, Biomembr.*, 2016, **1858**, 2421–2430.
53. N. A. Peppas and J. J. Sahlin, *Biomaterials*, 1996, **17**, 1553–1561.
54. C. E. Brubaker and P. B. Messersmith, *Langmuir*, 2012, **28**, 2200–2205.
55. P. Flammang and R. Santos, *Interface Focus*, 2014, **5**, 1–3.
56. T. R. Garrett, M. Bhakoo and Z. Zhang, *Prog. Nat. Sci.*, 2008, **18**, 1049–1056.
57. N. C. Schellmann, *Stud. Conserv.*, 2007, **52**, 55–66.
58. J. Eisenhaure and S. Kim, *Micromachines*, 2017, **8**, 1–38.
59. C. Heinzmann, C. Weder and L. M. De Espinosa, *Chem. Soc. Rev.*, 2016, **45**, 342–358.
60. J. Gomes, A. Grunau, A. K. Lawrence, L. Eberl and K. Gademann, *CHIMIA Int. J. Chem.*, 2013, **67**, 275–278.
61. S. Amini, S. Kolle, L. Petrone, O. Ahanotu, S. Sunny, C. N. Sutanto, S. Hoon, L. Cohen, J. C. Weaver, J. Aizenberg, N. Vogel and A. Miserez, *Science (80-.)*., 2017, **357**, 668–673.
62. D. R. Miller, J. E. Spahn and J. H. Waite, *J. R. Soc., Interface*, 2015, **12**, 20150614.
63. M. Donkerwolcke, F. Burny and D. Muster, in *Biomaterials*, 1998, vol. 19, pp. 1461–1466.
64. J. E. Murphy-Ullrich, *J. Clin. Invest.*, 2001, **107**, 785–790.
65. D. N. Woolfson, *Biopolymers*, 2010, **94**, 118–127.
66. D. N. Woolfson and M. G. Ryadnov, *Curr. Opin. Chem. Biol.*, 2006, **10**, 559–567.
67. J. M. Fletcher, R. L. Harniman, F. R. H. Barnes, A. L. Boyle, A. Collins, J. Mantell, T. H. Sharp, M. Antognozzi, P. J. Booth, N. Linden, M. J. Miles, R. B. Sessions, P. Verkade and D. N. Woolfson, *Science (80-.)*., 2013, **340**, 595–599.
68. S. Mann, *Nat. Mater.*, 2009, **8**, 781–792.
69. J.-P. Douliez, N. Martin, C. Gaillard, T. Beneyton, J.-C. Baret, S. Mann and L. Beven, *Angew. Chem. Int. Ed.*, 2017, **56**, 13689–13693.

70. S. Mann, *Angew. Chem. Int. Ed.*, 2008, **47**, 5306–5320.
71. Royal Society of Chemistry, Developing a glue, http://www.rsc.org/learn-chemistry/resource/res00000459/developing-a-glue?cmpid=CMP00005014.
72. A. Kumar, S. Grover, J. Sharma and V. K. Batish, *Crit. Rev. Biotechnol.*, 2010, **30**, 243–258.
73. M. Kieliszek and A. Misiewicz, *Folia Microbiol.*, 2014, **59**, 241–250.
74. F. Jin, X. Li and S. Park, *J. Ind. Eng. Chem.*, 2015, **29**, 1–11.
75. R. D. Adams, A. Collins, D. Cooper, M. W. Digby, A. W. Farmer, A. Laurence, K. Patel, M. Stevens and R. Watkins, in *Structural Integrity and Durability of Advanced Composites: Innovative Modelling Methods and Intelligent Design*, Woodhead Publishing, 2015, vol. 21, pp. 763–792.
76. U. Ali, K. J. B. A. Karim and N. A. Buang, *Polym. Rev.*, 2015, **55**, 678–705.
77. Y. Taira and Y. Imai, *Dent. Mater. J.*, 2014, **33**, 291–304.
78. P. A. Leggat and U. Kedjarune, *Int. Dent. J.*, 2003, **53**, 126–131.
79. K. P. Davis and R. W. Derlet, *Wilderness Environ. Med.*, 2013, **24**, 67–74.
80. H. Yuk, T. Zhang, S. Lin, G. A. Parada and X. Zhao, *Nat. Mater.*, 2015, **15**, 190.
81. M. M. Feldstein, K. A. Bovaldinova, E. V. Bermesheva, A. P. Moscalets, E. E. Dormidontova, V. Y. Grinberg and A. R. Khokhlov, *Macromolecules*, 2014, **47**, 5759–5767.
82. A. Colas and J. Curtis, in *Biomaterials Science*, Elsevier Academic Press, 2004, pp. 80–86.
83. F. Abbasi, H. Mirzadeh and A. A. Katbab, *Polym. Int.*, 2001, **50**, 1279–1287.
84. D. A. Armitage, M. N. Hughes and A. W. Sinden, *J. Chem. Educ.*, 1973, **50**, 434.
85. Y. Xia and G. M. Whitesides, *Angew. Chem. Int. Ed.*, 1998, **37**, 550–575.
86. D. Qin, Y. Xia and G. M. Whitesides, *Nat. Protoc.*, 2010, **5**, 491.
87. A. M. Collins, in *Nanotechnology Cookbook*, Elsevier, 2012, pp. 35–204.
88. N. H. Mohd Kaus, A. M. Collins, O. Bikondoa, P. T. Cresswell, J. M. Bulpett, W. H. Briscoe and S. Mann, *J. Mater. Chem. C*, 2014, **2**, 5447–5452.
89. A. M. Collins, N. H. Mohd Kaus, F. Speranza, W. H. Briscoe, D. Rhinow, N. Hampp and S. Mann, *J. Mater. Chem.*, 2010, **20**, 9037–9041.
90. K. Koch and H.-J. Ensikat, *Micron*, 2008, **39**, 759–772.
91. K. Koch, B. Bhushan and W. Barthlott, *Soft Matter*, 2008, **4**, 1943–1963.
92. K. Koch, A. Dommisse, W. Barthlott and S. N. Gorb, *Acta Biomater.*, 2007, **3**, 905–909.
93. A. Julia, K. Koch, M. Spaeth and W. Barthlott, *Acta Biomater.*, 2009, **5**, 1848–1854.
94. S. H. White and W. C. Wimley, *Annu. Rev. Biophys. Biomol. Struct.*, 1999, **28**, 319–365.
95. A. M. Seddon, P. Curnow and P. J. Booth, *Biochim. Biophys. Acta, Biomembr.*, 2004, **1666**, 105–117.
96. V. V. Kumar, *Proc. Natl. Acad. Sci. U. S. A.*, 1991, **88**, 444–448.

97. A. M. Collins, in *Materials Science and Technology 2010 Conference and Exhibition*, The Mineral, Metals & Materials Society, 2010, vol. 1.
98. D. McNaught and A. Wilkinson, *UPAC. Compendium of Chemical Terminology*, 2nd edn, Blackwell Scientific Publications, Oxford, 1997.
99. D. M. Raymond and B. L. Nilsson, *Chem. Soc. Rev.*, 2018, **47**, 3659–3720.
100. E. N. Trifonov and I. N. Berezovsky, *Curr. Opin. Struct. Biol.*, 2003, **13**, 110–114.
101. R.-P. Baumann, M. Schranz and N. Hampp, *Phys. Chem. Chem. Phys.*, 2010, **12**, 4329–4335.
102. J. J. Ramsden, *Chem. Soc. Rev.*, 1995, **24**, 73.
103. V. Vogel and G. Baneyx, *Annu. Rev. Biomed. Eng.*, 2003, **5**, 441–463.
104. S. Chen, J. Zheng, L. Li and S. Jiang, *J. Am. Chem. Soc.*, 2005, **127**, 14473–14478.
105. C. Leng, H.-C. Hung, S. Sun, D. Wang, Y. Li, S. Jiang and Z. Chen, *ACS Appl. Mater. Interfaces*, 2015, 7, 16881–16888.
106. S. I. Jeon, J. H. Lee, J. D. Andrade and P. G. De Gennes, *J. Colloid Interface Sci.*, 1991, **142**, 149–158.
107. I. M. Kuznetsova, B. Y. Zaslavsky, L. Breydo, K. K. Turoverov and V. N. Uversky, *Molecules*, 2015, **20**, 1377–1409.
108. A. M. Hyde, S. L. Zultanski, J. H. Waldman, Y. L. Zhong, M. Shevlin and F. Peng, *Org. Process Res. Dev.*, 2017, **21**, 1355–1370.
109. P. Vilaseca, K. A. Dawson and G. Franzese, *Soft Matter*, 2013, 1–9.
110. A. W. Perriman and S. Mann, *ACS Nano*, 2011, **5**, 6085–6091.
111. K. P. Sharma, A. M. Collins, A. W. Perriman and S. Mann, *Adv. Mater.*, 2013, **25**, 2005–2010.
112. J. P. K. Armstrong, R. Shakur, J. P. Horne, S. C. Dickinson, C. T. Armstrong, K. Lau, J. Kadiwala, R. Lowe, A. Seddon, S. Mann, J. L. R. Anderson, A. W. Perriman and A. P. Hollander, *Nat. Commun.*, 2015, **6**, 1–6.
113. T. Farrugia, A. W. Perriman, K. P. Sharma and S. Mann, *Chem. Commun.*, 2017, **53**, 2094–2097.
114. K. A. Woodhouse, J. I. Weitz and J. L. Brash, *J. Biomed. Mater. Res.*, 1994, **28**, 407–415.
115. K. R. Davey, S. Chandrakash and B. K. O'Neill, *Food Control*, 2013, **29**, 248–253.
116. R. Schlögl, *Angew. Chem. Int. Ed.*, 2015, **54**, 3465–3520.
117. K. Hiti, J. Walochnik, C. Faschinger, E.-M. Haller-Schober and H. Aspöck, *Eye*, 2004, **19**, 1301.
118. D. M. Lehmann, M. E. Cavet and M. E. Richardson, *Contact Lens Anterior Eye*, 2010, **33**, S24–S32.
119. N. J. Wood, S. E. Maddocks, H. J. Grady, A. M. Collins and M. E. Barbour, *Int. J. Nanomedicine*, 2014, **9**, 4145–4152.
120. M. E. Barbour, S. E. Maddocks, H. J. Grady, J. A. Roper, M. D. Bass, A. M. Collins, R. M. Dommett and M. Saunders, *Nanomedicine*, 2016, **11**, 2049–2057.

121. N. Maccallum, C. Howell, P. Kim, D. Sun, R. Friedlander, J. Ranisau, O. Ahanotu, J. J. Lin, A. Vena, B. Hatton, T. S. Wong and J. Aizenberg, *ACS Biomater. Sci. Eng.*, 2015, **1**, 43–51.
122. A. Tripathy, P. Sen, B. Su and W. H. Briscoe, *Adv. Colloid Interface Sci.*, 2017, **248**, 85–104.
123. A. M. Collins, J. Zabkiewicz, C. Ghiggi, J. C. Hauser, A. K. Burnett and S. Mann, *Small*, 2011, 7, 1635–1640.
124. M. E. Barbour, S. E. Maddocks, N. J. Wood and A. M. Collins, *Int. J. Nanomedicine*, 2013, **8**, 3507–3519.
125. A. E. L. Roberts, S. E. Maddocks and R. A. Cooper, *Microbiol.*, 2012, **158**, 3005–3013.
126. S. Pogodin, J. Hasan, V. A. Baulin, H. K. Webb, V. K. Truong, T. H. Phong Nguyen, V. Boshkovikj, C. J. Fluke, G. S. Watson, J. A. Watson, R. J. Crawford and E. P. Ivanova, *Biophys. J.*, 2013, **104**, 835–840.
127. G. Binnig, H. Rohrer, C. Gerber and E. Weibel, *Appl. Phys. Lett.*, 1982, **40**, 178–180.
128. C. Toumey, *Nat. Nanotechnol.*, 2010, **5**, 239–241.
129. Y. Martin, C. C. Williams and H. K. Wickramasinghe, *J. Appl. Phys.*, 1987, **61**, 4723–4729.
130. Z. Shao, J. Yang and A. P. Somlyo, *Annu. Rev. Cell Dev. Biol.*, 1995, **11**, 241–265.
131. P. G. Arscott, G. Lee, V. A. Bloomfield and D. F. Evans, *Nature*, 1989, **339**, 484–486.
132. A. Cricenti, S. Selci, A. C. Felici, R. Generosi, E. Gori, W. Djaczenko and G. Chiarotti, *Science (80-.)*, 1989, **245**, 1226–1227.
133. D. D. Dunlap and C. Bustamante, *Nature*, 1989, **342**, 204–206.
134. Y. L. Lyubchenko, B. L. Jacobs and S. M. Lindsay, *Nucleic Acids Res.*, 1992, **20**, 3983–3986.
135. W. M. Heckl and G. Binnig, *Ultramicroscopy*, 1992, **42–44**, 1073–1078.
136. C. R. Clemmer and T. P. Beebe, *Science (80-.)*, 1991, **251**, 640.
137. C. Bustamante, A. Vesenka, C. T. Lin, W. Rees, M. Guthold and R. Keller, *Biochemistry*, 1992, **31**, 22–26.
138. N. H. Thomson, S. Kasas, B. L. Smith, H. G. Hansma and P. K. Hansma, *Langmuir*, 1996, **12**, 5905–5908.
139. E. Henderson, *Nucleic Acids Res.*, 1992, **20**, 445–447.
140. J. Yang, K. Takeyasu and Z. Shao, *FEBS Lett.*, 1992, **301**, 173–176.
141. F. Zenhausern, M. Adrian, B. Ten Heggeler-Bordier, R. Emch, M. Jobin, M. Taborelli and P. Descouts, *J. Struct. Biol.*, 1992, **108**, 69–73.
142. Y. Bustanji, C. R. Arciola, M. Conti, E. Mandello, L. Montanaro and B. Samori, *Proc. Natl. Acad. Sci.*, 2003, **100**, 13292–13297.
143. E. L. Florin, V. T. Moy and H. E. Gaub, *Science (80-.)*, 1994, **264**, 415–417.
144. C. Verbelen and Y. F. Dufrêne, *Integr. Biol.*, 2009, **1**, 296–300.
145. C. K. Riener, C. M. Stroh, A. Ebner, C. Klampfl, A. A. Gall, C. Romanin, Y. L. Lyubchenko, P. Hinterdorfer and H. J. Gruber, *Anal. Chim. Acta*, 2004, **506**, 115.

146. J. M. Fernandez and H. Li, *Science (80-.)*., 2004, **303**, 1674–1679.
147. M. Rief, M. Rief, M. Gautel, F. Oesterhelt, J. M. Fernandez and H. E. Gaub, *Science (80-.)*., 2013, **1109**, 1109–1112.
148. X. Chen, M. C. Davies, C. J. Roberts, K. M. Shakesheff, S. J. B. Tendler and P. M. Williams, *Anal. Chem.*, 1996, **68**, 1451–1455.
149. R. N. Jagtap and A. H. Ambre, *Indian J. Eng. Mater. Sci.*, 2006, **13**, 368–384.
150. M. Born and E. Wolf, in *Principles of Optics*, Elsevier, 1980, pp. 1–70.
151. B. C. Wilson and S. L. Jacques, *IEEE J. Quantum Electron.*, 1990, **26**, 2186–2199.
152. K. Koch, C. Neinhuis, H. J. Ensikat and W. Barthlott, *J. Exp. Bot.*, 2004, **55**, 711–718.
153. T. R. Albrecht, S. Akamine, T. E. Carver and C. F. Quate, *J. Vac. Sci. Technol., A*, 1990, **8**, 3386–3396.
154. E. Meyer, H. Heinzelmann, P. Grutter, T. Jung, T. Weisskopf, H.-R. Hidber, R. Lapka, H. Rudin and H.-J. Güntherodt, *J. Microsc.*, 1988, **152**, 269.
155. A. Boisen, O. Hansen and S. Bouwstra, *J. Micromech. Microeng.*, 1996, **6**, 58–62.
156. D. Saya, K. Fukushima, H. Toshiyoshi, G. Hashiguchi, H. Fujita and H. Kawakatsu, *Sensors Actuators, A*, 2002, **95**, 281–287.
157. W. A. Ducker, T. J. Senden and P. A. Pashley, *Nature*, 1991, **353**, 2239.
158. M. Kappl and H. J. Butt, *Part. Part. Syst. Charact.*, 2002, **19**, 129–143.
159. H. J. Butt, B. Cappella and M. Kappl, *Surf. Sci. Rep.*, 2005, **59**, 1–152.
160. P. Dörig, D. Ossola, A. M. Truong, M. Graf, F. Stauffer, J. Vörös and T. Zambelli, *Biophys. J.*, 2013, **105**, 463–472.
161. D. Croft, G. Shed and S. Devasia, *J. Dyn. Syst. Meas. Control*, 2001, **123**, 35.
162. P. Eaton and P. West, *Atomic Force Microscopy*, Oxford University Press, 2010.
163. D. J. Müller, D. Fotiadis, S. Scheuring, S. A. Müller and A. Engel, *Biophys. J.*, 1999, **76**, 1101–1111.
164. S. Belaidi, P. Girard and G. Leveque, *J. Appl. Phys.*, 1997, **81**, 1023–1030.
165. J. N. Israelachvili, *Intermolecular and Surface Forces*, 3rd edn, Academic Press, San Diego, 2011, pp. 341–380.
166. W. T. Johnson, *Imaging DNA in Solution with the AFM*, Agilent Technologies, 2008, pp. 1–4.
167. Y. L. Lyubchenko, *Micron*, 2012, **42**, 196–206.
168. J. Goldstein and E. Johnson, *Theory, Cult. Soc.*, 2015, **32**, 61–81.
169. H. Irschik, *Eng. Struct.*, 2002, **24**, 5–11.
170. D. Evangelista, S. Hotton and J. Dumais, *J. Exp. Biol.*, 2011, **214**, 521–529.
171. A. R. Studart and R. M. Erb, *Soft Matter*, 2014, **10**, 1284–1294.
172. A. Rafsanjani, V. Brulé, T. L. Western and D. Pasini, *Sci. Rep.*, 2015, **5**, 8064.
173. S. Alben, *Adv. Comput. Math.*, 2014, **41**, 1–32.

174. E. Reyssat and L. Mahadevan, *J. R .Soc., Interface*, 2009, **6**, 951–957.
175. S. Reichert, A. Menges and D. Correa, *CAD Comput. Aided Des.*, 2015, **60**, 50–69.
176. J. Agbenyega, *Mater. Today*, 2009, **12**, 12.
177. K. Kuribayashi, K. Tsuchiya, Z. You, D. Tomus, M. Umemoto, T. Ito and M. Sasaki, *Mater. Sci. Eng., A*, 2006, **419**, 131–137.
178. N. S. Gettys and E. K. Jacobsen, *J. Chem. Educ.*, 2001, **78**, 2–3.
179. M. Mariano, N. El Kissi and A. Dufresne, *J. Polym. Sci., Part B: Polym. Phys.*, 2014, **52**, 791–806.
180. M. C. Mulakkal, A. M. Seddon, G. Whittell, I. Manners and R. S. Trask, *Smart Mater. Struct.*, 2016, **25**, 095052.

CHAPTER 3

Energy Conversion and Storage

BIRGIT SCHWENZER

Division of Materials Research, National Science Foundation, Alexandria, Virginia 22314, USA
Email: bschwenz@nsf.gov

3.1 Introduction

Research on energy storage and conversion materials has increased sharply over the last 20 years. Between 1998 and 2017 the annual number of research publications relating to, *i.e.* using the term, 'energy storage' multiplied by more than a factor of 13 (see Figure 3.1). By summer 2018, more than 200 new research articles on 'energy storage' were published every month. Querying scientific databases for publications with the keywords 'energy conversion', 'battery' or 'photovoltaic' show slightly lower overall numbers, but similar trends, with an annual increase in research articles on these topics of ~800–900% since 1998. For all these search criteria, the number of publications in 2017 that include the words 'bioinspired' or 'biomimetic' is equal to or less than 1% each. Acknowledging that the subgroup of bioinspired research cannot easily be described and binned by using just one or two search terms, the numbers seem vanishingly small right now. Nevertheless, it is encouraging to see – as the inset in Figure 3.1 shows – that the scientific publication output on 'bioinspired energy storage' research has steadily increased over recent years and even outpaces the rate of increase compared to that of research on 'energy storage' overall.[1] However, of this small but non-negligible number of studies, many of the publications that indicate that the research is bioinspired or contains biomimicking aspects

Inorganic Materials Series No. 4
Bioinspired Inorganic Materials: Structure and Function
Edited by Simon R. Hall

Published by the Royal Society of Chemistry, www.rsc.org

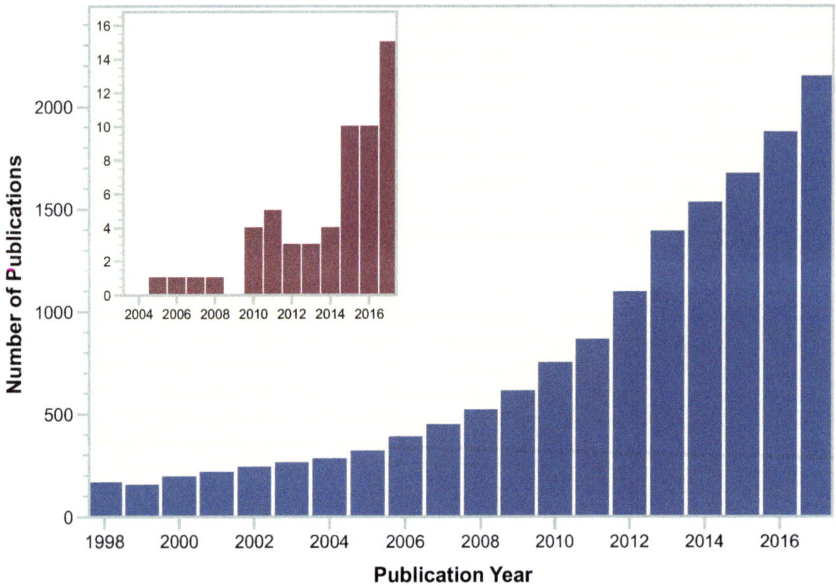

Figure 3.1 Graph depicting the results of a literature search (SciFinder; 12 November 2018) for publications on 'energy storage' between 1998 and 2017; inset shows the subset of this search for publications with the term 'bioinspired'. [Note: all values on the *y*-axis plotted in the main graph were given as minimum values in the literature search result, *i.e.* ≥2145 for 2017; the *y*-values in the inset were stated as actual numbers, *i.e.* 15 for 2017.].

do not provide any details justifying or explaining why the authors categorise their research as bioinspired.

In general, however, bioinspiration can come from many sources. For example, the eureka moment regarding new functional inorganic materials for energy conversion and storage applications may have been triggered by observations or studies of marine sponge skeletons,[2] butterflies,[3] anatomy,[4] soil,[5] or even bacteria.[6,7] The list includes living systems, as well as inanimate objects found in nature, which themselves often do not possess any energy conversion or storage functionalities.

As Valentine Vullev, University of California Riverside, eloquently explained in a perspective in 2011 'Biomimesis and bioinspiration carry different denotations and connotations, and as such, their implications for science and engineering also differ'.[8] Functional synthetic organic materials are often biomimetic, as they usually mimic, *i.e.* closely imitate, biological systems in their design and functionality, and sometimes even include biological molecules of lower complexity than the original molecular function. As an example for biomimetics, Vullev details the photosynthesis process observed in nature and argues that '[M]imicking the processes of water-splitting and carbon dioxide fixation provides this missing link in solar energy conversion'.[8]

Bioinspired materials on the other hand, especially inorganic materials as discussed in this chapter, rely on the translation of functionalities observed in the biological/natural organism or system on which they are based. They no longer contain biological components, nor do they, in most cases, closely resemble the original biomolecules. The difference between bioinspired and biomimetic approaches to energy conversion and storage materials affects the feasibility to scale up materials synthesis, fabricate proof-of-concept devices, and subsequently produce them at larger scale. While both avenues are interesting from an academic point of view and can provide valuable insights, bioinspired synthesis methods to prepare inorganic materials typically have the advantage in that they can be modified to fit existing manufacturing processes as long as the required chemical and physical conditions are met. Biomimetic pathways that include biological molecules, or fragments thereof, might be confined to certain narrow ranges of temperature and pH making them incompatible with, for example, many semiconductor processing conditions. Overall, Vullev rightly cautions that premature transition from biomimetic to bioinspired methodologies might hamper technological progress by steering scientific efforts in the wrong direction if the underlying biomimetic concepts are not yet properly understood.[8]

The bioinspiration aspect, as it pertains to inorganic materials for energy conversion and storage applications, can generally be grouped into two categories: (1) bioinspired synthesis/approach to obtain the active, *e.g.* photo- or electron-conducting, material, or (2) bioinspired design/functionality. Employing biomineralisation principles that have been translated to aid the formation of silica,[9–11] iron oxides,[7] or calcium carbonate[12,13] '*ex vivo*', or even to prepare ferroelectrics[14] and high-temperature superconductors,[15,16] among other inorganic compounds, falls into the first category of bioinspiration.

The concept of bioinspired design and functionality refers to atomic-level structure or macroscopic assembly.[17–19] In the context of this chapter, the layered crystal structure of inorganic clays is an example of an atomic-level structure type, which is desirable in battery electrode materials. Although it is possible to construct a battery directly from soil,[20] bioinspired approaches that employ different materials with a similarly layered atomic-level structure are, for the most part, more successful because electronic and ionic conductivities can be tuned in these artificial materials to achieve superior battery performance.[18,21–23]

Focusing on the commercialised and most commonly used methods to either convert or store energy, this chapter is structured into sections on photovoltaics (Section 3.2), thermal energy storage systems (TES) and phase change materials (PCMs) (Section 3.3), batteries (Section 3.4) and supercapacitors (Section 3.5). Each section first describes examples of bioinspired syntheses of functional inorganic materials relevant to the specific topic, or the fabrication of devices that contain active inorganic or hybrid materials that were prepared employing a bioinspired synthesis method. Subsequently, each section goes into more detail on some novel device designs translated from biology or nature for each energy storage or conversion system. In the case of supercapacitors, where no examples for the use of bioinspired functional inorganic materials in devices were readily accessible in the literature, approaches based on traditionally prepared inorganic materials are briefly described and parallels are drawn to existing bioinspired materials that could potentially be incorporated into supercapacitors.

The overarching aim of this chapter on bioinspired functional inorganic materials is to showcase the progress as well as the opportunities for how bioinspiration can contribute to existing energy conversion and storage solutions, or even help to overcome challenges regarding current approaches. Although by no means comprehensive, the following sections highlight some of the most innovative and creative approaches that have been reported so far for bioinspired inorganic materials in the area of energy-related applications.

3.2 Photovoltaics

With regard to synthesis of bioinspired materials and employing them for energy conversion and storage purposes, one of the earliest examples comes from Daniel E. Morse's laboratory at the University of California Santa Barbara.[2] Over several years the Morse research group went from isolating and identifying the enzyme silicatein α, responsible for the formation of the skeletal spicules around a protein filament in *Tethya aurantia*, an orange puffball sponge (both shown in Figure 3.2, upper left panel),[9,11] to using its catalytic and structure-directing hydrolysis and condensation cascade to synthesise different inorganic materials *in vitro* from silicatein filaments, such as SiO_2,[10] anatase TiO_2,[24] and Ga_2O_3 (see Figure 3.2, lower left panel)[25] under benign conditions.

The research sequence can be considered a textbook case for the transition from biomimetic research to bioinspired approaches. Early in the research, genetic engineering by site-directed mutagenesis revealed the essential roles of two specific amino acids, serine and histidine. More precisely, it illuminated the importance of their close spatial proximity. This particular interaction between the two amino acids enables the formation of the needle-like siliceous sponge skeleton. Through hydrogen bonding between the hydroxyl group of the serine and an imidazole side chain of the histidine, a catalytic reaction centre is created in the enzyme by increasing the nucleophilicity of the oxygen atom, which is part of the serine hydroxyl group. Under these conditions hydrolysis and condensation reactions can take place under benign reaction conditions (and in the ocean) yielding SiO_2 and directing the formation of the sponge skeleton (see Figure 3.2).

Subsequently, the Morse group mimicked this catalytic reaction centre: first with the help of small molecules as part of self-assembled monolayer devices,[26] and later through a kinetically controlled vapour-diffusion synthesis mechanism at the meniscus of aqueous solutions.[27,28] These reactions yield crystalline materials at room temperature or slightly above. At this point in the research, any organic reagents had been eliminated from the reaction conditions, making this a truly bioinspired rather than a

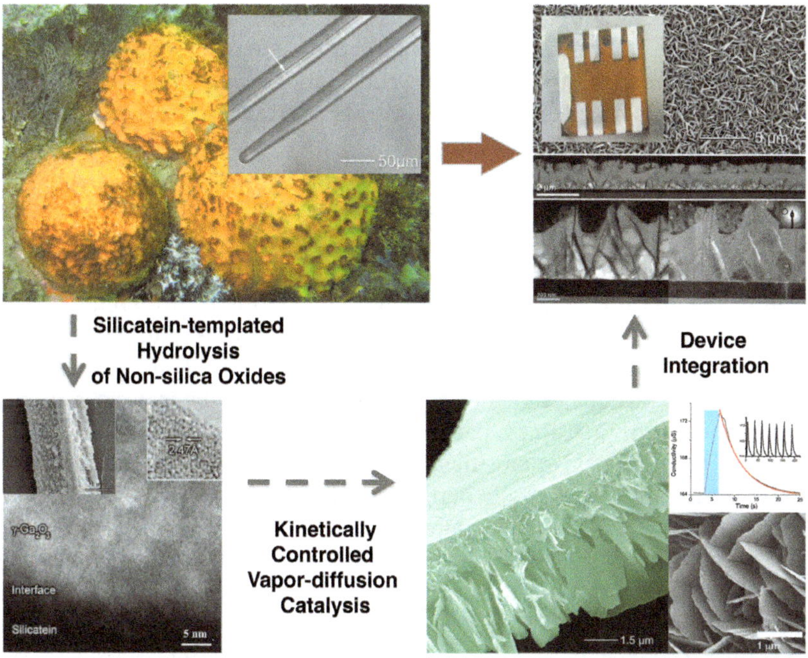

Figure 3.2 Panel top left: *Tethya aurantia*, a marine sponge (photo credit: Birgit Schwenzer), inset: optical image of silica needles that form the sponge skeleton, arrow indicates the axial silicatein filament. (Reprinted with permission from *Microsc. Res. Technol.* 2003, **62**, 4, 356–367. © 2003 Wiley-VCH Verlag GmbH & Co. KGaA, Weinheim.) Panel bottom left: high-resolution TEM image of Ga_2O_3 catalysed and templated by native silicatein filament from $Ga(NO_3)_3$ solution at room temperature, inset upper right: SEM image of the product on silicatein filament, inset upper left: lattice fringes of Ga_2O_3 product indicated by arrows. (Reprinted and adapted with permission from *Adv. Mater.* 2005, **17**, 3, 314–318. © 2005 Wiley-VCH Verlag GmbH & Co. KGaA, Weinheim.) Panel bottom right: SEM images showing side view (false colour) and bottom view of $Co_5(OH)_8(NO_3)_2{\cdot}2H_2O$ film synthesised by kinetically controlled vapour-diffusion catalysis approach, graph illustrating photoconductive behaviour of $Co_5(OH)_8(NO_3)_2{\cdot}2H_2O$ on a platinum interdigitated micro-electrode (upper right).[27] Panel top right (top to bottom): SEM image of photovoltaically active $Co_5(OH)_8(NO_3)_2{\cdot}2H_2O$/P3BT layer in the solar cell, inset: entire solar cell (photo credit: Birgit Schwenzer), overview cross-section TEM image of the active device below the Al electrode, higher resolution cross-section TEM images, bright and dark field condition, inset: SAED. Reprinted and adapted from *Thin Solid Films,* **517**, B. Schwenzer, J. R. Neilson, K. Sivula, C. Woo, J. M. J. Frechet and D. E. Morse, Nanostructured p-type cobalt layered double hydroxide/n-type polymer bulk heterojunction yields an inexpensive photovoltaic cell, 5722–5727. Copyright 2009, with permission from Elsevier.

biomimetic synthesis pathway to prepare functional inorganic materials. In the case of the kinetically controlled vapour-diffusion synthesis approach, first reported by Schwenzer *et al.*,[27] low concentrations of ammonia vapour are diffused into aqueous metal salt solutions. The formation of the inorganic material and the structure-directing aspects are governed by the change in pH (increasingly basic) and the spatial/vectorial control imparted by the directional diffusion of the ammonia reactant through the meniscus of the solution. By comparison, in the growth of the sponge skeleton/siliceous spicules, from which the process was translated, the formation of inorganic material and the structure-directing aspects both originate from the silicatein filament. For more details on the translation of the molecular mechanism of biosilification that controls the skeleton formation of *Tethya aurantia*, the marine puffball sponge, to soft chemistry principles, readers are referred to a 2008 review article by Brutchey and Morse.[29] In the years after the publication of this review article, Ould-Ely *et al.* demonstrated that the bioinspired, kinetically controlled, catalytic hydrolysis and polycondensation approach to preparing $BaTiO_3$ and other metal oxides can be scaled up to industrial quantities, enabling the synthesis of these nanocrystals under environmentally benign conditions.[30,31]

The achievement of synthesising metal oxides, hydroxides and some metal phosphates from over 30 different transition and main group metals using this bioinspired, kinetically controlled vapour-diffusion synthesis approach is impressive in itself (not all results published), but perhaps even more astonishing was the observation that, for example, $Co_5(OH)_8(NO_3)_2{\cdot}2H_2O$, one of the first layered hydroxide salts prepared by kinetically controlled vapour-diffusion catalysis in Morse's research group, is a photovoltaically active p-type semiconductor (shown in Figure 3.2, lower right panel).[27] Inherently p-type semiconductors are rare. In this case the p-type conductivity is the result of the cationic brucite-type layers within the crystal structure of $Co_5(OH)_8(NO_3)_2{\cdot}2H_2O$. Intrigued by this observation, Schwenzer *et al.* teamed up with experts in semiconductor polymer research at the Molecular Foundry, a Department of Energy user facility at Lawrence Berkeley National Laboratory. They demonstrated a functional bulk heterojunction solar cell (shown in Figure 3.2, upper right panel).[2] Poly(3-butylthiophene) (P3BT), which exhibits reasonable hole- as well as electron-conducting properties, was used in conjunction with the nano- and micro-structured film of the aforementioned $Co_5(OH)_8(NO_3)_2{\cdot}2H_2O$. Hampered by several issues, not least being the non-availability of a well-matched, chemically compatible n-type polymer and the unoptimised overall device thickness, the bulk heterojunction solar cell exhibited an extremely low conversion efficiency despite an open circuit voltage (V_{OC}) of 1.38 V, which is close to the theoretical maximum of this cell. Leaving room for improvement, this result was nevertheless a first step towards optoelectronic devices containing a functional inorganic material based on bioinspiration obtained from a marine sponge (*Tethya aurantia*) skeleton.[2]

Around the same time, Angela M. Belcher (now at Massachusetts Institute of Technology (MIT)), started her work on employing M13 phage display libraries to screen peptides for their ability to bind to semiconductor crystals and to mimic their organisation and nucleation of inorganic materials[32] as is seen by proteins in natural systems, for example, magnetic bacteria.

Today, the use of filamentous M13 bacteriophage in Belcher's work might still prove problematic in a potential commercialisation step for larger scale preparation of inorganic materials, compared to the inexpensive, low-technology approach Morse *et al.* developed. However, since bacteriophage materials are extensively used in other fields, scientists have already thought about technical solutions to increase M13 bacteriophage production capabilities.[33] Phage display technologies have been developed and are widely used in certain subfields of biology and medicine. With the 2018 Nobel Prize in Chemistry this technology has been further thrust into the limelight related to its use in pharmaceuticals development. Frances H. Arnold (California Institute of Technology, USA), George P. Smith (University of Missouri, USA) and Sir Gregory P. Winter (MRC Laboratory of Molecular Biology, Cambridge, UK) were jointly awarded the Nobel Prize 'for the directed evolution of enzymes'[34] (Arnold, 50%) and 'for the phage display of peptides and antibodies'[34] (Smith and Winter, 50%), respectively. The abstract of the press release issued by the Royal Swedish Academy of Sciences reads: 'The power of evolution is revealed through the diversity of life. The 2018 Nobel Laureates in Chemistry have taken control of evolution and used it for purposes that bring the greatest benefit to humankind. Enzymes produced through directed evolution are used to manufacture everything from biofuels to pharmaceuticals. Antibodies evolved using a method called phage display can combat autoimmune diseases and in some cases cure metastatic cancer.'[34] What the abstract (and the entire statement) does not address is the impact the development of phage display libraries has had on materials science. In particular, it is a powerful tool in the area of biomimetic and also bioinspired inorganic materials research.

Readers interested in this aspect of enzyme evolution and its use are referred to previously published overview articles[35–38] for more general information about phage display libraries, and in particular a review article by Sarikaya *et al.* from 2003: 'Molecular biomimetics: nanotechnology through biology'[39] and the book chapter titled 'Learning from Biology: Viral-Templated Materials and Devices' by Elaine D. Haberer in *Microelectronics to Nanoelectronics: Materials, Devices & Manufacturability*.[35] Haberer's chapter describes the role of viruses as templates to achieve long-range order, hierarchical templates, and specific chemical binding. She likens the biology-based assembly, which constitutes a virus, to organic nanoparticles with highly uniform physical dimensions and surface chemistry. Through the introduction of binding sites for inorganic materials into the proteins that make up the virus, inorganic materials can be synthesised, and this approach can subsequently lead to virus-templated devices. Examples of nanostructured photovoltaic devices, battery electrodes and photocatalytic water-splitting systems are among the applications Haberer highlights as achievements in virus-templated nanotechnology. This reference provides more details about the biological aspects behind the virus-enabled materials synthesis and complements the chapter you are currently reading, which focuses on the structure and properties of the resulting inorganic materials.

Belcher's group at first used selective peptide binding to pattern arrays of inorganic nanoparticles onto substrates[32] and to spatially arrange preformed nanoparticles with the help of M13 bacteriophage.[40] The M13 bacteriophage was modified by fusing peptides, which are specific to the inorganic material studied, into the virus capsid. Subsequently, Belcher's research group proceeded with synthesising inorganic nanoparticles *via* virus templating (Figure 3.3).[41] The researchers specifically note 'that peptide structure, not only sequence, is important in determining crystal phase' of the thus-prepared ZnS and CdS nanoparticles.[42] This notion is similar to what Morse *et al.* observed regarding the formation of the catalytic reaction centre of silicatein α.

The observed control and homogeneity over nanoparticle nucleation soon gave rise to forming nanowires through M13 bacteriophage-directed nucleation and templating.[43,44] As in the case of the Morse group's research, not only inorganic materials found in living organisms can be prepared by this virus-assisted nucleation and templating approach. Many technologically relevant semiconductor materials are accessible through this bacteriophage-templated approach as well. However, unlike in the previously described approach, the bacteriophage method to prepare functional inorganic materials is not strictly a low-temperature synthesis. While the actual nanomaterial formation and orientation takes place under ambient conditions, the

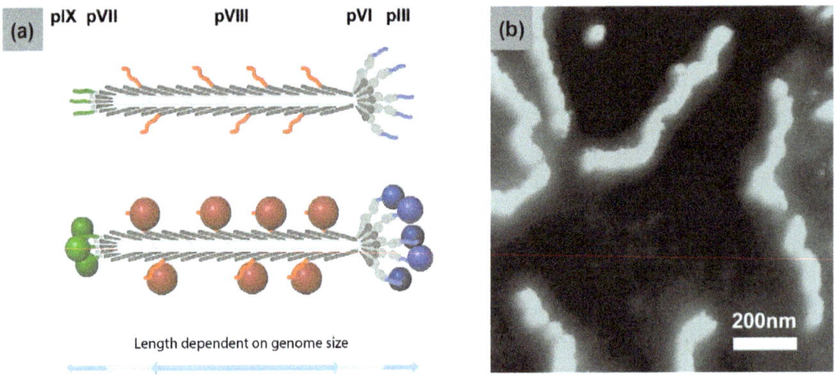

Figure 3.3 (a) Top: schematic of a M13 virus with different peptides (shown in different colours) fused to its natural coat proteins pIX (green peptide attached), pVIII (orange peptide attached), and pIII (blue peptide attached); bottom: schematic of a M13 virus with nanoparticles, represented by spheres, surrounding the peptide strands to illustrate the materials engineering potential to form and template nanostructures using the virus. (b) High-angle annular dark field scanning transmission electron microscopy (HAADF STEM) image depicting M13-templated ZnS nanowires, ZnS nanoparticles are nucleated by a specifically engineered peptide on the M13 virus protein.
Reprinted from *Acta Materialia*, **51**, C. E. Flynn, S.-W. Lee, B. R. Peelle and A. M. Belcher, Viruses as vehicles for growth, organisation and assembly of materials, 5867–5880. Copyright 2003, with permission from Elsevier.

nanoparticle assemblies and nanowires are usually annealed at moderate temperatures of 350 °C to improve the crystallinity of the inorganic material and, just as importantly, to remove the bacteriophage template.[44]

Table 3.1 provides an overview of the different solar cell types that have been prepared to date with M13 bacteriophage-templated photovoltaically active materials. For example, in 2011 Dang *et al.* reported the synthesis of single-walled carbon nanotube (SWNT)-TiO_2 nanocrystal core–shell nanocomposites using the said bacteriophage-templating technique, and used the composite as photoanodes in dye-sensitised solar cells (DSSCs).[45] After annealing the composite photoanode at 600 °C to enhance the contact between the SWNTs and TiO_2, the biomineralised TiO_2 was identified as anatase. The best-performing DSSC in this study yielded a conversion efficiency of 10.6% (0.1 wt% semiconducting SWNT; virus-to-SWNT ratio 1 : 2.5), making this a promising device design. The same device type with only TiO_2 nanoparticles yielded 8.3% conversion efficiency. The highest confirmed '1-sun' DSSC cell results are currently 11.9 ± 0.4% conversion efficiency,[46]

Table 3.1 Photovoltaic device performance metrics from representative research studies in which the photovoltaically active material was prepared using a bacteriophage-templating approach. The term 'virus' in the first column indicates that the formation of the inorganic material was assisted and templated by M13 bacteriophage.

Photoactive material	PCE (%)	V_{OC} (mV)	J_{SC} (mA cm^{-2})	Fill factor	Device type (ref.)
Virus/SWNT-TiO_2[a]	10.6	~780	~21	~0.7	DSSC (45)
Virus-$BiFeO_3$[b]	0.17	578	0.735	0.40	Solid–liquid junction (47)
Mesoporous virus-TiO_2 nanowire network[c]	7.1	750	14.0	0.67	Liquid-state DSSC (48)
Virus-TiO_2/$CH_3NH_3PbI_3$[c]	7.5	910	17.8	0.46	Solid-state hybrid (48)
Nanostructured virus-TiO_2/PbS[d]	2.93[e]	510 ± 40	13.2 ± 2.0	0.32 ± 0.03	Thin film BHJ (49)
Nanostructured virus-TiO_2/PbS with plasmonic enhancement[d,f]	3.96[e]	550 ± 40	15.5 ± 1.2	0.35 ± 0.04	Thin film BHJ (49)
Graphene/virus-WO_3/PEDOT:PSS	5.30[eg]	700 ± 10	15.1 ± 0.2[h]	0.49 ± 0.01[h]	Bilayer-structured OPV (50)

[a]Contains 0.1 wt% SWNT.
[b]Electrolyte: 1-butyl-3-methylimidazolium iodide, I_2, guanidinium thiocyanate and 4-*tert*-butyl pyridine in acetonitrile/valeronitrile (volume ratio, 85 : 15).
[c]30 nm TiO_2 nanowires.
[d]PbS nanoparticles 2.9 nm in diameter.
[e]Best device performance.
[f]Triangular Ag nanoplates.
[g]Average device performance: 5.19 ± 0.07%.
[h]Rounded values.

and this virus/SWNT-TiO_2-containing DSSC was measured under AM1.5 solar simulation at 100 mW cm^{-2}.[47]

The following year, in 2012, Belcher's group published the M13 virus-assisted synthesis of perovskite nanomaterials, specifically the M13 bacteriophage had been genetically engineered to mineralise strontium titanate, $SrTiO_3$, and bismuth ferrite, $BiFeO_3$.[47] The synthesis of ternary metal oxides by this approach was achieved *via* functionalising the virus appropriately and tuning it to interact with strontium titanium ethylene glycolate precursors, in the case of $SrTiO_3$, and bismuth iron ethylene glycolate for preparing $BiFeO_3$. Again, mild temperature elevation ($SrTiO_3$) or annealing ($BiFeO_3$) completed the reaction to yield highly crystalline $SrTiO_3$ nanowires and small $BiFeO_3$ nanoparticles, respectively. Both materials proved to be photoactive in nanostructure morphology, and Nuraje *et al.* demonstrated photocatalytic water-splitting for $SrTiO_3$ nanowires deposited with Pt nanoparticles as co-catalyst and a liquid junction solar cell made of $BiFeO_3$ nanoparticles with a conversion efficiency of 0.17% under AM1.5 solar simulation at 100 mW cm^{-2}. Unlike other perovskites, $BiFeO_3$ has a direct band gap in the visible range of the light spectrum, not in the UV, yet this was the first time the photovoltaic effect of $BiFeO_3$ nanoparticles had been investigated.

As indicated in Table 3.1, subsequent research on solar energy conversion based on M13 bacteriophage-templated inorganic materials involves, among other approaches, DSSCs with tuneable mesoporous semiconducting networks of TiO_2 nanowires.[48] The nanowire network morphology and architecture resulted in higher electron diffusion lengths compared to traditional TiO_2 nanoparticle photoanodes, and consequently higher power conversion efficiency. The thus-prepared nanowire networks could also be filled with organolead iodide perovskite material through first impregnating the porous network with PbI, then treating it with a CH_3NH_3I solution. This yielded an even more efficient solid-state hybrid solar cell than the DSSC employing the pure TiO_2 network, with power conversion efficiencies of 7.5% *versus* 7.1%, respectively (see Table 3.1).[48] Based on the same TiO_2 network, Dorval Courchesne *et al.* reported a similar hybrid bulk heterojunction thin-film solar cell architecture in 2015.[49] They incorporated PbS quantum dots (2.9 nm in diameter) into the continuous nanoporous M13 bacteriophage-templated TiO_2 network. Compared to similar thin-film bulk heterojunction solar cells, which utilise a mesoporous TiO_2 nanoparticulate network, the bacteriophage-templated architecture provides more direct pathways to conduct electrons, which leads to higher carrier transport, resulting in a more than twofold increase in power conversion efficiency compared to mesoporous TiO_2 nanoparticulate network/PbS quantum dot devices. Additionally, the authors fabricated another thin-film solar cell based on M13 bacteriophage-templated TiO_2/PbS quantum dots that incorporated various plasmonic metal nanoparticles to increase light harvesting *via* localised surface plasmon resonance. They state that by "binding metal nanoparticles to the virus prior to film assembly, the particles could be evenly distributed throughout the final nanoporous anatase film without aggregation".[49]

However, there are no details in the experimental section of the publication of how this was achieved. The best-performing cell with silver nanoplates had a power conversion efficiency of 3.96% (see Table 3.1; typical device performance of $2.98 \pm 0.58\%$).

All these examples come from a collaboration at MIT between Belcher and Paula T. Hammond's laboratories. Recently, researchers in South Korea also started exploring the use of M13 bacteriophage for solar applications.[50] Transition metal oxide thin films are often used as interlayers for charge extraction between the active layers and electrodes in organic photovoltaic devices. For graphene electrode-based organic photovoltaic devices this approach to deposit transition metal oxide films is challenging because pristine graphene is hydrophobic. Lee *et al.* report a bilayer-structured organic photovoltaic device, WO_3/PEDOT:PSS ((3,4-ethylenedioxythiophene):poly(styrenesulfonate)), for which the WO_3 thin film was prepared by engineering the M13 interface to react with sodium tungstate, the precursor to biotemplate WO_3 nanowires (see Figure 3.4a). Graphene was coated with a dilute aqueous solution of M13 bacteriophage and the liquid crystalline behaviour of M13 resulted in an ordered monolayer of the virus, which then reacted with sodium tungstate. The device preparation is shown in Figure 3.4a and b. Figure 3.4c depicts the current density–voltage plot of this biotemplated WO_3 device, as well as several graphene electrode-based organic photovoltaic control devices. As can be seen, I_{SC} (short circuit current), V_{OC} and the fill factor of the WO_3/PEDOT:PSS device are superior to the control devices, resulting in a power conversion efficiency of $5.19 \pm 0.07\%$ (see Table 3.1).[50]

After this extensive foray into photovoltaic devices that incorporate photoactive layers prepared by bioinspired synthesis approaches, bioinspired design principles for photovoltaic devices are discussed in the following paragraphs.

Butterflies and other insects use a mix of pigmentation and structural colour, the interaction between light and the structure of their wings, for a variety of purposes but predominantly for camouflage[51] and communication.[52] It is well-known that in many butterfly and moth wings, as well as beetle shells, photonic crystal-like structures are the origin of dazzling colour displays.[53] This understanding has been used to develop new bioinspired architectures and to design novel photovoltaically useful materials for solar energy conversion.

Advances in optical and electron microscopy have greatly aided the investigation of butterfly wing architectures, especially over the last two decades and much has been written about the scales, which make up the butterfly wing, and their periodic microstructure skeletons. A 2015 review article by Zhang *et al.* provides an overview of (1) structural characterisation and optical property analysis of butterfly and moth wings; (2) modelling and simulation of the optical properties and microstructure of the scales that form the wing surface; and (3) the fabrication of artificial structures inspired by these wing structures.[54] Zhang *et al.* highlight examples of butterfly wings with different microstructures, as they call them, but to list all of them goes

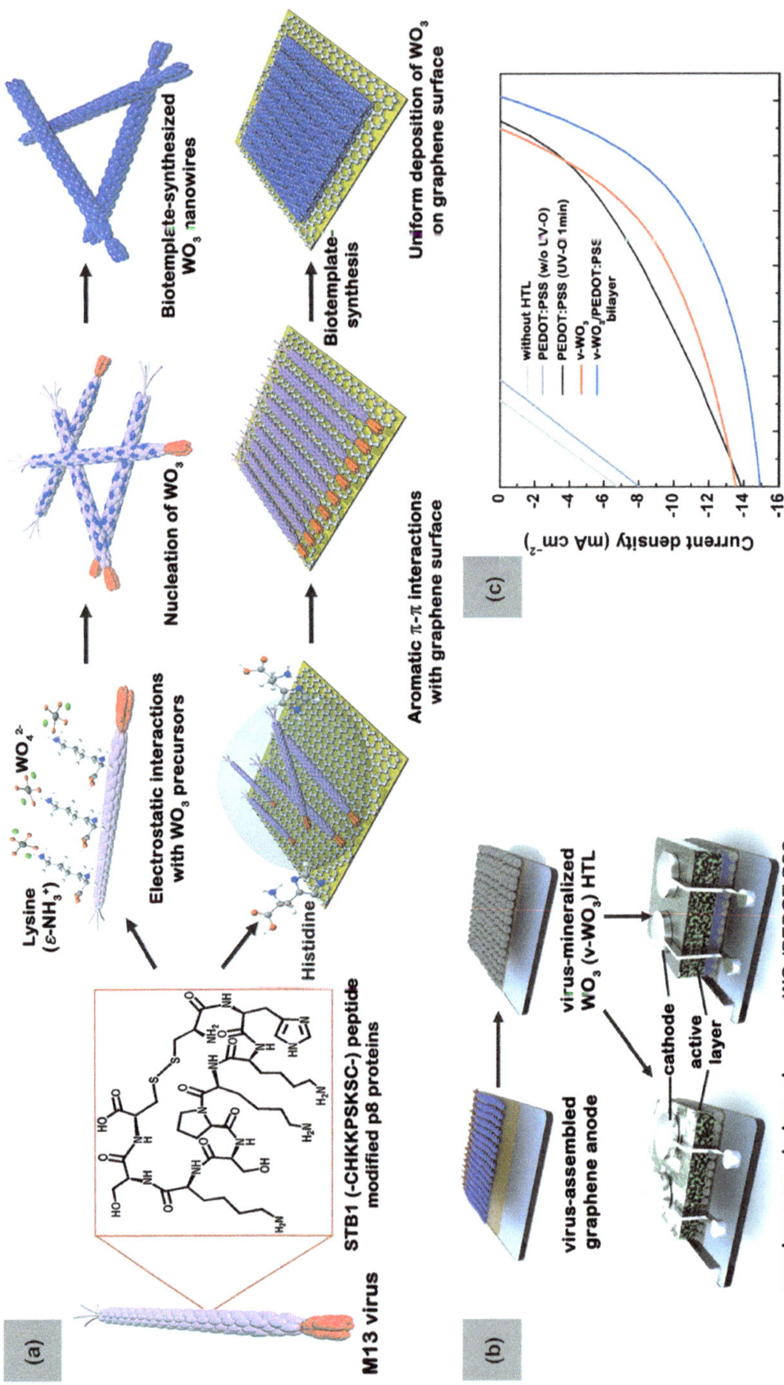
(a)
M13 virus
STB1 (-CHKKPSKSC-) peptide modified p8 proteins
Lysine (ε-NH3+)
WO4 2-
Electrostatic interactions with WO3 precursors
Nucleation of WO3
Biotemplate-synthesized WO3 nanowires
Histidine
Aromatic π-π interactions with graphene surface
Biotemplate-synthesis
Uniform deposition of WO3 on graphene surface
(b)
virus-assembled graphene anode
virus-mineralized WO3 (v-WO3) HTL
cathode
active layer
graphene anode-based OPV with v-WO3 HTL
v-WO3/PEDOT:PSS bilayer-structured HTL
(c)
without HTL
PEDOT:PSS (w/o UV-O)
PEDOT:PSS (UV-O 1 min)
v-WO3
v-WO3/PEDOT:PSS bilayer
Current density (mA cm-2)
Potential (V)

beyond the scope of this chapter. An example of the correlation between light interaction and wing design is the microstructure exhibited by many *Morpho* butterflies, which are native to South and Central America. It is a 'ridge-specialised' structure,[54] meaning that the scales exhibit ridges with multiple lamellae. Thereby, a multilayered reflector structure is created. The larger the number of lamellae on the ridges or the smaller the spacing between them, the brighter the colour of the butterfly wing.[55] More relevant to bioinspired optoelectronic energy conversion devices are highly ordered three-dimensional (3D) photonic crystal microstructures, which can be found in several lycaenid and papilionid butterfly scales. The constructive interference of light that results in a photonic band gap originates from chiral gyroid structures, and wing scales of these species have been used for biotemplating to create inorganic versions of these 3D photonic crystals.[53,56] The hybrid bulk heterojunction solar cell reported by Snaith and co-workers in 2009, with an ordered bicontinuous gyroid semiconducting network as the active material, is not based on biotemplated fabrication of the microstructure, but it nicely demonstrates how the architecture observed in butterfly wings can be translated and how the resulting 3D photonic crystal can be used for optoelectronic energy conversion devices (see Figure 3.5).[57] According to the authors, this was the first functioning electronic device of any type to make use of a gyroid structure. The preparation of the freestanding gyroid network involved a block copolymer template (poly(4-fluorostyrene)-*b*-poly(D,L-lactide); PFS-*b*-PLA), dissolution of the PLA network from the block copolymer, and infilling of the now absent PLA network with anatase TiO_2 *via* electrochemical deposition. The subsequent annealing not only led to the crystallisation and densification of the TiO_2 gyroid network, but also removed the PSF. Figure 3.5a–c show a scanning electron microscopy (SEM) image, a high-resolution transition electron microscopy (HRTEM) image and a low-magnification cross-section SEM image, respectively, of a thus-prepared TiO_2 bicontinuous gyroid network.[57] After impregnation of the TiO_2 gyroid array with either Z907, a bipyridyl ruthenium dye, or D149, an indoline-based dye, the polymer spiro-MeOTAD was infiltrated into the porous architecture by spin-coating. DSSCs with active layers prepared in this way exhibited overall power conversion efficiencies of 0.7% and 1.7%, respectively, providing another example of proof-of-concept for bioinspired solar cells.[57]

Figure 3.4 (a) Schematic of the synthesis steps to prepare a functionalised graphene surface with a M13 virus-templated and -mineralised WO_3 layer; the chemical structure of the disulfide bond-constrained peptide used to interact with the inorganic precursor ions is shown on the left. (b) Schematic illustrating the device structure for an organic photovoltaic (OPV) device incorporating the virus-mineralised WO_3 layer. (c) Current–voltage measurements comparing different OPV devices with and without the virus-mineralised WO_3 layer.
Reprinted with permission from *ChemSusChem* 2015, **8**, 14, 2385–2391.

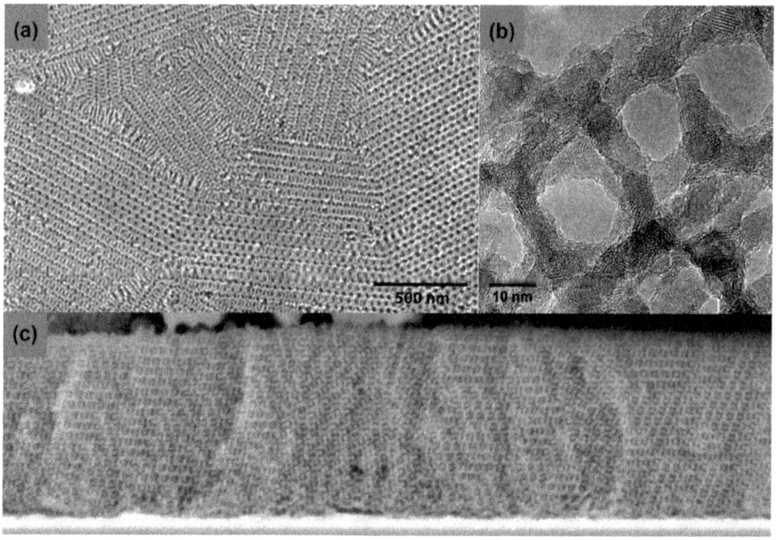

Figure 3.5 An anatase TiO_2 structure prepared in the laboratory replicates the concept of a bicontinuous gyroid network, such as that observed in butterfly wings, which results in a photonic structure. (a) SEM image of the surface of the anatase TiO_2 network prepared from infilling a titania sol–gel into a block copolymer matrix with one component removed (image after calcination shows empty array). (b) HRTEM image of the same polycrystalline TiO_2 structure, individual grains are approximately 10 nm in size. (c) Overview cross-section SEM image of a TiO_2 gyroid network shows the quality of the synthesis process over 3 μm (entire length of the image). Reprinted with permission from *Nano Lett.* 2009, **9**, 8, 2807–2812. Copyright 2009 American Chemical Society.

Yet another example of a bioinspired photovoltaic device goes beyond a mere proof-of-concept demonstration. It also originated in Henry J. Snaith's laboratory at the University of Oxford, UK, in collaboration with Hernán Míguez's group from Científicas-Universidad de Sevilla, Spain. In 2015 they reported highly efficient perovskite solar cells (see schematic in Figure 3.6a), which are based on the principles of tuneable structural colour.[58] By integrating a porous photonic crystal microstructure in the photoactive layer of a perovskite-based photovoltaic device the researchers succeeded in increasing the light absorption of the perovskite solar cell in the green-to-blue region of the visible spectrum, which yielded iridescent solar cells with a bluish colour and an efficiency of up to 8.8%. Preparation of the photonic microstructure, which needs to meet strict requirements of maximum thickness and high reflectance, was achieved by spin-coating SiO_2 nanoparticles and TiO_2 sol–gel precursor solutions to create low and high refractive index layers. Depositing the layer of SiO_2 nanoparticles as a polymer-SiO_2 composite, helped circumvent problems of TiO_2 sol–gel precursor solutions bleeding into the other layers. As in the case of the TiO_2 gyroid arrays described above, annealing of the entire photonic crystal

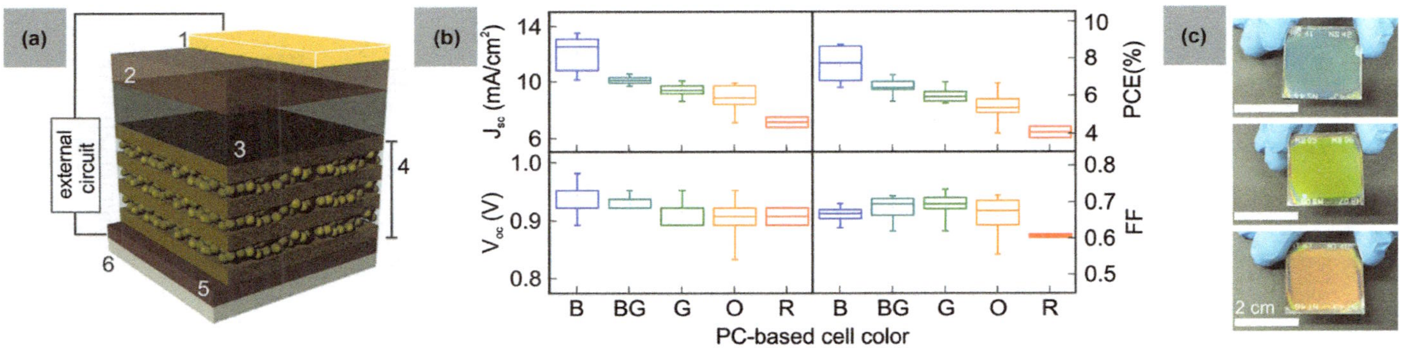

Figure 3.6 (a) A schematic of the photonic crystal-based perovskite solar cell with the different components of the device labelled 1–6 (1 = metallic contact; 2 = hole transport material, spiro-OMeTAD; 3 = perovskite overlayer, 4 = photonic crystal; 5 = transparent conducting oxide, FTO; 6 = glass substrate). (b) Box plots of the device performance parameters for different photonic crystal-based solar cells (clockwise from the upper right quadrant): PCE, fill factor (FF), open circuit voltage (V_{oc}) and short circuit current density (J_{sc}), labels along the x-axis indicate the colour of the solar cell, blue, blue-green, green, orange and red, respectively. (c) Pictures of a blue, green and orange photonic crystal-based perovskite solar cell (top to bottom). Reprinted with permission from *Nano Lett.* 2015, **15**, 8, 1698–1702. Copyright 2015 American Chemical Society.

increases the density of the TiO_2 layer and decomposes the sacrificial polymer content in the SiO_2 nanoparticle layer. Depending on the precise photonic microstructure that is incorporated into the photonic crystal-based perovskite solar cell, the hue of the device can be changed over an impressive range of the CIE 1931 *xy* chromaticity space.[58] Four different parameters are plotted in Figure 3.6b, highlighting trends in power conversion efficiency (PCE), fill factor (FF), open circuit voltage (V_{oc}) and short circuit current density (J_{sc}) for different photonic crystal-based perovskite solar cells. From a scientific point of view it is noteworthy that the colour of the device can be tuned by employing this concept, which originates from the phenomenon of structural colour in butterfly wings. From a technological and architectural point of view this means that solar cells with good efficiencies that are visually appealing and can be colour-coordinated to blend in with their area of deployment are within reach (examples shown in Figure 3.6c).

3.3 Thermal Energy Storage Systems and Phase Change Materials

Thermal energy storage (TES) systems, for example heat pumps, are more ubiquitous and well-known than the materials enabling the technology, such as polyethylene glycol in the case of smaller heat pumps or molten salt TES for industrial systems.[59,60] TES materials and systems are not as heavily researched as battery, supercapacitor, or photovoltaic materials. However, over the last few years a research emphasis on functional TES materials has emerged to accommodate a wider range of energy input, including solar, wind, or chemical energy.[61] More importantly in the context of this book, TES materials and systems can also be found in nature, and researchers are starting to take notice.

One example is the elaborate heat regulation in butterflies and other insects.[3,62] In addition to the inspiration butterflies have lent to photovoltaic cell design (see Section 3.2), researchers are also looking at the photonic structures of butterfly wings to model thermal management. With the extension into the mid- and near-infrared, the utilisation of solar energy in TES systems encompasses a larger range of the solar spectrum than photovoltaic applications, which opens up different avenues of research.

Butterflies do not possess internal thermal regulatory systems. Instead, they heat their bodies by exposing themselves to solar radiation. For example, a study of a lycaenid butterfly sister species pair by Biró *et al.* revealed that the brown colour of the species living at altitudes of 2000–2500 m in mountainous areas is not solely due to pigmentation (Figure 3.7). Instead, unlike in the related blue-winged species living at lower altitude, the nanostructure of the wing skeleton of the high-altitude butterfly does not act as a natural photonic band gap, which means it can absorb energy more efficiently.[63]

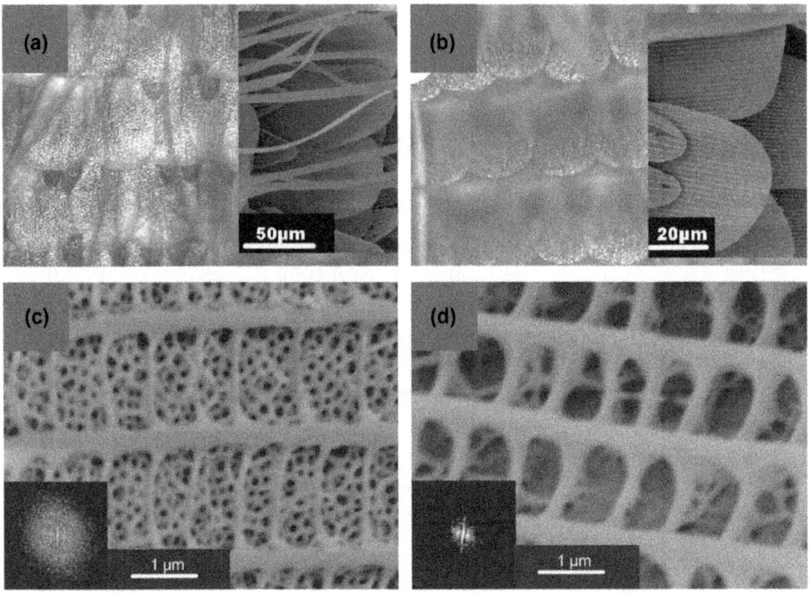

Figure 3.7 Comparison of the wing structure of a blue (BL) male lycaenid butterfly (panels a and c) and a brown (BR) male lycaenid butterfly (panels b and d). Seen through a low-magnification optical microscope (panels a and b) or studied by SEM (insets in panels a and b) at low resolution the morphology of the scales seems identical. High-resolution SEM images (panels c and d) show the fine structure of the scales with rounded ends: this reveals a distinct sponge-like structure, called 'pepper-pot' structure by entomologists, for the BL scales in panel c, but for the most part cannot be detected in the BR scales (panel d). The insets in the lower left-hand corner show the two-dimensional, logarithmic Fourier power spectra of square areas selected from the SEM images. The structural differences between the BL and BR scales are reflected here by the absence of strong features in the Fourier spectrum in panel d. Adapted with permission from L.P Biró, Z. Balint, K. Kertesz, Z. Vertsesy, G. I. Mark, Z. E. Horvath, J. Balazs, D. Mehn, I. Kiricsi, V. Lousse and J. P. Vigneron, *Phys. Rev. E*, **67**, 021907, 2003. Copyright 2003 by the American Physical Society.

Over the last two decades, *Troides helena*, a *butterfly* belonging to the family *Papilionidae*, has been studied by biologists[64,65] as well as by other scientists.[66] In 2015 *T. helena* served as the bioinspiration for Tian and co-workers in their quest for new approaches to energy harvesting.[3] They used a *T. helena* forewing, black in colour, as the scaffold to template and to fabricate novel Au–CuS-coupled photothermal materials. The chitin-based biomimetic template, a *T. helena* forewing, was functionalised with amine moieties and Au nanoparticles were grown on the aminated surface before CuS seeds, which subsequently grow into CuS nanoparticles, were deposited hydrothermally. This metal-semiconductor nanoparticle combination, in conjunction with the sub-micrometre periodic architecture of the underlying butterfly wing scaffold, led to a sub-micrometre, antireflection, quasi-photonic structure that displays enhanced broad band infrared absorption and lowered reflectance compared to other multicomponent systems.[67–71]

The SEM images in Figure 3.8 illustrate the structure of the *T. helena* forewing scaffold before (panel a) and after (panel b) coating with Au–CuS nanoparticles. A SEM cross-section image of the quasi-periodic roof-type ridges of the Au–CuS-coated *T. helena* forewing structure (panel c) highlights the sub-micrometre, antireflection, quasi-photonic structure further. According to the authors 'The triangular roof-type ridges can focus light into the scale interior *via* multiple antireflections and form the antireflection quasi-photonic structure that facilitates the capture of the light. Declining

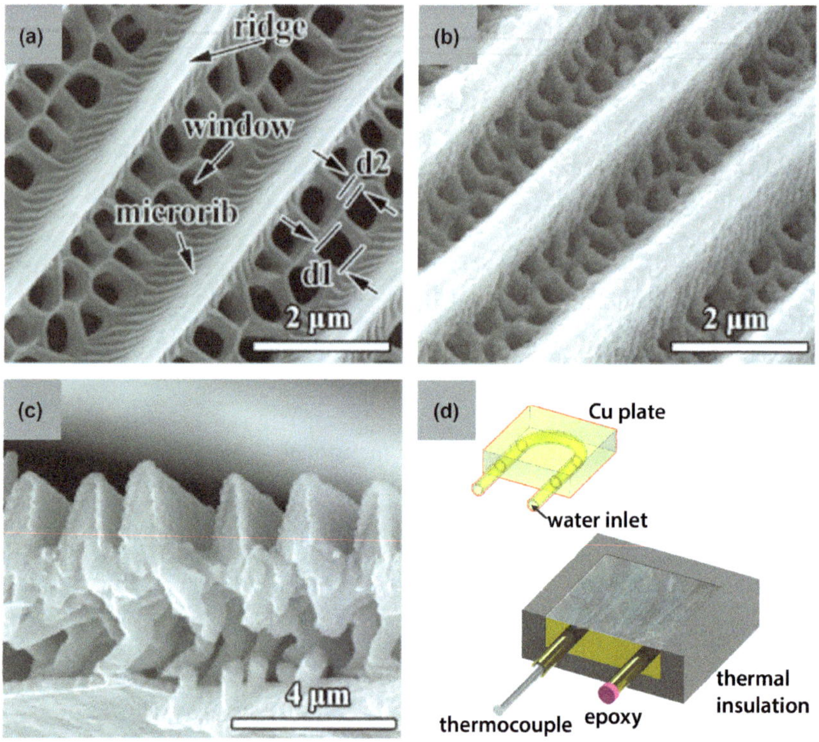

Figure 3.8 SEM images of (a) the structure of the *T. helena* forewing scaffold; (b) the same scaffold after coating with Au–CuS nanoparticles with the skeletal structure still recognisable; (c) a cross-section of the quasi-periodic roof-type ridges of the Au–CuS-coated *T. helena* forewing structure; and (d) a design schematic of the flat plate solar collector with the sub-micrometre antireflection quasi-photonic structure on top of a Cu plate; the other sides of the Cu plate are encapsulated in a thermal insulator.
Adapted from *Nano Energy*, **17**, J. Tian, W. Zhang, J. Gu, T. Deng and D. Zhang, Bioinspired Au–CuS coupled photothermal materials: enhanced infrared absorption and photothermal conversion from butterfly wings, 52–62. Copyright 2015, with permission from Elsevier.

microribs run down the sides of the ridge, which induces internal light scattering and assists in trapping light. Double-row staggered quasi-periodic lattice works are present between every two ridges, forming a series of windows. The windows elongate the effective light path and the energy density distribution interspace, implying a potentially important linkage between this feature and the light-harvesting capacity.'[3] Panel (d) in Figure 3.8 shows a schematic illustration of the simple flat plate solar collector incorporating the Au–CuS-coated *T. helena* forewing structure on a Cu plate. The authors claim this material exhibits a solar absorbance of up to 98% over a wavelength range of 300 to 25 000 nm and is effective for low-temperature ($T < 60$ °C) solar photothermal conversion. Further examples of bioinspired thermal materials were recently summarised in a review article by Tao Deng's group in Shanghai,[72] and it is easily imaginable that current work on infrared detection sensors, also based on the structure of butterfly wings,[73] can be adapted to further improve bioinspired TES systems.

Solid-state phase change materials (PCMs) exhibit different properties based on external triggers and their congruent changes in crystal structure, hence their name.[74] For the field of TES, solid-state PCMs that reversibly absorb, store and release heat based on their crystal structure transformations have tremendous potential to advance the technology.[75,76] Many PCMs exhibit a high heat storage density and a huge latent heat storage capacity, but, currently, low thermal conductivity of PCMs inhibits higher heat transfer rates and hampers their widespread commercial application as TES materials.[75,76] Polyethylene glycol (PEG) has been established as a versatile PCM for conversion of solar energy to thermal energy. It has favourable thermal properties, including high latent enthalpy, a wide transition temperature range, as well as thermal and chemical stability. Although, just like other organic PCMs, PEG also has a low thermal conductivity.[60] It is therefore commonly used in composite TES materials.[77–79]

While quite a number of different bio-based PCMs are being investigated for energy storage and conversion applications,[80–82] to date only one research article has been published on a bioinspired PCM. In 2016, Yang *et al.* from Sichuan University reported a PEG-based composite PCM with a mussel-adhesion-inspired modification of boron nitride (BN) for solar–thermal energy conversion and storage (Figure 3.9).[83] Mussels and other sessile organisms adhere to their wet, slippery and wave-pounded environment through an assortment of proteins, including the *Mytilus* foot protein 5 (mfp-5). Thirty percent of the residues in the mfp-5 sequence are 3,4-dihydroxyphenyl-L-alanine (DOPA) and 15% are lysine.[84] The catechol feature of DOPA and the primary amine of lysine, respectively, are mimicked by dopamine, a small molecule (see inset in Figure 3.9a).

In 2007, Lee *et al.* had demonstrated that a coating of polymerised dopamine, polydopamine (PDA), adheres to all types of different surfaces.[85] Yang *et al.* used this finding to prepare PDA-functionalised BN microplatelets (Figure 3.9a). By blending PEG with PDA-functionalised BN microplatelets the researchers were able to improve the dispersibility of BN in

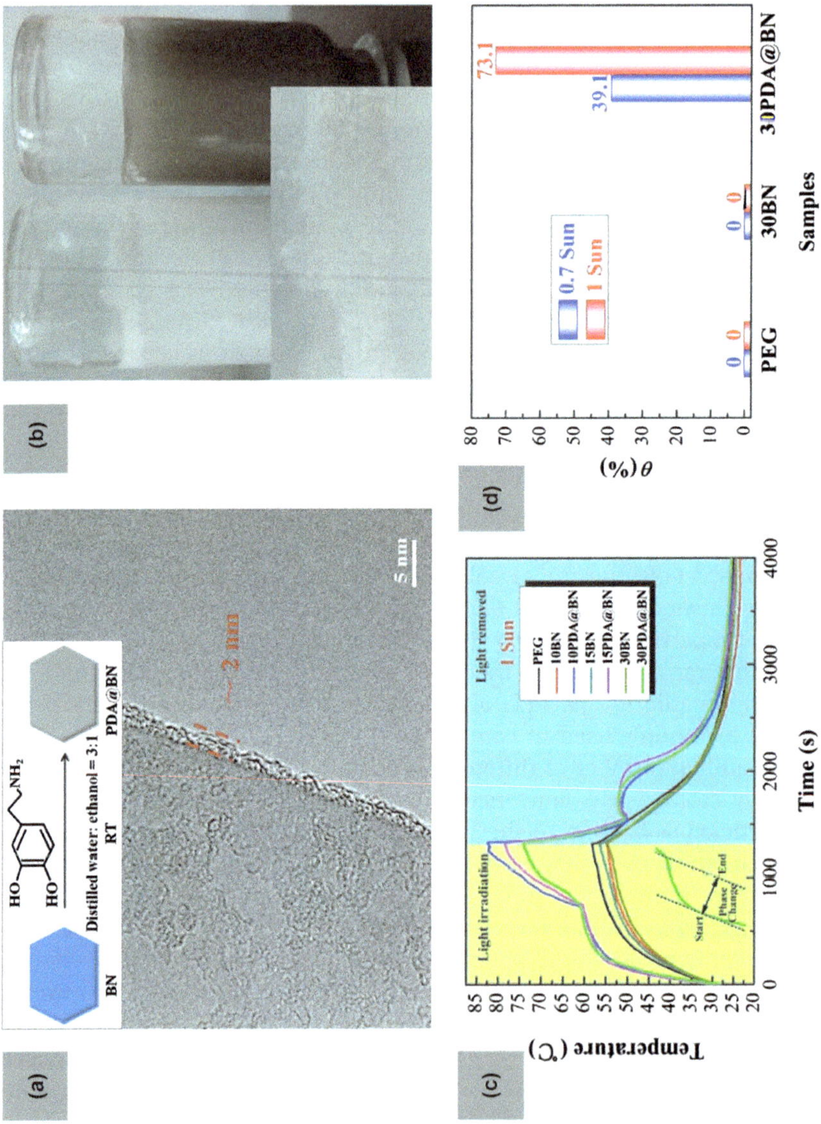
(a)
(b)
(c)
(d)
BN
HO
HO
NH2
Distilled water: ethanol = 3:1
RT
PDA@BN
~2 nm
5 nm
Light irradiation
Light removed
1 Sun
PEG
10BN
10PDA@BN
15BN
15PDA@BN
30BN
30PDA@BN
Start
Phase Change
End
Temperature (°C)
Time (s)
0.7 Sun
1 Sun
39.1
73.1
30BN
30PDA@BN
Samples
θ (%)

water (Figure 3.9b) and the treatment also leads to superior thermal conductivities and energy storage efficiencies compared to pure PEG and PEG/BN composites without PDA coating.[83] Figure 3.9c illustrates the differences in thermal conductivities between non-functionalised PEG/BN and PDA-functionalised PEG/PDA@BN composites of different BN microplatelet content. Thermogravimetric analysis (TGA) measurements revealed that approximately 4 wt% PDA was non-covalently bound to the BN microplatelets. The authors credit the improved dispersion of BN platelets in the PEG following PDA surface modification to the higher thermal conductivity of the composite PCM. PDA coating leaves the BN platelets less hydrophobic and, therefore, the miscibility with PEG increases, distributing the thermally conductive filler material more evenly throughout the PCM composite.

As can be seen in Figure 3.9d, this leads to an impressive increase in light-to-heat and energy storage efficiency. Additionally, the enhancement of absorption over the entire UV and visible light range due to PDA, which is a component of eumelanin, contributes a stronger heat absorption ability to PEG/PDA@BN (30PDA@BN in Figure 3.9d indicates a 30% mass content of BN), which results in a faster thermal response during heating and cooling. Consequently, the enhancement in light-to-heat and energy storage efficiency is more pronounced for illumination under 1 sun (100 $mW\,cm^{-2}$) *versus* 0.7 sun (70 $mW\,cm^{-2}$) with $\theta = 73.1\%$ and 39.1%, respectively.[83] Although the authors do not make this connection, one could also describe the system, perhaps even more accurately, as bioinspired with regard to the PDA eumelanin connection. Superior shape stability, as well as heating cycle reliability and stability tests, was also encouraging, which indicates that the bioinspired design of PEG/PDA@BN composite PCMs may indeed be a viable candidate for improving solar–thermal energy storage.

3.4 Batteries

Batteries *per se* do not exist in nature, even though living organisms have found numerous ways to store energy. Their energy storage capabilities are based on organic molecules and/or enzymatic mechanisms, organisms or natural building blocks, which are sometimes used *ex situ* to mineralise

Figure 3.9 (a) High-resolution TEM image depicting the edge of a polydopamine-functionalised boron nitride (PDA@BN) microplatelet (inset: a reaction schematic of the surface modification of BN microplatelets with dopamine, which polymerises to form PDA under the indicated conditions). (b) Photographs illustrating the differences between BN (left) and PDA@BN (right) microplatelets dispersed in water. (c) Temperature evolution curves of pure PEG, PEG/BN, and PEG/PDA@BN composite PCMs with different BN mass contents under 100 $mW\,cm^{-2}$ solar illumination intensity (inset: tangential method for determining the start and end points of the phase change). (d) Light-to-heat conversion and energy storage efficiency (θ) of PEG, 30BN, and 30PDA@BN.
Adapted from ref. 83 with permission from the Royal Society of Chemistry.

inorganic materials with desirable compositions for energy storage. A 2013 review article by Kim and Park[86] highlights some of the most notable examples of biomimetic mineralisation of inorganic materials for lithium-ion battery applications. While the use of microbes[87,88] or bacteria[89,90] as part of the actual materials synthesis is intriguing from a basic science point of view and has yielded remarkable results, at this time it seems doubtful that these biomimetic growth techniques can overcome the challenges of mass production, such as quantity and cost. The downside of using M13 bacteriophage as a nucleation and templating medium is similar and has already been discussed in Section 3.2. Nevertheless, the work by Belcher and her group is included in this section of bioinspired inorganic materials syntheses and designs for battery applications.

As mentioned previously, inorganic clays are suitable electrode materials in their own right, but they are by no means efficient electrode materials. Recently, layered transition metal carbides (MXenes)[18,21,22] have been explored as electrode materials, inspired by their layered crystal structure, which is similar to the atomic structure of natural clays and promotes intercalation of small molecules and ions. Kalaga *et al.*, however, did not pursue the approach of using synthetic layered materials, instead they employed microflakes of naturally occurring clay particles drenched in a solution of lithiated, room temperature ionic liquid to form a quasi-solid-state electrolyte for lithium-ion batteries.[5]

The potential of lithium-ion batteries for high-temperature applications is still not fully realised. Solid-state electrolyte research has gained momentum after some high-profile incidents of laptops catching fire, but steadily operating lithium-ion batteries at temperatures in the range of 100–150 °C poses different challenges in addition to minimising (among other aspects) the flammability of battery components or implementing features to prevent runaway reactions. Currently, the thermal and mechanical properties of some ceramic electrolytes make this class of materials one of only a few viable options for lithium-ion batteries operating in this temperature range, but their ionic conductivities are still several orders of magnitude lower than those of traditional liquid organic electrolytes. A different approach is to use room temperature ionic liquids (RTILs) as electrolytes.[91] These exhibit low vapour pressures, high thermal stability, Li^+ ion conductivity comparable to organic electrolytes and a large electrochemical potential window. Additionally, they are not flammable. At temperatures below ~60 °C, RTILs have been used extensively as electrolytes by themselves; at elevated temperatures, where RTILs exhibit lower viscosity, they have been used in composite electrolytes in combination with, for example, ceramic materials or gel polymer electrolytes. Inspired by the ability of clay to intercalate different molecules and ions, two research groups at Rice University and Wayne State University, respectively, jointly studied an RTIL/clay composite in 2015. This bioinspired hybrid electrolyte separator can be employed in ambient- and high-temperature batteries according to the authors of the study.[5] The composite material, which replaces not only the electrolyte but also the separator in traditional lithium-ion batteries, consists of micrometre-sized clay flakes that have been impregnated with lithiated RTIL. Bentonite, an aluminium phyllosilicate clay and known absorbent material, was used in this study, in combination with a $Li_4Ti_5O_{12}$ (LTO) electrode. Just as RTILs can be used to exfoliate graphene and numerous other layered materials,[92,93] they also intercalate into clays and separate the anionically charged oxide layers.[94,95] For the study in question, however, intercalated water molecules first needed to be removed from the clay structure. The presence of water molecules would result in stability and device performance issues, as Li^+ ions get introduced into the system. Therefore, Kalaga *et al.* heated the bentonite clay flakes at 650 °C under vacuum to remove both absorbed and lattice water molecules. In their thorough, multiparameter study the researchers determined the optimum ratio of RTIL/bentonite flakes:lithium salt. At the targeted operating temperature for the battery of 120 °C, the best ionic conductivity, 3.35 mS cm^{-1}, was observed for a composition of 1-methyl-1-propylpiperidinium bis(trifluoromethylsulfonyl)imide [PPMI]/bentonite and a 1 M concentration of bis(trifluoromethane)sulfonimide lithium salt [LiTFSI] (see Figure 3.10a for a schematic representation of the battery assembly and its components).[5] Galvanostatic charge–discharge measurements on this new class of quasi-solid-state electrolyte system were also conducted in a half-cell configuration, to test the capacity and cyclability at 120 °C. A stable capacity of 65 mA g^{-1} was observed at a C-rate of C/3 (Figure 3.10b)

As described, the research groups of Arava and Ajayan used naturally occurring bentonite clay in a solid-state electrolyte battery.[5] Yury Gogotsi's research group at Drexel University has been working on a different class of materials for several years: MXenes.[96] These two-dimensional (2D) metal carbides and nitrides exhibit a similar layered structure to natural clays, and Gogotsi's group and other researchers have therefore investigated them as anode and as cathode materials in batteries. MXenes are derived from layered ternary or quaternary carbides and combine metallic conductivity in the metal carbide layers with ion exchange capability between the layers, just like certain clays might. Two examples of MXene materials for which their feasibility as anode materials has been demonstrated are oxidised Mo_2CT_x MXene[22] and V_2CT_x,[21] with T_x denoting anionic surface functional groups such as T = O, OH, or F, in both cases. After investigating the energetics of general alkali ion exchange (K^+, Na^+, Li^+) in aqueous solutions for $Ti_3C_2T_x$, a different clay-like MXene, in an earlier study in 2017,[97] Gogotsi's research group subsequently reported a MXene-derived molybdenum oxide (MoO_2)-disordered carbon hybrid material as being a suitable anode material for Li-ion batteries and V_2CT_x as an anode material for Na-ion batteries. In the first study, the researchers used mild oxidation with CO_2[98] to controllably convert nanometre-thin MXene flakes into a MoO_2-disordered carbon hybrid material, which they then employed as an anode material (see schematic in Figure 3.11a).[22] When the MoO_2-disordered carbon hybrid material was tested in a standard coin-type half-cell against Li-foil, they observed an irreversible charge loss during the first discharge/charge cycle, but nevertheless a high

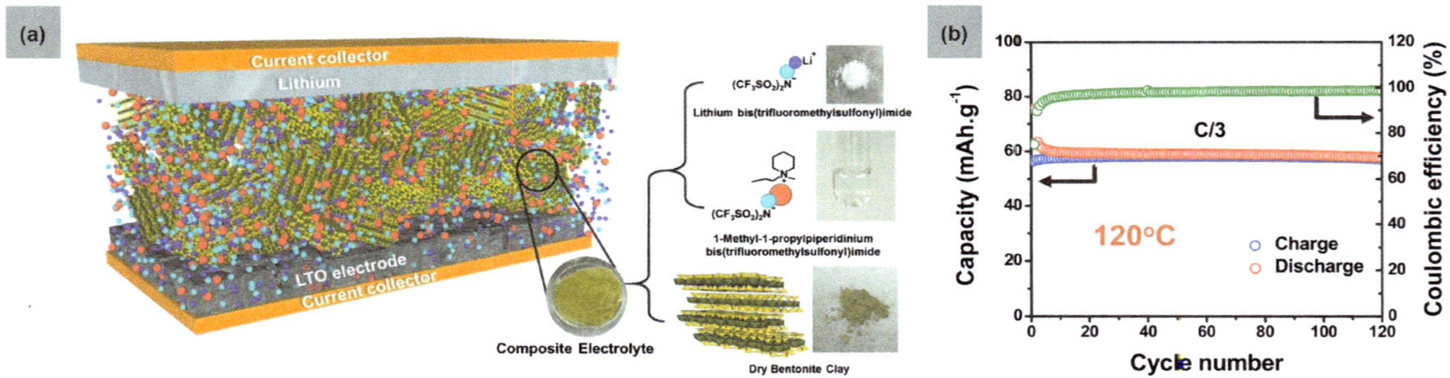

Figure 3.10 (a) Structure of the clay–RTIL quasi-solid-state electrolyte lithium-ion battery as well as schematics, formulae and images of the individual components. (b) Capacity and coulombic efficiency of this half-cell composed of $Li_4Ti_5O_{12}$ as the electrode and clay–RTIL as the quasi-solid-state electrolyte over 120 cycles at a C-rate of C/3.
Reprinted with permission from *ACS Appl. Mater. Interfaces* 2015, 7, 46, 25777–25783. Copyright 2015 American Chemical Society.

capacity value of 332 mAh g^{-1} at 50 mA g^{-1} (0.1 C) and 180 mAh g^{-1} at 1 A g^{-1} (4.0 C) and a coulombic efficiency of 99% after 280 cycles (Figure 3.11b) were obtained. By comparison, untreated Mo_2CT_x exhibited a lower capacity than did increasingly oxidised Mo_2CT_x (Figure 3.11c). According to Byeon *et al.* an increase in Li^+ ion intercalation/deintercalation for the molybdenum oxide (MoO_2)-disordered carbon hybrid material explains this increase in observed capacity.[22] For untreated Mo_2CT_x, Li^+ ions are only able to intercalate between the layers of the MXene, Mo_2CT_x. Once MoO_2 particles form during the CO_2-induced oxidation of Mo_2CT_x (see Figure 3.11a), two different intercalation/deintercalation possibilities are available for Li^+ ions during discharge: (1) the same lithiation/delithiation between the layers of the residual Mo_2CT_x and (2) through reversible phase transformation of MoO_2 to Li_xMoO_2 and back.

In the second study, on the other hand, V_2CT_x was used directly as an anode material for Na-ion batteries, which is possible because the charge storage mechanism is different.[21] Bak *et al.* collected hard X-ray absorption near edge structure (XANES) spectra for V_2CT_x at several different discharge–charge voltages during which Na^+ intercalation (discharge) and deintercalation

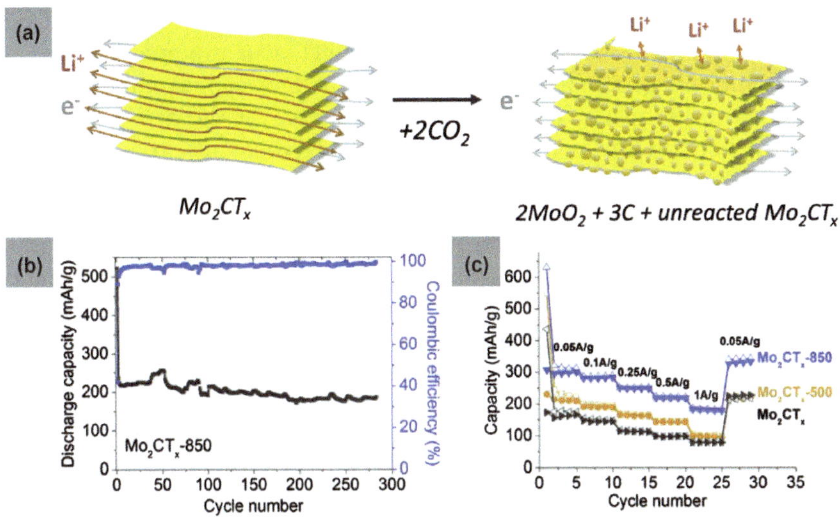

Figure 3.11 (a) Schematic of MoO_2–carbon hybrid material synthesis from Mo_2CT_x MXene using CO_2 oxidation yielding MoO_2 nanoparticles on the surface of residual Mo_2CT_x. (b) Cycle stability of a half-cell with MoO_2-disordered carbon hybrid anode material synthesised by oxidation at 850 °C, Mo_2CT_x-850. (c) Comparison of rate capability data from tests with MoO_2-disordered carbon hybrid anode material in a half-cell against Li-foil for differently treated Mo_2CT_x materials.
Reprinted from *Electrochimica Acta*, **258**, A. Byeon, C. B. hatter, J. H. Park, C. W. Ahn, Y. Gogotsi and J. W. Lee, Molybdenum oxide/carbon composites derived from the CO_2 oxidation of Mo_2CT_x (MXene) for lithium-ion battery anodes, 979–987. Copyright 2017, with permission from Elsevier.

(charge) take place. They were able to confirm that V_2CT_x undergoes a redox reaction at the transition metal site during the Na^+ intercalation/deintercalation cycle. The observed V K-edge energies, at half height of normalised XANES spectra, shift to lower energies during sodiation, which indicates a reduction of vanadium, and subsequently shifts back during deintercalation of Na^+, a shift that stems from the oxidation of vanadium. In this redox reaction the average oxidation state change of vanadium is, according to the authors, 'about the charge of 0.2 electron (*i.e.* by ~0.2 ē per V atom) over the voltage range from 0.1 to 3 V'[21] and determines the electrochemical charge storage. Additional information about MXenes and their use for energy storage can be found in a 2017 review article by Anasori *et al.*[18]

In 2018, VahidMohammadi *et al.* extended the utility of MXenes even further into the realm of next-generation battery technologies.[23] The researchers at Auburn University demonstrated that V_2CT_x used as a cathode can accommodate the intercalation of Al^{3+} ions between the MXene layers. Even though further studies into the exact charge storage mechanism are still pending, an observed relatively high discharge potential, combined with the high specific capacity of more than 300 mAh g^{-1}, makes V_2CT_x one of the best cathode materials for rechargeable Al-ion batteries to date.

In addition to the extensive research on photovoltaics described in detail in Section 3.2, Angela Belcher's group at MIT also explored the use of M13 bacteriophage-templated materials for battery applications. The synthesis of the inorganic materials is the same as for the active inorganic materials used in solar cells. Among other achievements, the group's research includes the synthesis of M13 virus-templated electrode materials for Li-ion[99–101] as well as Na-ion batteries.[102] In particular, the work on lithium–oxygen batteries, a collaboration in 2013 between Angela Belcher's group and Yang Shao-Horn's group, also in the Department of Materials Science and Engineering at MIT, created a lot of interest in and beyond the research community.[103,104] In their study they used Belcher's go-to virus, M13, to prepare biotemplated manganese oxide nanowires (bio MnO nanowires), in which the average oxidation state of manganese was determined to be between 3+ and 4+. The researchers incorporated 1–3 wt% Pd nanoparticles, a known oxygen reduction reaction (ORR) catalyst, into these porous, 3D-connected bio MnO nanowires. By comparison, more conventionally prepared lithium–oxygen battery electrode materials at the time contained ≥40 wt% of Pd.[105] The high surface area of the biotemplated MnO material, in combination with the low noble metal loading, resulted in a new catalyst electrode with increased capacity of 7340 mAh $g^{-1}{}_{c+catalyst}$ and a good cycle life (50 cycles with 400 mAh $g^{-1}{}_{c+catalyst}$) of lithium–oxygen batteries at the highest gravimetric current density compared to other transition metal oxide-based ORR electrodes in 2013.[103] The following year, the researchers were successful in engineering bio $Co_3O_4/Co(OH)_x$ nanowires with thermal treatments and incorporation of Ni nanoparticles, and they demonstrated that high-performing ORR electrodes do not require noble metal catalysts.[104]

In an entirely different approach, although also a biomimetic one, *i.e.* involving biological material, researchers from Okayama University in Japan employed a species of aquatic iron-oxidising bacteria, *Leptothrix ochracea*, to synthesise iron-based oxide.[6,7] The bacteriogenic iron-based oxide material exhibited an X-ray diffraction pattern similar to that of two-line ferrihydrite, a nanocrystalline iron oxyhydroxide, and the 3 nm amorphous nanoparticles self-assembled into microtubules of approximately 1 μm in diameter. First, they reported a discharge capacity for this bacteriogenic material that exceeded 70 $mAh\,g^{-1}$ at a very high current rate of 1670 $mA\,g^{-1}$.[6] In a subsequent publication they doped the two-line ferrihydrite nanoparticles (1.5–3 nm in diameter) with varying amounts of Si. The crystallinity of the material decreased with increased doping of the crystal structure: Si–O–Fe bonds were observed within the nanocrystalline structures for Si doping of less than 0.30 mole, and an amorphous structure was detected when the Si molar ratio surpassed 0.30. For iron-based oxide cathodes with ~0.30 molar Si doping the half cells exhibited a discharge capacity of approximately 100 $mAh\,g^{-1}$ after 20 cycles at a high current rate of 500 $mA\,g^{-1}$.[7] In both cases, using undoped and low-concentration Si-doped two-line ferrihydrite, the authors credited the size of the nanoparticles and their amorphous nature with the observed high Li-ion diffusion in and out of the cathode. Additionally, Mössbauer spectroscopy revealed the potential existence of ultrafine Fe^0 nanoparticles and/or a very complex coordination environment, which could lead to conversion reactions involving Fe^0 and suggest the suitability of the material as a high capacity anode material as well.

Moving away from using biological organisms for the synthesis of electrode materials as described in the previous paragraphs, Zhang and Morse modified the bioinspired, kinetically controlled vapour-diffusion method (described earlier and pioneered by Daniel E. Morse's group) to synthesise 'tin-in-graphite' composites, Sn@NG, as anode materials for Li-ion batteries.[106] This was the first time that homogeneous dispersion of metal particles (200–500 nm-sized Sn particles) with concurrent retention of the crystal structure of the graphite matrix was achieved. Previously reported methods to prepare such composites by ball milling led to acceptable Sn particle distribution but destroyed the crystal structure of the graphene matrix.[107] Other approaches using chemical syntheses to form such composites retained the crystal structure of graphite but resulted in precipitation of heterogeneous mixtures of Sn particles during the reduction of the Sn-containing precursors.[108] The researchers coupled the bioinspired, kinetically controlled vapour-diffusion hydrolysis of $SnCl_2$, the tin precursor, in aqueous solution with a subsequent carbothermal reaction. First, ammonia vapour was introduced into the aqueous $SnCl_2$/natural graphite solution as a catalyst, which led to the formation of a finely dispersed homogeneous mixture of SnO and SnO_2 nanoparticles within the suspension. The dried and rinsed product then underwent an *in situ* carbothermal reduction at 950 °C. As demonstrated with composites containing 5 wt%, 15 wt% and 25 wt% of tin nanoparticles, respectively, the tin-loading of the composite material can be tuned by changing the ammonia concentration or the reaction temperature. Raman spectroscopy confirmed that the observed G- and D-bands, and their ratio relative to each other, for the natural graphite starting material and the Sn@NG composite with 15 wt% Sn are very similar, which indicates that the crystallinity of the graphite matrix remains intact. Only the electrochemical performance of the 15 wt% Sn@NG composite material has been reported, with a reversible capacity of approximately 450 mAh g^{-1}, an initial coulombic efficiency of >80% and excellent cyclability, with a capacity retention of 97% over 50 cycles. All these measurements indicate a superior performance of Sn@NG compared to a composite material with 15 wt% Sn synthesised by dropwise addition of ammonia followed by carbothermal reduction, instead of kinetically controlled ammonia vapour diffusion.[106]

Shortly after the publication of this novel composite anode material, Zhang *et al.* reported a lithium vanadium oxide cathode material, LiV_3O_8, prepared by the same bioinspired ammonia vapour-diffusion synthesis approach.[109] The material itself was not new, but the flaky morphology of the 120 nm-thick sheets was unique compared to the agglomeration of irregular particles resulting from conventional methods to prepare LiV_3O_8 (see Figure 3.12a and b).

As can be seen in Figure 3.12c, d and e, this difference in morphology strongly influences the electrochemical properties of LiV_3O_8. The data titled 'Vapour diffusion' refer to the LiV_3O_8 nanoflakes (Figure 3.12a) prepared using the bioinspired synthesis method, the data labelled 'Dropwise addition' refer to LiV_3O_8 particles (Figure 3.12b) prepared by conventional sol–gel synthesis through dropwise addition of $NH_3{\cdot}H_2O$ into a metal salt precursor solution. A half-cell with a LiV_3O_8 nanoflake cathode (vapour-diffusion sample) and a lithium foil counter electrode exhibits higher reversible electrochemical capacity over 20 cycles compared to a half-cell containing a LiV_3O_8 sol–gel-synthesised cathode (Figure 3.12c). Most notably, the initial discharge capacity of the vapour-diffusion sample is ~210 mAh g^{-1}, then improves to ~250 mAh g^{-1} at the second cycle and remains relatively stable at a higher level for the rest of the cycling period of 20 cycles at a rate of 0.1 C. This increase of ~40 mAh g^{-1} from the first to the second discharge indicates, according to Zhang *et al.*, that 0.4 Li^+ ions per formula initially occupy the tetrahedral sites in the layered structure.[109] As these Li^+ ions are extracted during the first charging process vacancies result that are then available for reversible insertion/deinsertion during the following cycles, explaining the observed increase in discharge capacity. The discharge capacity for the sol–gel sample (dropwise addition), on the other hand, decreases continuously from ~195 mAh g^{-1} to ~170 mAh g^{-1} over the same period.

Results from rate capability tests carried out on the two samples are shown in Figure 3.12c (middle). The observed rate capability of the half-cell containing the vapour diffusion-prepared LiV_3O_8 is also notably superior at all rates tested to the rate performance seen for the comparison sample. At the rate of 2 C the vapour-diffusion sample exhibits a capacity that is nearly

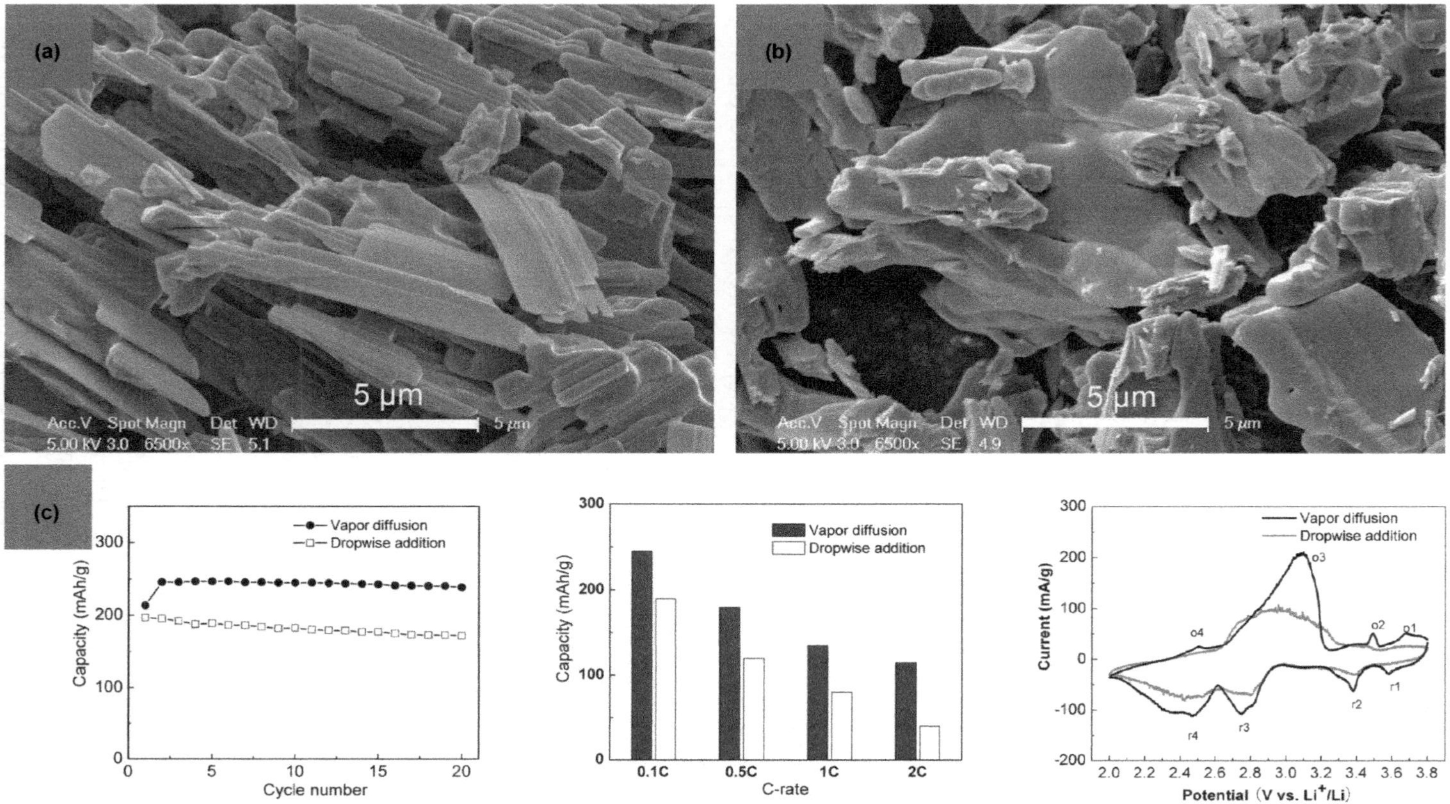

Figure 3.12 SEM images of lithium vanadium oxide (predominantly LiV_3O_8 phase) prepared by (a) bioinspired vapour-diffusion sol–gel synthesis and (b) dropwise addition of $NH_3{\cdot}H_2O$ into the same metal salt precursor solution. (c) Comparison of electrochemical properties for lithium-ion batteries containing the two different cathode materials (from left to right): cyclability, capacity at different cycle rates, and cyclic voltammetry data.
Reprinted with permission from *J. Phys. Chem. C* 2010, **114**, 19550–19555. Copyright 2010 American Chemical Society.

300% (~115 mA g^{-1}) higher than the one observed for the sol–gel sample (~40 mAh g^{-1}) prepared by dropwise addition of $NH_3{\cdot}H_2O$. The authors attribute this advantage to the shorter diffusion pathway of Li^+ ions through the thinner nanoflakes compared to the aggregate morphology of the other LiV_3O_8 material, stating that the former facilitates faster kinetics of Li^+ ion insertion/deinsertion. This assertion is supported by the sharper, better-resolved redox peaks observed for the vapour diffusion-synthesised LiV_3O_8 in a comparison of cyclic voltammetry curves (not shown here). In summary, the bioinspired, kinetically controlled vapour-diffusion synthesis method, which traces back to the formation of a marine sponge skeleton and employs no organic surfactants, yields LiV_3O_8 in a nanoflake morphology that proves to be more effective for Li^+ ion insertion/deinsertion and thereby energy storage applications. This cathode material exhibits good structural and morphological stability and outperforms the corresponding sol–gel-fabricated material by a statistically significant margin with regard to cycling stability as well as rate capability.[109]

Yuan Yang, along with his group at Columbia University and collaborators, addressed a completely different aspect of bioinspiration to develop Li-ion batteries, namely the design.[4] As society moves further in the direction of wearable electronics, along with smaller flexible electronics in general, the need for flexible power sources becomes more urgent. Conventional batteries such as coin cells or cylindrical batteries will not be able to change their shape along with the flexible device. The researchers at Columbia University developed a spine-like flexible Li-ion battery with high energy density. In their 2018 *Advanced Materials* communication the authors describe the battery architecture as: 'A thick, rigid segment to store energy through winding the electrodes corresponds to the vertebra of animals, while a thin, unwound, and flexible part acts as 'marrow' to interconnect all vertebra-like stacks together, providing excellent flexibility for the whole battery.' Figure 3.13a illustrates this comparison of the seven-layer anode/separator/cathode stack architecture to a spine.[4]

The battery architecture is very flexible and mechanically stable. In addition to electrochemical cycling (Figure 3.13b and c) under stress, the authors carried out mechanical load tests, numerical simulations and post-cycling SEM imaging of the $LiCoO_2$ cathode. The latter did not display any obvious cracking or peeling from the Al foil. Instead, while deforming the full cell from flat to flexed and back to flat and then to twisted, the initial discharge capacity of 151 mAh g^{-1} showed a retention of >94.3% after 100 cycles (Figure 3.13b and c). A stable coulombic efficiency of >99.9% was observed. Vacuum-sealed in aluminised polyethylene packing material the spine-like battery had the following characteristics and dimensions: area of active materials ($LiCoO_2$), 69.30 cm^2; mass of active materials ($LiCoO_2$), 0.852 g; entire length of the cell, 69 mm; length of each hard stack component (vertebra), 5 mm; length of interconnected spaces, 1 mm; thickness of the hard stack component (vertebra) including the packaging material, 2.4 mm. This unoptimised flexible battery exhibited an energy density of

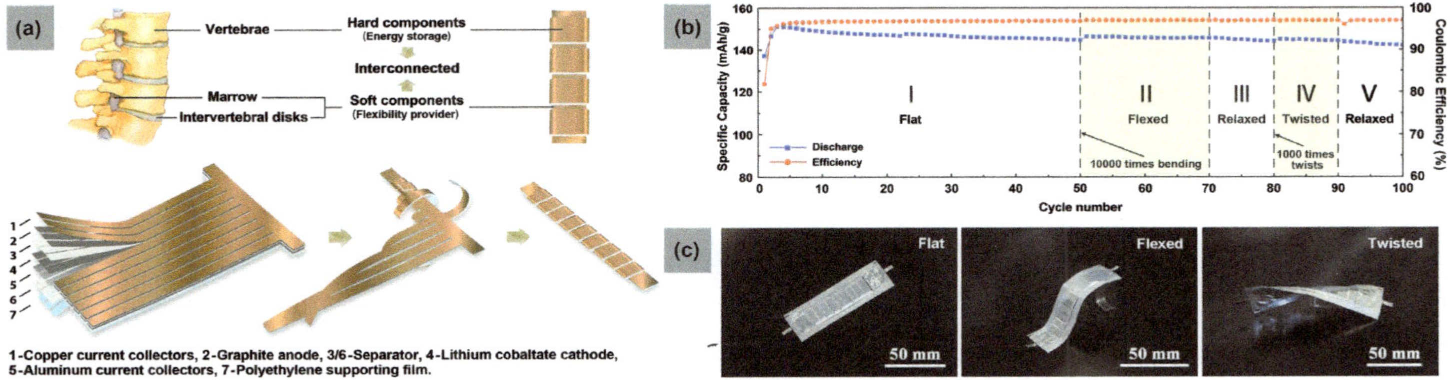

Figure 3.13 (a) A schematic of the bioinspired spine-like battery (right) compared to the section of a spine (left) above an illustrated description of the battery fabrication and design. (b) Charge/discharge cycling data of a spine-like battery with the different sections indicating what type of mechanical deformation, if any, was performed on the battery. (c) Optical images of a spine-like battery in the flat, flexed and twisted state during which the battery performance was evaluated.
Reprinted with permission from *Advanced Materials* 2018, **30**, 12, 1704947. © 2018 Wiley-VCH Verlag GmbH & Co. KGaA, Weinheim.

>85% of a similar one in conventional packing.[4] This novel battery design is a great example of a biological/natural design feature being translated to benefit an entirely unrelated application.

Similarly bioinspired, although by comparison almost a more obvious translation of design feature, are approaches to improve battery performance by taking cues from botany.[19,110] In their *Nature Communications* article, Xianfeng Zheng and co-workers observed that: 'Both plants and animals possess analogous tissues containing hierarchical networks of pores, with pore size ratios that have evolved to maximise mass transport and rates of reactions. The underlying physical principles of this optimised hierarchical design are embodied in Murray's law. However, we are yet to realise the benefit of mimicking nature's Murray networks in synthetic materials due to the challenges in fabricating vascularised structures.'[19] In 2017, the international research team from China, Belgium, USA and the UK reported the preparation of a bioinspired inorganic Murray material, a network with decreasing pore sizes across multiple scales, and its use as Li-ion battery electrodes (among other demonstrated applications). The concept of a Murray network is schematically explained in Figure 3.14a–c. The panel on the right in Figure 3.14 outlines the connections between the leaf veins, a Murray network that evolved in nature (top two images), and a hierarchical structure with macro-meso-micropores (M-M-M) self-assembled from microporous ZnO nanoparticles (bottom image).

The scheme in Figure 3.14d illustrates the fabrication of such an artificial Murray network through bottom-up assembly of microporous nanoparticles. The process utilises evaporation-induced, layer-by-layer self-assembly of microporous ZnO nanoparticles. Typically, sacrificial templating material (often polymer or silica spheres) is used to create macropores. In this study, however, the authors employ the process of solvent evaporation to form macropores, an approach driven by the change in meniscus of the evaporating solvent. As a result of self-assembling the microporous ZnO nanoparticle in this way mesopores are formed during the nanoparticle close-packing, yielding the M-M-M Murray network. Overall, not only is the newly formed Murray material based on bioinspiration, but the process to make it is also environmentally benign.[19]

This type of functional inorganic, bioinspired Murray network can be utilised for many applications, and the authors list and demonstrate several. In the context of this book, the most interesting one is its incorporation into Li-ion batteries as an electrode material. Zheng *et al.* deposited ZnO films of bioinspired Murray networks on Cu foil and incorporated them as the cathode material in a coin cell with 1 M solution of $LiPF_6$ in ethylene carbon/diethyl carbonate (1 : 1 wt%) as electrolyte and a lithium metal anode. Compared to other architectures made from either microporous or non-porous ZnO nanoparticles, batteries with ZnO-based Murray network cathodes exhibited much higher rate capabilities, outstanding cycle stability and a high reversible capacity of ~1660 mAh g^{-1} after 50 cycles at 0.05 A g^{-1}. As the authors explain, if the Li-ion transfer process in an electrode is dominated by ion transfer in

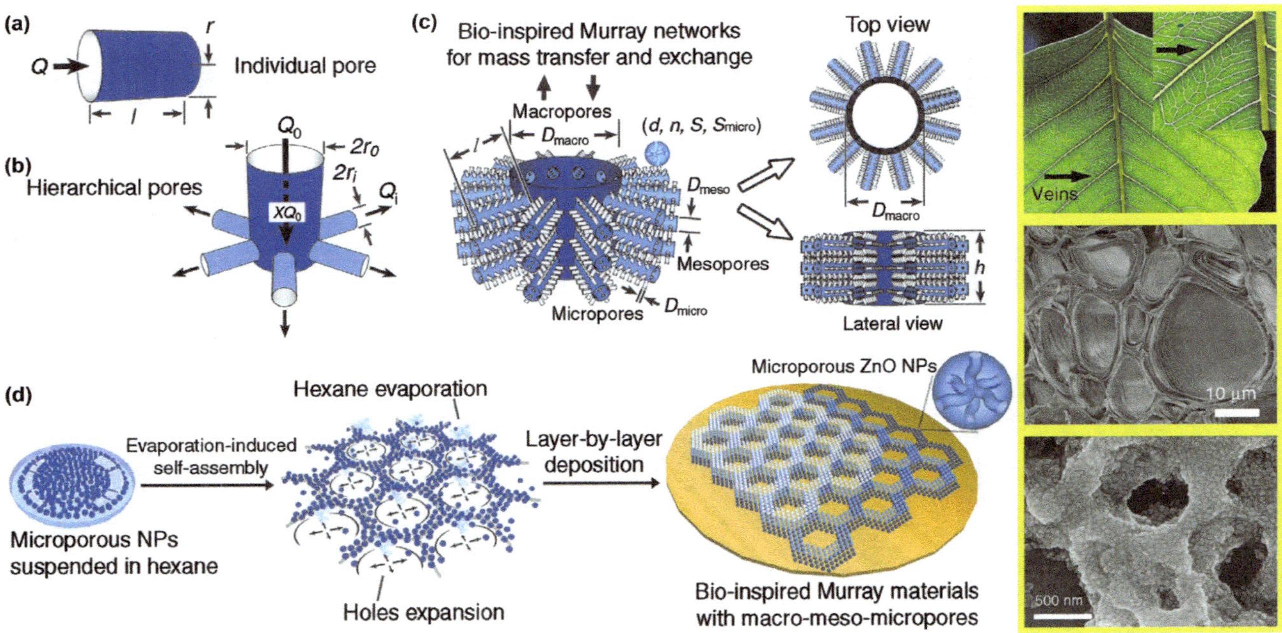

Figure 3.14 Schematic models of (a) an individual pore (with Q = laminar flow rate, r = radius, l = length). (b) A parent pore connecting to smaller pores (with X = the factor by which Q changes while moving through the pore). (c) A conceptualised hierarchically porous network based on the synthesised hierarchically macro-meso-microporous (M-M-M) materials seen from different directions (with d = diameter of nanoparticles (NPs), n = average number of micropores within a single NP, S = specific surface area of all NPs, S_{micro} = surface area of the micropores, D_{micro} and D_{meso} = diameters of micropores and mesopores, respectively). (d) The preparation of bioinspired Murray materials *via* layer-by-layer evaporation-driven self-assembly of ZnO NPs. The panel on the right shows (top to bottom): photograph of a living Murray network, namely leaf veins; cross-section SEM image of such leaf veins; and high magnification SEM image of microporous ZnO nanoparticles self-assembled into a macro-meso-microporous (M-M-M) network on a Si wafer.

pores filled with electrolyte instead of slower solid-state ion diffusion, cycling stability and rate capability increase. Additionally, high surface area is desirable in electrode materials for Li-ion batteries to enhance interfacial Li storage at the nanoscale. The bioinspired ZnO-based Murray material fulfils both requirements. The space-filling pores enable ultra-short solid-phase Li diffusion, and the optimised porous network filled with electrolyte leads to full and rapid Li-ion transfer with the electrolyte.[19]

Plant roots, which incidentally are an example of a Murray network, have also inspired a battery design. The year after Zheng *et al.* reported their findings, researchers from Huazhong University of Science and Technology and Tongji University, both in China, realised a high-area sulfur-loading cathode based on a TiN/C matrix.[110] The 3D porous interconnected structure of the root-like TiN/C matrix provides an environment for highly efficient transport of polysulfide, electrons and liquid electrolyte, thereby facilitating the redox reaction inherent to Li–S battery operation. By electrospinning a reaction solution containing polyvinyl pyrrolidone (PVP), titanium(IV) butoxide and paraffin oil, and subsequently annealing the fibre mat at 500 °C under argon, the researchers synthesised a TiO_2/C precursor material in the desired macroscopic fibre structure with mesopores resulting from calcination of the organic precursors. This material was then converted to TiN/C, because a research team lead by John B. Goodenough at the University of Texas at Austin, USA, who reported the first Li–S batteries with mesoporous TiN cathodes in 2016, highlighted the benefit that TiN is able to better bind polysulfides (through polar Ti–S and also N–S bonds) compared to TiO_2, and also possesses better electrical conductivity.[111] Hence, the approach of Liao *et al.*, of converting the TiO_2/C fibres into TiN/C, offers the same beneficial properties regarding Li–S battery performance.

The bioinspired 3D porous interconnected structure of TiN particles within the composite TiN/C fibres (average diameter of the fibre ~500 nm) delivers a slightly lower, although overall comparable performance to other Li–S batteries with TiN cathodes. Liao *et al.* did not explicitly state if the reported 56 wt% sulfur loading for their cathode material refers to TGA data;[110] assuming this is the case, battery performance data from the first reported mesoporous TiN cathodes (58.8 wt% sulfur)[111] lends itself for comparison. Li–S batteries containing mesoporous TiN cathodes exhibited an initial discharge capacity of 988 $mAh\,g^{-1}$ at 0.5 C current density and a discharge capacity of 644 $mAh\,g^{-1}$ after 500 charge/discharge cycles, which equals a capacity decay rate of 0.07% per cycle.[111] Incorporation of the root-like TiN/C matrix into Li-S batteries as the cathode material yielded an initial discharge capacity of 983 $mAh\,g^{-1}$ at 0.2 C current density, with a discharge capacity of 685 $mAh\,g^{-1}$ remaining after 300 cycles (decay rate of 0.1% per cycle).[110]

In conclusion it stands to reason that (1) employing a less energy intense synthesis approach to creating hierarchical networks of pores with varying sizes and (2) focusing on optimising transport through these pores, as in the research described above for the Murray materials,[19] might be a strategy to further enhance the impact of bioinspired inorganic materials for a variety of battery chemistries.

3.5 Supercapacitors

In supercapacitors, energy is stored electrostatically. Hence, they can be charged and discharged very quickly because, unlike in batteries, no chemical reactions take place during charge and discharge. To date, not many reports describe the incorporation of functional inorganic, bioinspired materials into supercapacitors. This is distinctly different from the trend observed for the area of battery research described in Section 3.4, where infiltration of bioinspiration is slow but noticeable. On the other hand, many of the materials and/or synthesis techniques described in Sections 3.2–3.4 could potentially be beneficial in the context of supercapacitor research as well. It is possible that these or similar approaches are already being used, but that in the field of supercapacitor research they are simply not indicated as originating from bioinspiration. The use of MXenes for supercapacitor applications and their incorporation into devices, for example, has already been reported by several research groups.[112–114]

As Oh *et al.* stated in their 2012 publication[101] on using a genetically engineered M13 virus to synthesise intimately intertwined graphene/metal oxide nanocomposites by employing graphene/virus nanotemplates: 'This study demonstrates the importance of well-dispersed graphene in aqueous media for synthesising composite materials and this general method could be extended to other materials for applications including biosensors, supercapacitors, catalysts and energy conversion applications.' Another obvious connection between supercapacitors and the research described here is that, overall, synthetic layered double hydroxide (LDH) materials, *i.e.* cationic clays, have been studied more frequently as electrode materials for solid-state hybrid energy storage devices[115] and supercapacitors[116] than for batteries. The bioinspired approach to synthetic LDHs[27] and their use in photovoltaic devices[2] has been described in Section 3.2. While LDH materials prepared by this particular bioinspired method have not yet been used to prepare supercapacitors, several examples of LDH-based energy storage devices are included in this section to illustrate that bioinspired material could compete in this field of energy application technologies.

A review article by Zhao *et al.* from 2017, titled 'Recent progress in layered double hydroxide-based materials for electrochemical capacitors: design, synthesis and performance' provides a good overview over the development and current state of this subfield.[116] Although pure LDH materials with different morphologies have been used for supercapacitor applications, their various hybrid composites, *e.g.* with carbon or polymer materials, usually exhibit better electrochemical performance. In 2010 Malak-Polaczyk *et al.* published one such study on LDH/activated carbon composite electrodes.[117] They prepared a known Co-Al-CO_3 LDH compound,[118] which has a similar layered structure to the naturally occurring mineral hydrotalcite

($Mg_6Al_2(OH)_{16}CO_3{\cdot}4H_2O$). To achieve homogeneous mixing of the LDH and the activated carbon material, activated carbon was already added during synthesis of the LDH from $Co(NO_3)_2{\cdot}6H_2O$, $Al(NO_3)_3{\cdot}9H_2O$ and urea. The authors credit "the controlled supply of carbonate and hydroxide by the decomposition of urea" in the aqueous M^{2+} and M^{3+} salt solution for the formation of monodisperse hydrotalcite particles. SEM images confirm a rather narrow particle size distribution (of the order of a few micrometres). X-ray diffraction clearly showed that these microplatelets are not single crystalline, but that the observed Scherrer broadening of the peaks indicates that even after annealing the composite material at up to 500 °C the individual crystals are rather small.[117] Unfortunately, the study did not include characterisation of the LDH microplatelets by dark field TEM, and it is impossible to assess whether the microplatelets consist of highly aligned individual nanoparticles. Based on the SEM images shown in the publication,[117] the habit that the Co–Al–CO_3 LDH microplatelets exhibit is very similar to the structure of the LDH material that Schwenzer *et al.* synthesised[27] and subsequently incorporated into a proof-of-principle solar cell.[2] Given the suggested similarities between the LDH materials synthesised employing the decomposition of urea and the one prepared using the bioinspired synthesis method, it seems justified to conclude that an LDH material, prepared *via* a bioinspired synthesis pathway, would also make a suitable active material in a supercapacitor. To the best of my knowledge though, nothing in this area has been reported yet.

3.6 Outlook

Along with the increasing public awareness that fossil fuel, natural oil and coal resources are limited, the need for new, transformative approaches to satisfy our dependence on a constant availability of electricity becomes clear. In several countries this has already lead to more efforts to advance renewable energies and implement less energy-intensive manufacturing approaches.[119] Future research along these lines will most likely include an increased focus on learning from nature and translating reaction mechanisms observed under benign conditions, energy storage and conversion principles, or the design of structures to benefit human needs. There are many examples of bioinspired synthesis of inorganic materials and design of inorganic structures already, some of which have been discussed in this chapter. So far, only a few approaches have led to proof-of-principle energy conversion and storage devices, and cases beyond a mere conceptual demonstration are still rare in this area of research and technology. Nevertheless, examples of viable bioinspired technologies like the perovskite solar cells with competitive power conversion efficiencies, which are based on the principles of tuneable structural colour,[58] or the idea of modelling a battery design on the spine-structure of vertebrates,[4] are very promising. They are bound to inspire more researchers to forego synthesis approaches with a large environmental footprint in favour of environmentally benign methods, such as the kinetically controlled vapour-diffusion synthesis described in this chapter.[27]

Acknowledgements

The author thanks D. E. Morse and her past collaborators for hours of constructive discussions about bioinspiration. The author's time to write this chapter was supported by (and while serving at) the National Science Foundation. Any opinion, findings, and conclusions or recommendations expressed in this material are those of the author and do not necessarily reflect the views of the National Science Foundation.

Birgit Schwenzer, past affiliation: Physical and Computational Sciences Directorate, Pacific Northwest National Laboratory, Richland, Washington 99352, United States. Pacific Northwest National Laboratory is operated for the United States Department of Energy by Battelle Memorial Institute under contract DE-AC05-76RL01830.

References

1. *SciFinder*; Chemical Abstracts Service: Columbus, OH; energy storage; bioinspired; https://scifinder.cas.org (accessed November 12, 2018).
2. B. Schwenzer, J. R. Neilson, K. Sivula, C. Woo, J. M. J. Fréchet and D. E. Morse, *Thin Solid Films*, 2009, **517**, 5722–5727.
3. J. Tian, W. Zhang, J. Gu, T. Deng and D. Zhang, *Nano Energy*, 2015, **17**, 52–62.
4. G. Qian, B. Zhu, X. Liao, H. Zhai, A. Srinivasan, N. J. Fritz, Q. Cheng, M. Ning, B. Qie, Y. Li, S. Yuan, J. Zhu, X. Chen and Y. Yang, *Adv. Mater.*, 2018, **30**, 1704947.
5. K. Kalaga, M.-T. F. Rodrigues, H. Gullapalli, G. Babu, L. M. Reddy Arava and P. M. Ajayan, *ACS Appl. Mater. Interfaces*, 2015, **7**, 25777–25783.
6. R. Sakuma, H. Hashimoto, G. Kobayashi, T. Fujii, M. Nakanishi, R. Kanno, M. Takano and J. Takada, *Mater. Lett.*, 2015, **139**, 414–417.
7. H. Hashimoto, Y. Nishiyama, M. Ukita, R. Sakuma, M. Nakanishi, T. Fujii and J. Takada, *Inorg. Chem.*, 2015, **54**, 7593–7599.
8. V. I. Vullev, *J. Phys. Chem. Lett.*, 2011, **2**, 503–508.
9. Y. Zhou, K. Shimizu, J. N. Cha, G. D. Stucky and D. E. Morse, *Angew. Chem., Int. Ed.*, 1999, **38**, 780–782.
10. J. N. Cha, K. Shimizu, Y. Zhou, S. C. Christiansen, B. F. Chmelka, G. D. Stucky and D. E. Morse, *Proc. Natl. Acad. Sci. U. S. A.*, 1999, **96**, 361–365.
11. J. C. Weaver and D. E. Morse, *Microsc. Res. Tech.*, 2003, **62**, 356–367.
12. E. Loste, R. J. Park, J. Warren and F. C. Meldrum, *Adv. Funct. Mater.*, 2004, **14**, 1211–1220.
13. S. Ludwigs, U. Steiner, A. N. Kulak, R. Lam and F. C. Meldrum, *Adv. Mater.*, 2006, **18**, 2270–2273.
14. R. L. Brutchey and D. E. Morse, *Angew. Chem., Int. Ed.*, 2006, **45**, 6564–6566.
15. D. Walsh, S. C. Wimbush and S. R. Hall, *Supercond. Sci. Technol.*, 2009, **22**, 015026.

16. S. R. Hall, S. C. Wimbush, Y. Shida and W. Ogasawara, *Chem. Phys. Lett.*, 2011, **507**, 144–150.
17. J. Wang, J. Harris, M. Li, D. Plana, M. E. Barbour, D. J. Fermin and S. Mann, *CrystEngComm*, 2013, **15**, 7152–7156.
18. B. Anasori, M. R. Lukatskaya and Y. Gogotsi, *Nat. Rev. Mater.*, 2017, **2**, 16098.
19. X. Zheng, G. Shen, C. Wang, Y. Li, D. Dunphy, T. Hasan, C. J. Brinker and B.-L. Su, *Nat. Commun.*, 2017, **8**, 14921.
20. https://www.youtube.com/watch?v=aCCK132OIGA (accessed November 3, 2018).
21. S.-M. Bak, R. Qiao, W. Yang, S. Lee, X. Yu, B. Anasori, H. Lee, Y. Gogotsi and X.-Q. Yang, *Adv. Energy Mater.*, 2017, 7, 1700959.
22. A. Byeon, C. B. Hatter, J. H. Park, C. W. Ahn, Y. Gogotsi and J. W. Lee, *Electrochim. Acta*, 2017, **258**, 979–987.
23. A. VahidMohammadi, A. Hadjikhani, S. Shahbazmohamadi and M. Beidaghi, *ACS Nano*, 2017, **11**, 11135–11144.
24. J. L. Sumerel, W. Yang, D. Kisailus, J. C. Weaver, J. H. Choi and D. E. Morse, *Chem. Mater.*, 2003, **15**, 4804–4809.
25. D. Kisailus, J. H. Choi, J. C. Weaver, W. Yang and D. E. Morse, *Adv. Mater.*, 2005, **17**, 314–318.
26. D. Kisailus, Q. Truong, Y. Amemiya, J. C. Weaver and D. E. Morse, *Proc. Natl. Acad. Sci. U. S. A.*, 2006, **103**, 5652–5657.
27. B. Schwenzer, K. M. Roth, J. R. Gomm, M. Murr and D. E. Morse, *J. Mater. Chem.*, 2006, **16**, 401–407.
28. J. R. Gomm, B. Schwenzer and D. E. Morse, *Solid State Sci.*, 2007, **9**, 429–431.
29. R. L. Brutchey and D. E. Morse, *Chem. Rev.*, 2008, **108**, 4915–4934.
30. T. Ould-Ely, M. Luger, L. Kaplan-Reinig, K. Niesz, M. Doherty and D. E. Morse, *Nat. Protoc.*, 2011, **6**, 97–104.
31. T. Ould-Ely, L. Kaplan-Reinig and D. E. Morse, *Adv. Funct. Mater.*, 2014, **24**, 1275–1282.
32. S. R. Whaley, D. S. English, E. L. Hu, P. F. Barbara and A. M. Belcher, *Nature*, 2000, **405**, 665–668.
33. C. M. Warner, N. Barker, S.-W. Lee and E. J. Perkins, *Bioprocess Biosyst. Eng.*, 2014, **37**, 2067–2072.
34. Press release: The Nobel Prize in Chemistry 2018; by The Royal Swedish Academy of Sciences, 3 October, 2018.
35. E. D. Haberer, in *Microelectronics to Nanoelectronics*, ed. A. B. Kaul, CRC Press Taylor & Francis Group, Boca Raton, FL, 2013, pp. 157–177.
36. J. K. Scott and G. P. Smith, *Science*, 1990, **249**, 386–390.
37. G. P. Smith and V. A. Petrenko, *Chem. Rev.*, 1997, **97**, 391–410.
38. T. Sawada and T. Serizawa, *Bull. Chem. Soc. Jpn.*, 2018, **91**, 455–466.
39. M. Sarikaya, C. Tamerler, A. K.-Y. Jen, K. Schulten and F. Baneyx, *Nat. Mater.*, 2003, **2**, 577–585.
40. S.-W. Lee, C. Mao, C. E. Flynn and A. M. Belcher, *Science*, 2002, **296**, 892–895.

41. C. E. Flynn, S.-W. Lee, B. R. Peelle and A. M. Belcher, *Acta Mater.*, 2003, **51**, 5867–5880.
42. C. E. Flynn, C. Mao, A. Hayhurst, J. L. Williams, G. Georgiou, B. Iverson and A. M. Belcher, *J. Mater. Chem.*, 2003, **13**, 2414–2421.
43. C. Mao, C. E. Flynn, A. Hayhurst, R. Sweeney, J. Qi, G. Georgiou, B. Iverson and A. M. Belcher, *Proc. Natl. Acad. Sci. U. S. A.*, 2003, **100**, 6946–6951.
44. C. Mao, D. J. Solis, B. D. Reiss, S. T. Kottmann, R. Y. Sweeney, A. Hayhurst, G. Georgiou, B. Iverson and A. M. Belcher, *Science*, 2004, **303**, 213–217.
45. X. Dang, H. Yi, M.-H. Ham, J. Qi, D. S. Yun, R. Ladewski, M. S. Strano, P. T. Hammond and A. M. Belcher, *Nat. Nanotechnol.*, 2011, **6**, 377–384.
46. M. A. Green, Y. Hishikawa, E. D. Dunlop, D. H. Levi, J. Hohl-Ebinger and A. W. Y. Ho-Baillie, *Prog. Photovolt. Res. Appl.*, 2018, **26**, 427–436.
47. N. Nuraje, X. Dang, J. Qi, M. A. Allen, Y. Lei and A. M. Belcher, *Adv. Mater.*, 2012, **24**, 2885–2889.
48. P.-Y. Chen, X. Dang, M. T. Klug, Ń.-M. Dorval Courchesne, J. Qi, M. N. Hyder, A. M. Belcher and P. T. Hammond, *Chem. Mater.*, 2015, **27**, 1531–1540.
49. N.-M. Dorval Courchesne, M. T. Klug, K. J. Huang, M. C. Weidman, V. J. Cantú, P.-Y. Chen, S. E. Kooi, D. S. Yun, W. A. Tisdale, N. X. Fang, A. M. Belcher and P. T. Hammond, *J. Phys. Chem. C*, 2015, **119**, 13987–14000.
50. Y. M. Lee, W. Kim, Y. H. Kim, J. K. Kim, J.-R. Jang, W.-S. Choe, J. H. Park and P. J. Yoo, *ChemSusChem*, 2015, **8**, 2385–2391.
51. B. D. Wilts, A. Matsushita, K. Arikawa and D. G. Stavenga, *J. R. Soc., Interface*, 2015, **12**, 20150717.
52. A. Sweeney, C. Jiggins and S. Johnsen, *Nature*, 2003, **423**, 31–32.
53. J. W. Galusha, L. R. Richey, M. R. Jorgensen, J. S. Gardner and M. H. Bartl, *J. Mater. Chem.*, 2010, **20**, 1277–1284.
54. D. Zhang, W. Zhang, J. Gu, T. Fan, Q. Liu, H. Su and S. Zhu, *Prog. Mater. Sci.*, 2015, **68**, 67–96.
55. P. Vukusic, in *Functional Surfaces in Biology, Little Structures with Big Effects*, ed. S. N. Gorb, Springer, Dordrecht, Heidelberg, London, New York, 2009, vol. 1, pp. 237–258.
56. C. Mille, E. C. Tyrode and R. W. Corkery, *Chem. Commun.*, 2011, **47**, 9873–9875.
57. E. J. W. Crossland, M. Kamperman, M. Nedelcu, C. Ducati, U. Wiesner, D.-M. Smilgies, G. E. S. Toombes, M. A. Hillmyer, S. Ludwigs, U. Steiner and H. J. Snaith, *Nano Lett.*, 2009, **9**, 2807–2812.
58. W. Zhang, M. Anaya, G. Lozano, M. E. Calvo, M. B. Johnston, H. Míguez and H. J. Snaith, *Nano Lett.*, 2015, **15**, 1698–1702.
59. Q. Peng, X. Yang, J. Ding, X. Wei and J. Yang, *Appl. Energy*, 2013, **112**, 682–689.

60. S. Sundararajan, A. B. Samui and P. S. Kulkarni, *J. Mater. Chem. A*, 2017, **5**, 18379–18396.
61. D. Lefebvre and F. H. Tezel, *Renewable Sustainable Energy Rev.*, 2017, **67**, 116–125.
62. L. P. Biró and J.-P. Vigneron, *Laser Photonics Rev.*, 2011, **5**, 27–51.
63. L. P. Biró, Z. Bálint, K. Kertész, Z. Vértesy, G. I. Márk, Z. E. Horváth, J. Balázs, D. Méhn, I. Kiricsi, V. Lousse and J.-P. Vigneron, *Phys. Rev. E*, 2003, **67**, 021907.
64. J. T. Herbeck and J. Novembre, *J. Mol. Evol.*, 2003, **56**, 691–701.
65. V. Nazari, E. V. Zakharov and F. A. H. Sperling, *Mol. Phylogenet. Evol.*, 2007, **42**, 131–156.
66. J. Jiang, H. Zhou, J. Ding, F. Zhang, T. Fan and D. Zhang, *Appl. Surf. Sci.*, 2015, **347**, 368–377.
67. Z. C. Orel and M. K. Gunde, *Sol. Energy Mater. Sol. Cells*, 2000, **61**, 445–450.
68. Y. Yin, Y. Pan, L. X. Hang, D. R. McKenzie and M. M. M. Bilek, *Thin Solid Films*, 2009, **517**, 1601–1606.
69. R.-C. Juang, Y.-C. Yeh, B.-H. Chang, W.-C. Chen and T.-W. Chung, *Thin Solid Films*, 2010, **518**, 5501–5504.
70. N. Selvakumar, S. Santhoshkumar, S. Basu, A. Biswas and H. C. Barshilia, *Sol. Energy Mater. Sol. Cells*, 2013, **109**, 97–103.
71. Q.-F. Geng, X. Zhao, X.-H. Gao and G. Liu, *J. Am. Ceram. Soc.*, 2011, **94**, 827–832.
72. P. Tao, W. Shang, C. Song, Q. Shen, F. Zhang, Z. Luo, N. Yi, D. Zhang and T. Deng, *Adv. Mater.*, 2015, **27**, 428–463.
73. F. Zhang, Q. Shen, X. Shi, S. Li, W. Wang, Z. Luo, G. He, P. Zhang, P. Tao, C. Song, W. Zhang, D. Zhang, T. Deng and W. Shang, *Adv. Mater.*, 2015, **27**, 1077–1082.
74. M. Wuttig and S. Raoux, *Z. Anorg. Allg. Chem.*, 2012, **638**, 2455–2465.
75. K. Kant, A. Shukla and A. Sharma, *Sol. Energy Mater. Sol. Cells*, 2017, **172**, 82–92.
76. Y. Lin, Y. Jia, G. Alva and G. Fang, *Renewable Sustainable Energy Rev.*, 2018, **82**, 2730–2742.
77. Y. Wang, B. Tang and S. Zhang, *RSC Adv.*, 2012, **2**, 5964–5967.
78. Y. Wang, B. Tang and S. Zhang, *Adv. Funct. Mater.*, 2013, **23**, 4354–4360.
79. R. Chen, R. Yao, W. Xia and R. Zou, *Appl. Energy*, 2015, **152**, 183–188.
80. S. Ebadi, S. H. Tasnim, A. A. Aliabadi and S. Mahmud, *Appl. Therm. Eng.*, 2018, **141**, 724–740.
81. Y. Kang, S.-G. Jeong, S. Wi and S. Kim, *Sol. Energy Mater. Sol. Cells*, 2015, **143**, 430–434.
82. W. Hu and X. Yu, *RSC Adv.*, 2012, **2**, 5580–5584.
83. J. Yang, G.-Q. Qi, L.-S. Tang, R.-Y. Bao, L. Bai, Z.-Y. Liu, W. Yang, B.-H. Xie and M.-B. Yang, *J. Mater. Chem. A*, 2016, **4**, 9625–9634.
84. H. Zhao and J. H. Waite, *J. Biol. Chem.*, 2006, **281**, 26150–26158.

85. H. Lee, S. M. Dellatore, W. M. Miller and P. B. Messersmith, *Science*, 2007, **318**, 426–430.
86. S. Kim and C. B. Park, *Adv. Funct. Mater.*, 2013, **23**, 10–25.
87. X. Xie, M. Ye, C. Liu, P.-C. Hsu, C. S. Criddle and Y. Cui, *Energy Environ. Sci.*, 2015, **8**, 546–551.
88. M. Parthasarathy, A. Narayanan, S. Narasimhan, S. A. Alex, V. Ogirala and V. Kalidasan, *Energy Technol.*, 2014, **2**, 625–633.
89. S. Suresh, R. W. Evitts and G. F. Kennell, *Adv. Chem. Eng. Sci.*, 2016, **6**, 49–61.
90. M. G. Kim, D.-H. Kim, T. Kim, S. Park, G. Kwon, M. S. Kim, T. J. Shin, H. Ahn and H.-G. Hur, *J. Mater. Chem. A*, 2015, **3**, 16978–16987.
91. D. R. MacFarlane, N. Tachikawa, M. Forsyth, J. M. Pringle, P. C. Howlett, G. D. Elliott, J. H. Davis, M. Watanabe, P. Simon and C. A. Angell, *Energy Environ. Sci.*, 2014, 7, 232–250.
92. M. Vijayakumar, B. Schwenzer, V. Shutthanandan, J. Z. Hu, J. Liu and I. A. Aksay, *Nano Energy*, 2014, **3**, 152–158.
93. S. Ravula, S. N. Baker, G. Kamath and G. A. Baker, *Nanoscale*, 2015, 7, 4338–4353.
94. S. Letaief and C. Detellier, *J. Mater. Chem.*, 2007, **17**, 1476–1484.
95. C. Takahashi, T. Shirai and M. Fuji, *Solid State Ionics*, 2014, **267**, 16–21.
96. M. Naguib, V. N. Mochalin, M. W. Barsoum and Y. Gogotsi, *Adv. Mater.*, 2014, **26**, 992–1005.
97. G. Sharma, E. Muthuswamy, M. Naguib, Y. Gogotsi, A. Navrotsky and D. Wu, *J. Phys. Chem. C*, 2017, **121**, 15145–15153.
98. C. Zhang, M. Beidaghi, M. Naguib, M. R. Lukatskaya, M.-Q. Zhao, B. Dyatkin, K. M. Cook, S. J. Kim, B. Eng, X. Xiao, D. Long, W. Qiao, B. Dunn and Y. Gogotsi, *Chem. Mater.*, 2016, **28**, 3937–3943.
99. Y. J. Lee, Y. Lee, D. Oh, T. Chen, G. Ceder and A. M. Belcher, *Nano Lett.*, 2010, **10**, 2433–2440.
100. Y. J. Lee and A. M. Belcher, *J. Mater. Chem.*, 2011, **21**, 1033–1039.
101. D. Oh, X. Dang, H. Yi, M. A. Allen, K. Xu, Y. J. Lee and A. M. Belcher, *Small*, 2012, **8**, 1006–1011.
102. G. Zhang, S. Wei and A. M. Belcher, *ACS Appl. Nano Mater.*, 2018, **1**, 5631–5639.
103. D. Oh, J. Qi, Y.-C. Lu, Y. Zhang, Y. Shao-Horn and A. M. Belcher, *Nat. Commun.*, 2013, **4**, 2756.
104. D. Oh, J. Qi, B. Han, G. Zhang, T. J. Carney, J. Ohmura, Y. Zhang, Y. Shao-Horn and A. M. Belcher, *Nano Lett.*, 2014, **14**, 4837–4845.
105. Y.-C. Lu, H. A. Gasteiger and Y. Shao-Horn, *J. Am. Chem. Soc.*, 2011, **133**, 19048–19051.
106. H.-L. Zhang and D. E. Morse, *J. Mater. Chem.*, 2009, **19**, 9006–9011.
107. G. X. Wang, J.-H. Ahn, M. J. Lindsay, L. Sun, D. H. Bradhurst, S. X. Dou and H. K. Liu, *J. Power Sources*, 2001, **97–98**, 211–215.
108. Y. Wang, J. Y. Lee and T. C. Deivaraj, *J. Electrochem. Soc.*, 2004, **151**, A1804–A1809.

109. H.-L. Zhang, J. R. Neilson and D. E. Morse, *J. Phys. Chem. C*, 2010, **114**, 19550–19555.
110. Y. Liao, J. Xiang, L. Yuan, Z. Hao, J. Gu, X. Chen, K. Yuan, P. K. Kalambate and Y. Huang, *ACS Appl. Mater. Interfaces*, 2018, **10**, 37955–37962.
111. Z. Cui, C. Zu, W. Zhou, A. Manthiram and J. B. Goodenough, *Adv. Mater.*, 2016, **18**, 6926–6931.
112. R. B. Rakhi, B. Ahmed, M. N. Hedhili, D. H. Anjum and H. N. Alshareef, *Chem. Mater.*, 2015, **27**, 5314–5323.
113. Y. Xin and Y.-X. Yu, *Mater. Des.*, 2017, **130**, 512–520.
114. J. Halim, S. Kota, M. R. Lukatskaya, M. Naguib, M.-Q. Zhao, E. J. Moon, J. Pitock, J. Nanda, S. J. May, Y. Gogotsi and M. W. Barsoum, *Adv. Funct. Mater.*, 2016, **26**, 3118–3127.
115. M. Padmini, S. K. Kiran, N. Lakshminarasimhan, M. Sathish and P. Elumalai, *Electrochim. Acta*, 2017, **236**, 359–370.
116. M. Zhao, Q. Zhao, B. Li, H. Xue, H. Pang and C. Chen, *Nanoscale*, 2017, **9**, 15206–15225.
117. A. Malak-Polaczyk, C. Vix-Guterl and E. Frackowiak, *Energy Fuels*, 2010, **24**, 3346–3351.
118. Z. Liu, R. Ma, M. Osada, N. Iyi, Y. Ebina, K. Takada and T. Sasaki, *J. Am. Chem. Soc.*, 2006, **128**, 4872–4880.
119. https://www.clickenergy.com.au/news-blog/12-countries-leading-the-way-in-renewable-energy/ (accessed November 26, 2018).

CHAPTER 4

Biomimetics of Structural Colours: Materials, Methods and Applications

AHU GÜMRAH DUMANLI[a] AND THIERRY SAVIN*[b]

[a] School of Materials, University of Manchester, Oxford Road, Manchester M13 9PL, United Kingdom; [b] Department of Engineering, University of Cambridge, Trumpington Street, Cambridge CB2 1PZ, United Kingdom
*Email: t.savin@eng.cam.ac.uk

4.1 Introduction

Bioinspiration and biomimicry have enabled revolutionary advances in the engineering of new materials. They still offer an attractive route for sustainable solutions for material design and fabrication. Among the many features and functionalities exhibited by natural materials, textures and colours are a striking display of engineering principles that have been precisely optimised by natural selection.

Indeed, colours play an essential role in the evolution and survival of plants and animals. To the best of our current knowledge, the colours of living creatures are produced either through pigments, bioluminescence, or spatial structures.[1] Pigmentary colours originate from the selective absorption of light by the electrons of molecules embedded in the materials. Bioluminescence is emitted by chemical reactions in the photophores of some organisms.[1,2] The third type of coloration is the result of the discrimination of wavelengths by the interaction of the incident light with structures on the reflective biomaterial. In the visible range of wavelengths, diffracting

Inorganic Materials Series No. 4
Bioinspired Inorganic Materials: Structure and Function
Edited by Simon R. Hall

Published by the Royal Society of Chemistry, www.rsc.org

structures must be in the submicrometre scale in order to interact with light, and such patterns are indeed quite common in biological assemblies.[3]

These so-called "structural colours" usually have a distinctive appearance (shiny, metallic aspect with often an angle-dependent apparent colour) that contrasts with the dull and diffuse aspect of more common pigmentary colours. Physically, structural colours are the result of the fundamental optical processes of diffraction, interference or scattering. They generally suffer little dissipative loss,[4] and do not fade like pigments, which are sensitive to chemical alterations and have relatively low environmental resistance. In nature, these structures have been optimised, after millions of years of evolutionary trial and error, to display the coloration most adapted to the various needs and purposes of living creatures, thus fulfilling important roles such as signalling (selection, advertising, *etc.*) or camouflage.

Structural colours in nature have fascinated scientists for centuries, but their accurate assessment was only achieved with the introduction of the scanning electron microscope (SEM) in the 1940s.[5] This discovery triggered an extraordinary new perspective on the description and understanding of the natural nanostructures found in living matter, and indeed those responsible for coloration. Today, a prodigious array of colour-producing architectures have been described by scientists in animals, plants and minerals, and new ones are still frequently discovered.

The engineering potentials of structural colours, for systematic usage in photonics (and possibly elsewhere), were first recognised about 30 years ago with the inception of so-called "photonic crystals".[6] These materials can "mould the flow of light",[7] in the manner that atomic crystals control the transport of electrons. Since then, intense research efforts have been carried out to overcome the two main and concomitant problems when building efficient photonic materials. First, how to best design the light-interfering structure for a given purpose. Second, how to actually build these nanostructured materials. As with many other examples of recent technological advances, biomimicry is an elegant shortcut towards the optimisation of design.[8] For this second challenge, scientists already benefit from a strong footing to establish micromanufacturing methods for photonic materials.

Until about two decades ago, progress in microfabrication technologies were mainly driven by the microelectronics industry, which began in the 1950s with the invention of the integrated circuit. Microfabrication techniques are now employed to manufacture a plethora of miniaturised devices, with scopes well beyond pure electronics. In fact, the implementation of microdevices established entirely new disciplines, depending on the function sought (sensor, actuator, imaging, communication, mechanics, *etc.*), the materials used (silicon, metals, polymer, ceramics, *etc.*), and the fabrication processes employed (patterning, etching, moulding, deposition, *etc.*). Microelectromechanical systems, in particular, are recent applications of microfabrication that have been booming in recent years. They have driven many innovations in micromachining and are, today, found in mundane consumer products.[9]

As a result of these advances, submicrometre features can currently be fabricated with, generally, high throughput and with a high degree of reproducibility. Building large assemblies, with the regular and periodic arrangements of microstructures required for photonic crystals, still remains a manufacturing challenge, particularly in three dimensions. The development of photonic nanomaterials, bioinspired or custom-designed, may now have become the main driving force behind contemporary endeavours to push progress in microfabrication. These current efforts attract the interest of a large body of scientists, across many fields: chemistry, engineering, physics, materials science, biology, *etc.*

There already exists a vast literature on structural coloration and photonic materials, their incidence in nature, and their physical origins and underlying optical mechanisms. Consequently, recent years have seen the publication of many comprehensive review articles[1,3,4,8,10–27] and books[2,28–30] on these subjects. However, reviews covering the fabrication of bioinspired photonic structures[31–33] are more rare, and do not specifically focus on the actual manufacturing techniques. We believe that the recent, and often remarkably ingenious, methods to reproduce natural structural colours deserve to be highlighted, and indeed have the potential to serve a wide spectrum of scientists.

This review therefore puts a strong emphasis on microfabrication techniques. It is organised as follows: we first give an overview of the physical principles behind the structural colorations of some natural systems. In particular, we highlight four typical structures found in nature, representing the canonical systems that, from our review of the latest literature, have attracted most studies in structural colour biomimicry. We next focus our attention on the recent techniques and attempts to mimick the structural colours found in nature. We distinguish two main strategies, "top-down" and "bottom-up", as this dicision of the technologies used is quite general when presenting routes for fabricating complex materials. The last part of this chapter is devoted to the most recent and promising applications of structural colours inspired by nature.

4.2 Natural Structural Colours

4.2.1 Physical Origins of Natural Structural Colours

Structural colours are produced by the microscopic patterns of a reflective material. Features with sizes comparable to the wavelength of the incident illumination diffract the light selectively to cause interference effects that can be constructive or destructive.[34] This interference thus only allows ranges of wavelengths to be reflected and produces, when visible, a particular coloration of the material.[2] These ranges of reflected wavelengths generally depend on the refracting angle, meaning that apparent colours change with the viewing orientation. Hence, structural colours often appear iridescent.

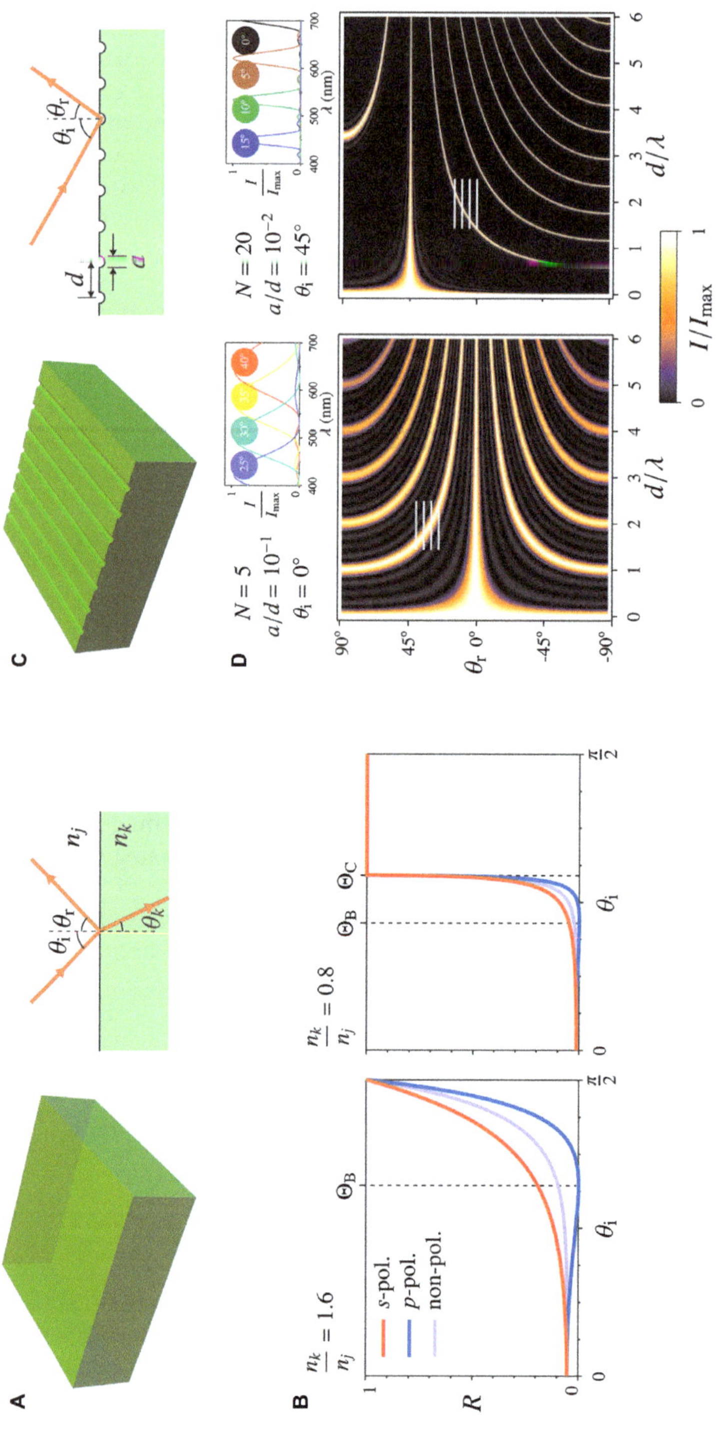
A
B
C
D
$\frac{n_k}{n_j} = 1.6$
s-pol.
p-pol.
non-pol.
$\frac{n_k}{n_j} = 0.8$
Θ_B
Θ_C
θ_i
θ_r
θ_k
n_j
n_k
R
d
a
N = 5
a/d = 10^{-1}
$\theta_i = 0°$
N = 20
a/d = 10^{-2}
$\theta_i = 45°$
d/λ
λ (nm)
I/I_{max}

To help understand in more detail the design criteria underlying the production of structural colours, we first review the physical principles behind the processes generating these phenomena.[3,28] The relevant quantity to study is the reflectance of the medium at its interface with the exterior, and for visible wavelengths of 400–700 nm. The calculation of reflectance is performed using physical optics principles.[2,4,34,35] We note that another way of producing colours *via* structure can be obtained with nanopatterned metallic films *via* their surface plasmon resonance.[36] This mechanism is not observed in natural systems, and will not be reviewed here.

The physical origins of structural colours have been summarised in several reviews.[2–4,10,11,18,19] Here we adopt an analytical approach, focusing on practical formulae for the reflectance of a slab of a structured material, supported by a substrate, and in contact with the incident medium (typically air or water).

4.2.1.1 *Diffraction Grating*

A diffraction grating is a common occurrence of structural coloration (*e.g.* the reflection from a compact disc). It typically involves a single substrate material, in contact with in incident medium containing a light source that is reflected on the substrate.

It is relevant at this point to recall the equations that dictate the propagation of light travelling in an incident medium labelled j of refractive index n_j, and encountering a planar interface separating j from another medium k with index n_k, at an angle $\theta_i = \theta_j$ from the normal to the interface. The law of reflection indicates that the light is reflected at the specular angle $\theta_r = \theta_i$ in medium j. The sign convention used here is to consider $\theta_i > 0$, and the reflected angle θ_r to be negative if it is in the same half-space as θ_i, and positive otherwise. In the specular configuration drawn in Figure 4.1A, both

Figure 4.1 (A) The specular reflectance of light incident from a medium j at the interface with a medium k is given in terms of the incident angle $\theta_i = \theta_r = \theta_j$ defined in the schematic. (B) Variation of the reflectance as a function of θ_j. At the Brewster's angle $\Theta_B = \arctan\left(\frac{n_k}{n_j}\right)$, p-polarised light is not reflected, while above the critical angle $\Theta_C = \arcsin\left(\frac{n_k}{n_j}\right)$, the light is entirely reflected in both polarisations. (C) A reflective diffraction grating is obtained when a reflecting surface displays periodic pattern: here, parallel grooves are etched at a regular interval; in the schematic, the light path has both incident and reflective angles, θ_i and θ_r respectively, positive according to our convention so that specular reflection is $\theta_r = \theta_i$ (see Figure 4.1A). (D) The reflectance for various values d/λ as a function of the reflected (or observer's) angle θ_r for a fixed incident (or illumination) angle θ_i; $\theta_i = 0°$ in the left panel, $\theta_i = 45°$ in the right panel. The insets show the λ-dependent reflectance for a typical patterning period $d = 1$ μm in the visible range, as taken along the white segments shown in the corresponding panel. The chips give the approximate perceived colour corresponding to the reflectance spectrum (we used the CIE 1931 colour space to convert the spectrum into colour[37]) at the indicated refracted angles.

θ_i and θ_r are positive angles. Snell's law further characterises the propagation of the light through the two refracting media,

$$n_j \sin\theta_j = n_k \sin\theta_k, \tag{4.1}$$

to relate the refractive angle in each medium with the index. The Fresnel equations describe the fraction of light reflected at the interface, $j|k$, through the reflectance R given by:

$$R = |\rho_{jk}|^2, \quad \text{with} \quad \rho_{jk} = \frac{\eta_j - \eta_k}{\eta_j + \eta_k} = -\rho_{kj} \tag{4.2}$$

the reflection coefficient of light in the medium j incident to medium k given in terms of the corresponding polarisation-dependent admittance η_j and η_k. The polarisation s corresponds to the component of the incident light with electric field perpendicular to the plane of incidence (the plane containing the interface's normal and the incident light's propagation vector), while the electric field of the polarisation p is parallel to the plane of incidence. Continuity of the electromagnetic fields at the interface gives:

$$\eta_j = \begin{cases} n_j \cos\theta_j & \text{for } s\text{-polarized,} \\ \dfrac{n_j}{\cos\theta_j} & \text{for } p\text{-polarised.} \end{cases} \tag{4.3}$$

The reflectance of a surface, and its dependency on the incident wavelength and the illumination/viewing angles, is the important observable for the characterisation of structural colours. In Figure 4.1B we plot the reflectance for various incident angles, and for s- and p-polarised incident light. Unpolarised light carries both s and p polarisation equally, and its reflectance is the average of both directions. Wavelengths are usually not separated in simple reflection, unless the medium is dispersive. Here, we will not consider dispersion, birefringence and absorption of materials, although separate results will be given for s- and p-polarised incident light when it is relevant. Figure 4.1B shows the existence of two particular values of the incident angle: at the Brewster's angle, Θ_B, p polarisation is not reflected; for incidence beyond the critical angle, Θ_C, the light is totally reflected ("total internal reflection") for any polarisation.

In a diffraction grating, the reflecting material is patterned with periodically arranged features. The size of the features is subwavelength, so each will diffract the light in all direction and interferences between the emerging rays lead to coloration. If we consider a one-dimensional (1D) grating pattern, for example, the parallel grooves etched on the reflecting substrate shown in Figure 4.1C, the reflected intensity of light rays with wavelength λ, incident at an angle θ_i and observed at a diffraction angle θ_r, is proportional to[11]

$$I \propto \left[\frac{\sin\left(\frac{a}{d}\frac{\phi}{2}\right)}{\frac{a}{d}\frac{\phi}{2}} \frac{\sin\left(N\frac{\phi}{2}\right)}{\sin\frac{\phi}{2}} \right]^2, \quad \text{with } \phi = \frac{2\pi}{\lambda} d(\sin\theta_i - \sin\theta_r), \tag{4.4}$$

for N periodic line features, each of size a and spaced by d ($a \leq d$). Here, we use the same convention as in Figure 4.1A to define θ_r, so that in the configuration presented in Figure 4.1C, both θ_i and θ_r are positive angles. Eqn (4.4) is obtained by summing the rays diffracted by each groove and accounting for the interferences between the fields emitted by them.

The maximum reflected intensity is observed at peaks centred at $\phi_m = m2\pi$ and of width $\Delta\phi_m = \phi_m \frac{2\pi}{N}$ with $m \in \mathbb{Z}$ an integer called the spectral order. According to eqn (4.4), these values, ϕ_m, correspond to a discrete set of wavelengths verifying

$$m\lambda_m = d(\sin\theta_i - \sin\theta_r). \tag{4.5}$$

The reflected spectrum given as a function of the wavelengths will have asymmetric peaks, and the reflected bands of wavelengths may be better described by their centres λ_m^c and widths $\Delta\lambda_m$ as:

$$\lambda_m^c = \lambda_m \left[1 - \left(\frac{\Delta\phi_m}{2\phi_m} \right)^2 \right]^{-1}, \tag{4.6a}$$

$$\Delta\lambda_m = \lambda_m^c \frac{\Delta\phi_m}{\phi_m}. \tag{4.6b}$$

The zero-order grating $m = 0$ corresponds to the specular reflection, $\theta_r = \theta_i$ and does not separate the wavelength of the incident light. The entire reflected spectrum may be factored by the reflectance between the incident and substrate medium, although the grating pattern may affect the light polarisation in a non-trivial way. Hence, when designing a diffraction grating for specific purposes, the geometry of the patterning (shape, size a, periodicity d, orientation with respect to polarisation) and the refractive index of the medium must be taken into account.[38]

In Figure 4.1D we show the variation of the reflected intensity with both the refracted angle θ_r and the inverse wavelength d/λ for a fixed incident angle ($\theta_i = 0°$ in the left panel, $\theta_i = 45°$ in the right panel). The graphs clearly show how the wavelengths are separated by different spectral order, except for the specular mode $m = 0$, and their dependency on the viewing angle. These variations give rise to the familiar iridescence that is characteristic of a diffraction grating.

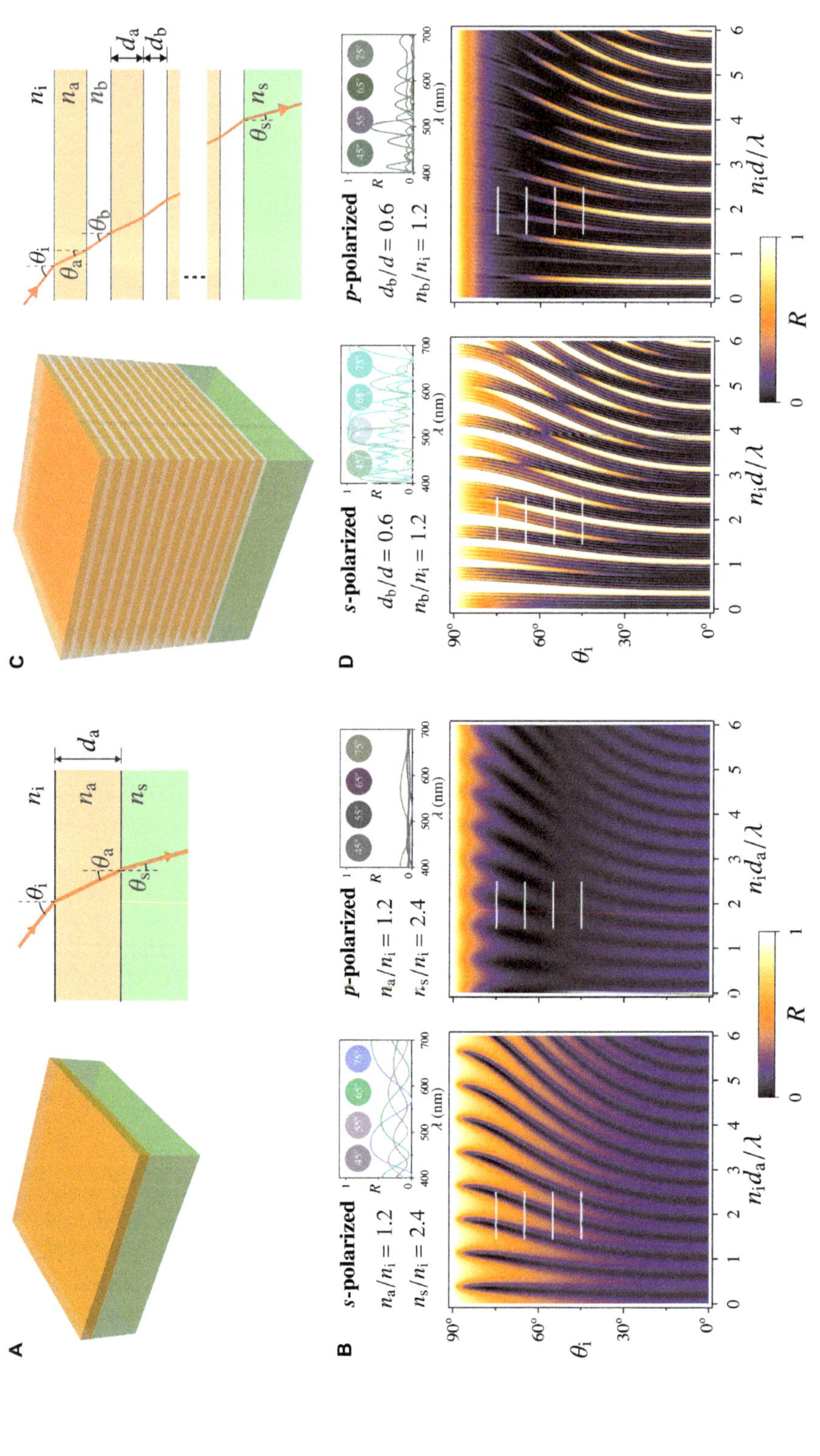
A
B
C
D
s-polarized
p-polarized
$n_a/n_i = 1.2$
$n_s/n_i = 2.4$
$d_b/d = 0.6$
$n_b/n_i = 1.2$
$n_i d_a/\lambda$
$n_i d/\lambda$
θ_i
R
λ (nm)

A two-dimensional (2D) grating, made with a periodic lattice of features on the reflecting surface, will also reflect the light according to eqn (4.4) but with $\phi_\alpha = \frac{2\pi}{\lambda}\mathbf{d}_\alpha \cdot (\hat{\mathbf{s}}_i - \hat{\mathbf{s}}_r)$, in each of the two directions $\alpha = 1,2$ of the lattice unit vectors $\mathbf{d}_\alpha$, for incident and reflected light parallel to the unit vectors $\hat{\mathbf{s}}_i$ and $\hat{\mathbf{s}}_r$, respectively.

Examples of structural colours obtained from diffraction gratings in nature include floral iridescence,[39] butterfly scales,[40] or mollusc shells.[41]

4.2.1.2 Layered Media

A simple structure that generates colour is a layered medium, where thin layers are stacked parallel to the reflecting interface. Layered media are very common in natural structural colours, and their fabrication based on thin-film deposition is routinely performed to produce optical components such as Bragg mirrors.[35,42]

The reflectance of a stack of layers results from the interference between the light reflected and transmitted multiple times at each layer boundary. It can be calculated with the transfer matrix formalism, and this calculation is sufficiently simple to be readily programmed on a computer. Consider N parallel layers, of refractive indices and thicknesses $\{(n_j, d_j)\}_{j=1...N}$ bounded by two media: on the side of the light source, an incident material of index n_i and on the other side a substrate of index n_s (see Figure 4.2C). Snell's law,

Figure 4.2 (A) Structural coloration can be obtained when a thin film of material "a" with thickness d_a is deposited on a substrate "s"; the relevant refractive indices are n_i, n_a and n_s of the incident medium, and of materials "a" and "s", respectively. (B) Density map of the reflectance as a function of the scaled inverse wavelength $n_i d_a/\lambda$ and the incidence angle θ_i, when the film is observed in specular reflection. The left and right panels are for s and p polarisations, respectively, and the insets show the λ-dependent reflectance in air ($n_i = 1$) for a film thickness $d_a = 1$ μm in the visible range, as taken along the white segments shown in the corresponding panel. The chips give the approximate perceived colour corresponding to the reflectance spectrum at the indicated angles of reflection. (C) Multilayers made of a periodic stacks of two alternating materials "a" and "b", with respective thicknesses d_a and d_b and refractive indices n_a and n_b, and deposited on a substrate "s" while immersed in the incident medium "i". (D) Density map of the reflectance as a function of the scaled inverse wavelength $n_i d/\lambda$, with $d = d_a + d_b$, and the incidence angle θ_i. All panels are obtained for $2N = 10$ layers of alternating materials "a" and "b", with $n_a/n_i = 1.8$, on a substrate with $n_s/n_i = 1.5$. The left and right panels are for s and p polarisations, respectively, and the insets show the λ-dependent reflectance in air ($n_i = 1$) for a period thickness $d = 1$ μm, in the visible range, as taken along the white segments shown in the corresponding panel. The chips give the approximate perceived colour corresponding to the reflectance spectrum at the indicated angles of reflection. The case shown, under normal incidence $\theta_i = 0°$, corresponds to an ideal multilayer (see text).

eqn (4.1), through the layers gives $n_j \sin\theta_j = n_i \sin\theta_i = n_s \sin\theta_s$ for $j = 1\ldots N$. The reflectance is then $R = \left|\dfrac{\eta_i - \eta_\Sigma}{\eta_i + \eta_\Sigma}\right|^2$, where, as before, η designates the polarisation-dependent admittance of a material.[43,44] The effective admittance η_Σ of the multilayer with substrate is obtained as

$$\eta_\Sigma = \frac{h}{e} \quad \text{with} \quad \begin{pmatrix} e \\ h \end{pmatrix} = \left[\prod_{j=1}^{N} \begin{pmatrix} \cos\phi_j & i\sin\phi_j/\eta_j \\ i\eta_j \sin\phi_j & \cos\phi_j \end{pmatrix}\right] \begin{pmatrix} 1 \\ \eta_s \end{pmatrix}, \tag{4.7}$$

in terms of the admittances $\{\eta_j\}_{j=1\ldots N}$, given by eqn (4.3), and the phase thicknesses $\left\{\phi_j = \dfrac{2\pi}{\lambda} n_j d_j \cos\theta_j\right\}_{j=1\ldots N}$, defined for each layer with an incident light at a wavelength λ. In all cases, the reflectance R can be expressed as

$$\frac{R - \rho_{is}^2}{1 - \rho_{is}^2} = \frac{\mathscr{F}(\phi_1, \ldots, \phi_N)}{1 + \mathscr{F}(\phi_1, \ldots, \phi_N)}, \tag{4.8}$$

where $\mathscr{F}$ is a function that depends on the structure of the stack.

4.2.1.2.1 Thin Film. For a single thin film (n_a, d_a), one may show that $\mathscr{F}(\phi_a) = F_{aa} \sin^2\phi_a$ and the reflectance follows

$$\frac{R - \rho_{is}^2}{1 - \rho_{is}^2} = \frac{F_{aa}\sin^2\phi_a}{1 + F_{aa}\sin^2\phi_a}, \tag{4.9}$$

where we have defined

$$F_{jk} = \frac{(\eta_j^2 - \eta_i^2)(\eta_k^2 - \eta_s^2)}{\eta_j \eta_k (\eta_i + \eta_s)^2} \neq F_{kj} \tag{4.10}$$

the "generalised coefficient of finesse" of an embedded bilayer $i|j|k|s$. Eqn (4.9) is a well-known formula encountered when studying the Fabry–Pérot interferometer, for example. The reflectance has extrema when $\phi_a = k\dfrac{\pi}{2}$ for $k \in \mathbb{Z}$ an integer, which will be either maxima or minima depending on the sign of F_{aa}. The results for the maximum peaks of reflecatnce $\phi_a = \phi_m$ and their width $\Delta\phi_m$ (at half-maximum) can be summarised as follows:

$$F_{aa} < 0: \ \phi_m = m\pi, \quad \Delta\phi_m = \pi - 2\arctan\left(\sqrt{1 + F_{aa}}\right); \tag{4.11a}$$

$$F_{aa} > 0: \ \phi_m = m\pi + \frac{\pi}{2}, \quad \Delta\phi_m = 2\arctan\left(\sqrt{1 + F_{aa}}\right); \tag{4.11b}$$

with $m \in \mathbb{Z}$. Higher values of $|F_{aa}|$ correspond to wider peaks (the term "finesse" is used for the Fabry–Pérot interferometer in transmission, which selects thinner ranges of wavelength as $|F_{aa}|$ increases). The corresponding bands in wavelengths can be calculated using eqn (4.6a) and (4.6b) with $\lambda_m = \phi_m{}^{-1} 2\pi\, n_a\, d_a \cos\theta_a$, that is:

$$F_{aa} < 0: \quad m\lambda_m = 2d_a\sqrt{n_a^2 - n_i^2 \sin^2\theta_i}, \tag{4.12a}$$

$$F_{aa} > 0: \quad \left(m + \frac{1}{2}\right)\lambda_m = 2d_a\sqrt{n_a^2 - n_i^2 \sin^2\theta_i}. \tag{4.12b}$$

The case $F_{aa} < 0$ is sometime called the "anti-reflection" setting (because it includes the case $n_s > n_a > n_i$ used in anti-reflective screens) and $F_{aa} > 0$ the "soap-bubble" setting (including the case $n_s = n_i < n_a$).[43] For p-polarised components, F_{aa} changes sign at the incident Brewster's angle $\Theta_B = \arctan\left(\frac{n_a}{n_i}\right)$.

In Figure 4.2B we show the reflectance of a single thin film on a substrate for various angles of incidence in specular reflection. In this figure, $n_i < n_a < n_s$ and we observe a strong separation of wavelength for s-polarised light, as characterised by the bright bands of reflectance in the left panel. Again, as in a diffraction grating, the locations of the bands on the spectrum vary with the incident angles, marking iridescence. All the wavelengths of the p polarisation are reflected equally at the Brewster's angle, with the reflectance ρ_{is}^2 calculated at this angle. The thin film reflects nearly all wavelengths of both polarisations above the critical angle $\Theta_C = \arcsin\left(\frac{n_a}{n_i}\right)$, since no light is transmitted through the "i|a" interface in that case (not shown in Figure 4.2B).

In nature, thin-film structural colours can be found, for example, in insect wings[45] or bird feathers.[46,47]

4.2.1.2.2 Multilayer and 1D Photonic Crystals. For a multilayer, the reflectance can be calculated numerically using eqn (4.7). In most applications, however, it is desirable to limit the number of materials used to build the multilayer. A particularly common configuration is a periodic multilayer made with two materials "a" and "b". In that case, one can show that the function $\mathcal{F}$ in eqn (4.8) has the following expression:

$$\begin{aligned}\mathcal{F}(\phi_a, \phi_b) = {} & F_{aa} S_N^2 \sin^2\phi_a + F_{bb} S_N^2 \sin^2\phi_b \\ & + (F_{ab} S_N S_{N+1} + F_{ba} S_{N-1} S_N) \sin\phi_a \cdot \sin\phi_b\end{aligned} \tag{4.13}$$

for the stack i|a|b|a|...|a|b|s made of a total of $2N$ alternating layers of materials "a" and "b" (N of type "a", N of type "b"), with respective thicknesses d_a and d_b, stacked between the incident medium i and the substrate s.†

†For the case i|a|b|a|...|b|a|s of $2N-1$ alternating layers (N of type "a", $N-1$ of type "b"), the function $\mathcal{F}$ is

$$\mathcal{F}(\phi_a, \phi_b) = F_{aa} S_N^2 \sin^2\phi_a + F_{bb} S_{N-1}^2 \sin^2\phi_b + (F_{ab} S_{N-1} S_N + F_{ba} S_{N-1} S_N) \sin\phi_a \cdot \sin\phi_b.$$

Here we have used the definition of the generalised coefficient of finesse, eqn (4.10), and defined $S_N = \dfrac{\sin(N\Phi)}{\sin\,\Phi}$ with

$$\cos\,\Phi = \frac{\cos(\phi_{\rm a}+\phi_{\rm b}) - \rho_{\rm ab}^2\,\cos(\phi_{\rm a}-\phi_{\rm b})}{1-\rho_{\rm ab}^2}. \tag{4.14}$$

The quantity Φ is the effective phase thickness of the three-layers symmetric system $\left(n_{\rm a}, \dfrac{d_{\rm a}}{2}\right)|(n_{\rm b}, d_{\rm b})|\left(n_{\rm a}, \dfrac{d_{\rm a}}{2}\right)$, which represents one period of the multilayer. If $|\cos\,\Phi| > 1$ (that is, Φ is a complex number), then the admittance of the period will be purely imaginary and the multilayer will reflect the incident wave. Inspection of this condition leads to the following results for the location of peaks of the reflectance in terms of $\phi = \phi_{\rm a} + \phi_{\rm b}$:[7]

$$\phi_m = m\pi + O(\rho_{\rm ab})^2, \tag{4.15a}$$

$$\Delta\phi_m = 4\left|\rho_{\rm ab}\,\sin\left[\frac{m\pi}{2}\left(1-\left|\frac{\phi_{\rm a}-\phi_{\rm b}}{\phi_{\rm a}+\phi_{\rm b}}\right|\right)\right]\right| + O(\rho_{\rm ab})^3, \tag{4.15b}$$

where $m \in \mathbb{Z}$. Note that if $\dfrac{\phi_{\rm a}}{\phi_{\rm b}} = k$ or k^{-1} with k an integer, then there are no reflectance peaks for m a multiple of $k+1$, since $\Delta\phi_m = 0$ in that case.

These equations can be expressed in terms of the wavelength using eqn (4.6a) and (4.6b) with $\lambda_m = \phi_m{}^{-1} 2\pi(n_{\rm a}d_{\rm a}\cos\theta_{\rm a} + n_{\rm b}d_{\rm b}\cos\theta_{\rm b})$. Within the approximation above,

$$m\lambda_m \approx 2d\sqrt{\bar{n}^2 - n_{\rm i}^2\,\sin^2\theta_{\rm i}} \tag{4.16}$$

where we defined $d = d_{\rm a} + d_{\rm b}$ the period, and $\bar{n}$ is an average refractive index for the multilayer. The effective index may be obtained by either the Bruggeman or Maxwell Garnett relations,[2]

$$-(1-f_{\rm a})\frac{\bar{n}^2 - n_{\rm b}^2}{L\bar{n}^2 + (1-L)n_{\rm b}^2} = f_{\rm a}\frac{\bar{n}^2 - n_{\rm a}^2}{L\bar{n}^2 + (1-L)n_{\rm a}^2}, \tag{4.17a}$$

$$\frac{\bar{n}^2 - n_{\rm b}^2}{L\bar{n}^2 + (1-L)n_{\rm b}^2} = f_{\rm a}\frac{n_{\rm a}^2 - n_{\rm b}^2}{Ln_{\rm a}^2 + (1-L)n_{\rm b}^2}, \tag{4.17b}$$

respectively, with the latter being more accurate when the volume fraction $f_{\rm a}$ of one material ("a", for example) is small. In these relations, L is the shape-dependent depolarisation factor, with $L = 1$ for layers for which $f_{\rm a} = d_{\rm a}/d$.

The instance $\phi_{\rm a} = \phi_{\rm b}$, called an ideal multilayer, is particularly simple to resolve. The bands of reflection are obtained for $\phi_m = m\pi$ with m odd, and their width $\Delta\phi_m = 4|\rho_{\rm ab}|$. Under normal incidence, ideal multilayers are obtained by ensuring $n_a d_a = n_b d_b$ in the design of the multilayer. In that case, the reflectance has minimum ρ_{is}^2 and maximum $\left(\frac{1-\gamma}{1+\gamma}\right)^2$, with $\gamma = \frac{n_s}{n_i}\left(\frac{n_a}{n_b}\right)^{2N}$.‡ This shows that for sufficiently large N, the maximum reflectance reaches 1.

In Figure 4.2D we show the reflectance for a multilayer designed to be ideal under normal incidence. The left panel for *s* polarisation shows that the peaks in reflectance are not only shifted with increasing incidence, but may also disappear (with others appearing). This implies a non-trivial pattern of colour formation, and iridescence for multilayer-based structural coloration. Also, there are no clear marks of a Brewster's angle with minimum reflected *p* polarisation (although reflection is still total for angles greater than $\Theta_C = \arcsin\left(\frac{n_a}{n_i}\right)$ under the choice of indices $n_a < n_i$ not shown in Figure 4.2B).

Note that eqn (4.16) is Bragg's law for constructive coherent scattering in a crystal lattice with constant d, often written $m\lambda_m = 2d\bar{n}\sin\bar{\theta}$ in a medium with effective refractive index $\bar{n}$, and where $\bar{\theta}$ is the angle of the incident and reflected beams as taken from the planar boundary of the medium. Combining Bragg's law with Snell's law, $n_i \sin\theta_i = \bar{n}\sin(\pi/2 - \bar{\theta})$ for the effective medium indeed returns eqn (4.16). This equation can also be used for the important case of helicoidal stacked polarising layers, similar to chiral nematic (or cholesteric) liquid crystals (see Figure 4.10 in Section 4.3.2.1), which are common in nature. In that case, the chiral nematic pitch substitutes the periodicity of the stack.[48]

Multilayer interference is very common in natural structural colours. It has been found and characterised in many organisms, including insects,[49,50] birds,[51] fish,[52] molluscs[53,54] and fruits.[55]

4.2.1.3 Photonic Crystal

Photonic crystals are materials having a spatially periodic refractive index, with the period near the wavelength of the incident light. In general, photonic crystals have dimensions much larger than their periodicity, so that they can be considered as infinite structures in calculations. The periodicity can be only in one dimension, in which case the crystal is the periodic multilayer described in the previous section. Examples of two-dimensional (2D) and three-dimensional (3D) crystals are shown in Figure 4.3A and B, respectively, where the pattern of varying refractive index is repeated on a 2D or 3D lattice. As such, photonic crystals may be regarded as a special case of composites, characterised by a refractive index invariant under the spatial translations of a crystalline lattice.

Photonic crystals can control the propagation of light in the manner that atomic crystals control electrons.[7] And these materials have indeed attracted

---

‡ $\gamma = \frac{n_a^2}{n_i n_s}\left(\frac{n_a}{n_b}\right)^{2N-2}$ for $2N$–1 alternating layers, with N of type "a".

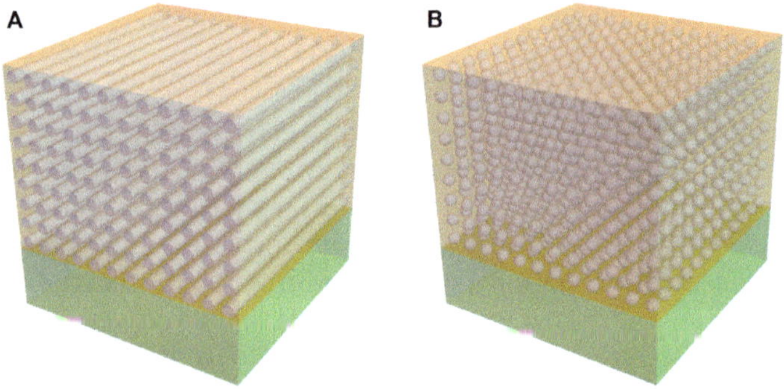

Figure 4.3 (A) 2D photonic crystals display invariance through translation in one direction. (B) 3D photonic crystals have periodicity in three directions. In principle, photonic crystals are considered infinite, but, in practice, they are bounded by a substrate and the exterior.

important research efforts since their inception about 30 years ago, particularly to overcome challenges in their manufacturing.[56–58]

The Bragg–Snell law, eqn (4.16), can be used to evaluate the approximate set of reflectance peak wavelengths in a given direction of the crystalline lattice. In most cases, however, there are no simple calculations to evaluate the width of these peaks, or even their existence. Indeed, we have seen in the previous section that in 1D photonic crystals, for example, some peaks given by eqn (4.16) in fact vanish when the layers of the material have commensurable optical thicknesses. Analytical expressions of reflectance from a photonic crystal slab can be obtained in some cases (patterned multilayer) *via* scattering-matrix calculations,[59] but the spectrum must generally be obtained *via* computational modelling relying on finite elements, plane-wave expansion, finite-difference time-domain method, to name a few examples. The focus of the calculations is to obtain the photonic band gaps in their optical modes, that is the frequency ranges in which propagation of light is prohibited by the crystal. Band gaps depend on the orientation of the propagation in the crystal, often quantified in the lattice reciprocal space, and correspond to iridescent reflectance peaks.[60]

Natural 2D photonic crystals provide the coloration of several marine animals[61] and birds.[62,63] Opals are archetypical of 3D photonic crystals,[64] which are also common in insects.[65–68]

Tuneable structural colours from photonic crystals are used by some organisms for camouflage, predation, communication, *etc.*[3] These rely on varying the various parameters setting the peak's wavelength of reflectance in eqn (4.16). Fudouzi,[21] Zhao *et al.*[22] and Xu and Guo[24] recently proposed reviews of these mechanisms in nature and recent efforts in their biomimicry.

4.2.1.4 Light Scattering

Light scattering is a fundamental process that describes the interactions of light with elementary particles of matter (scatterers). It is different from the processes previously described, although the laws of refraction are the results of the multiple light-scattering events occurring in a bulk material. In this regard, a classification for the production of structural colour *via* scattering qualifies the mechanism as either coherent or incoherent. In coherent scattering, there is a phase relationship between the scattered waves by ordered scatterers, and this usually yields iridescence. Incoherent scattering is the result of disordered dispersion of scatterers and has a diffuse appearance in reflection, with the spectrum being independent of the observation or illumination angle (except in amplitude). Incoherent light scattering is the origin of the white colour of milk or the blue colour of the sky. An intermediate state of coherent, but non-iridescent scattering has been observed, where the scatterers are in fact quasi-ordered (or "weakly localised").[69]

To describe light scattering, we consider the intensity propagating in the direction of the unit vector $\hat{\mathbf{s}}$, $I(\hat{\mathbf{s}}, \hat{\mathbf{s}}')$, as scattered by a single non-absorbing spherical particle under an incident light $I_0(\hat{\mathbf{s}}')$, directed parallel to $\hat{\mathbf{s}}'$. $I(\hat{\mathbf{s}}, \hat{\mathbf{s}}')$ is given by:

$$I(\hat{\mathbf{s}}, \hat{\mathbf{s}}') = I_0(\hat{\mathbf{s}}') \frac{\sigma}{4\pi r^2} p(\hat{\mathbf{s}} \cdot \hat{\mathbf{s}}'),$$

where r is the distance from the scatterer, σ is the scattering cross-section and $p(\hat{\mathbf{s}} \cdot \hat{\mathbf{s}}')$ is the phase function of the particle, which depends on $\cos\theta = \hat{\mathbf{s}} \cdot \hat{\mathbf{s}}'$, the cosine of the angle between the incident and scattering directions (see Figure 4.4B). The quantity p is normalised with respect to the average over all solid angles Ω': $\oiint p(\hat{\mathbf{s}} \cdot \hat{\mathbf{s}}') \frac{d\Omega'}{4\pi} = 1$. The scattering of light by a single spherical particle is called Mie scattering, and its analytical expression, as well as computing software[§] can be found in many textbooks.[70] In Figure 4.4B and C we show the variation of p with $\cos\theta$ and the particle's scaled diameter d/λ for an incident wavelength λ, which is conserved through scattering. In Figure 4.4C we also show the Mie scattering cross-section σ. The limit for small spheres with refractive index n_a, embedded in a substrate "s", and with diameter $d \ll \lambda$ is given by the Rayleigh scattering:

$$p(\hat{\mathbf{s}} \cdot \hat{\mathbf{s}}') = \frac{3}{4}(1 + \cos^2\theta) \quad \text{and} \quad \sigma = \frac{8}{3}\frac{\pi d^2}{4}\left(\frac{\pi d}{\lambda}\right)^4 \left(\frac{n_a^2 - n_s^2}{n_a^2 + 2n_s^2}\right)^2,$$

for non-polarised (with respect to the $(\hat{\mathbf{s}}, \hat{\mathbf{s}}')$-plane) incident light. The dependency of σ with λ^{-4} indicates that blue light is more scattered than red. This phenomenon extends to non-spherical particles with sizes near λ, in which case it is named the Tyndall effect, and explains the blue tint of milky opals or, possibly, of irises with low melanin content.[71] The scattering becomes increasingly anisotropic for larger scatterers, and we also show in

---

[§]See for example http://www.scattport.org

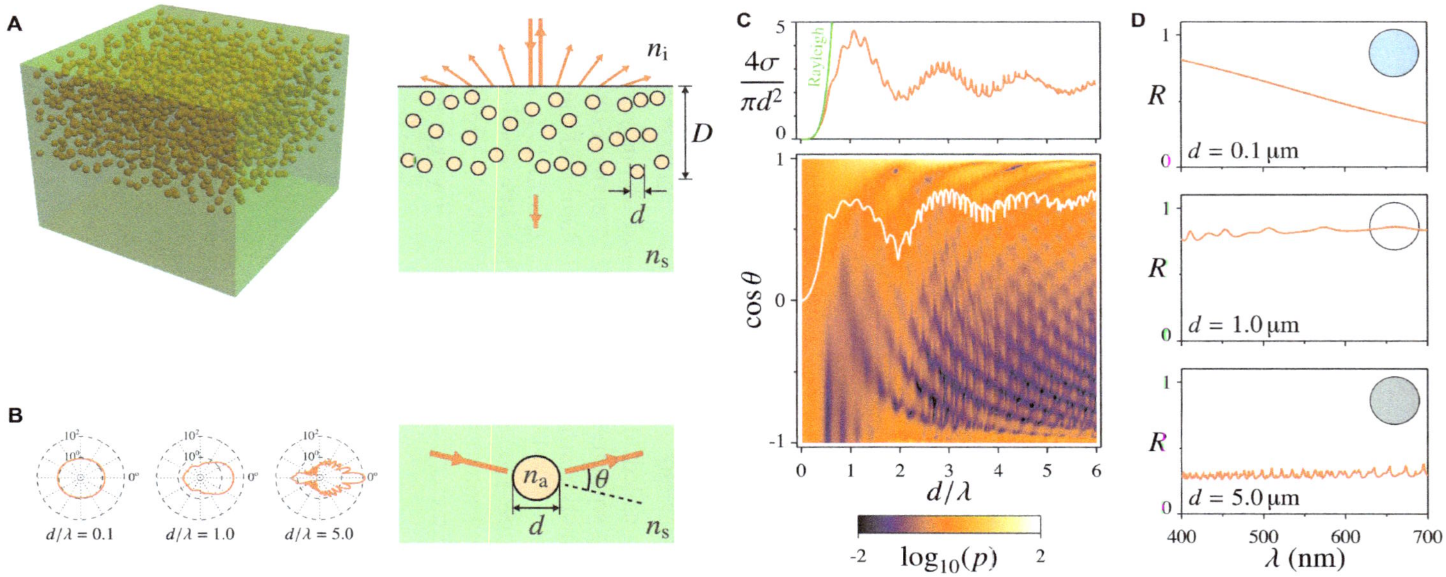

Figure 4.4 (A) Structural coloration can be obtained from the light scattered by a slab of a turbid material with thickness D, and made by spherical scatterers of size d randomly dispersed in a substrate material "s". (B) Angular dependence of the Mie scattering phase function for various scaled diameters d/λ. The scattering angle θ is relative to the incident light, as shown in the schematic to the right; the scattering polar diagrams have a log-scaled radial distance. (C) The scaled Mie scattering cross-section (top) and the average scattering angle's cosine g as a function of d/λ (bottom), over the density map of the phase function. (D) The reflectance of a turbid slab of thickness $D = 100$ µm for various sizes of scatterers (from eqn (4.18)). The chip gives the approximate perceived colour corresponding to the reflectance spectrum. All curves in this figure are for spherical scatterers with refractive index $n_a = 2.4$, in a substrate with index $n_s = 1.5$. The reflectance spectra are in air ($n_i = 1.0$) for a volume fraction of scatterers $f = 10\%$.

Figure 4.4C (white line) the average angle's cosine $g = \oiint \hat{\mathbf{s}} \cdot \hat{\mathbf{s}}' p(\hat{\mathbf{s}} \cdot \hat{\mathbf{s}}') \frac{d\Omega'}{4\pi}$, quantifying the scattering anisotropy and increasing with d/λ.

The radiative transfer equation was established from the energy balance to describe the light propagation resulting from the multiple scattering in a turbid medium made of randomly distributed particles at a volume fraction f. At steady state, the light intensity $I(\mathbf{r}, \hat{\mathbf{s}})$ at a position $\mathbf{r}$ propagating in the direction $\hat{\mathbf{s}}$ verifies $\ell \nabla \cdot \hat{\mathbf{s}} I(\mathbf{r}, \hat{\mathbf{s}}) = -I(\mathbf{r}, \hat{\mathbf{s}}) + p^* I(\mathbf{r}, \hat{\mathbf{s}})$, where $\ell = \frac{\pi d^3}{6 f \sigma}$ is the mean-free path, and the convolution $p^* I(\mathbf{r}, \hat{\mathbf{s}}) = \oiint I(\mathbf{r}, \hat{\mathbf{s}}') p(\hat{\mathbf{s}} \cdot \hat{\mathbf{s}}') \frac{d\Omega'}{4\pi}$ describes the effect of scattering.

A convenient approximation of the radiative transfer equation is diffusion¶, which, in the case of a parallel turbid slab under normal illumination presented in Figure 4.4A, allows one to solve for the reflectance:‖

$$\frac{R - \rho_{\mathrm{is}}^2}{1 - \rho_{\mathrm{is}}^2} = \frac{(1 - g) D/\ell - (1 - z_{\mathrm{s}})(1 - e^{-D/\ell})}{(1 - g) D/\ell + z_{\mathrm{s}} + z_{\mathrm{i}}}, \tag{4.18}$$

where z_k is an extrapolation length ratio given by

$$z_k = \frac{2}{3} \frac{1 + \int_0^{\pi/2} 3 \bar{\rho}_{k\mathrm{m}}^2(\theta) \cos^2\theta \sin\theta d\theta}{1 - \int_0^{\pi/2} 2 \bar{\rho}_{k\mathrm{m}}^2(\theta) \cos\theta \sin\theta d\theta} \tag{4.19}$$

for both materials, the incident material "i" and the substrate "s", relative to the turbid medium "m". Here $\bar{\rho}_{k\mathrm{m}}^2(\theta)$ is the reflectance of unpolarised light at the $k|\mathrm{m}$ interface with incidence θ (see eqn (4.2) and (4.3); unpolarised reflectance is the average of the s and p-polarised ones). The effective refractive index of the diffuse medium "m" can be calculated using, for example, the Maxwell Garnett mixing rule, eqn (4.17b) with depolarisation $L = 1/3$ for spheres[2] and host matrix material $n_{\mathrm{b}} = n_{\mathrm{s}}$.[72]

¶The diffusion approximation is valid if the dimensions of the medium are bigger than the mean-free path of the photon. In that case, the intensity I is split into a ballistic, non-scattered component I_0, that solves $\ell \nabla \cdot \hat{\mathbf{s}} I_0(\mathbf{r}, \hat{\mathbf{s}}) = -I_0(\mathbf{r}, \hat{\mathbf{s}})$, and a diffusing term $I_{\mathrm{d}} = I - I_0$ that is approximated to $I_{\mathrm{d}}(\mathbf{r}, \hat{\mathbf{s}}) \approx \bar{I}_{\mathrm{d}}(\mathbf{r}) + 3\hat{\mathbf{s}} \cdot \mathbf{J}_{\mathrm{d}}(\mathbf{r})$ in terms of its mean $\bar{I}_{\mathrm{d}}(\mathbf{r}) = \oiint I_{\mathrm{d}}(\mathbf{r}, \hat{\mathbf{s}}) \frac{d\Omega}{4\pi}$ and its flux $\mathbf{J}_{\mathrm{d}}(\mathbf{r}) = \oiint \hat{\mathbf{s}} I_{\mathrm{d}}(\mathbf{r}, \hat{\mathbf{s}}) \frac{d\Omega}{4\pi}$. One can then establish $\ell \nabla \cdot \mathbf{J}_{\mathrm{d}}(\mathbf{r}) = \bar{I}_0(\mathbf{r})$ with $(1 - g)\mathbf{J}_{\mathrm{d}}(\mathbf{r}) = -\ell \nabla \bar{I}_{\mathrm{d}}(\mathbf{r})/3 + \oiint \hat{\mathbf{s}} p^* I_0(\mathbf{r}, \hat{\mathbf{s}}) \frac{d\Omega}{4\pi}$, and the accompanying extrapolated boundary conditions $\bar{I}_{\mathrm{d}}(\mathbf{r}) = 3z\hat{\mathbf{n}} \cdot \mathbf{J}_{\mathrm{d}}(\mathbf{r})$ for $\mathbf{r}$ on the boundary, $\hat{\mathbf{n}}$ thc unit vector normal to the boundary pointing outside the diffuse medium, and z given by eqn (4.19).

‖If the turbid medium is made with particles in the incident medium, and in contact with "s", then one may use instead: $\frac{R - \rho_{\mathrm{is}}^2 e^{-2D/\ell}}{1 - \rho_{\mathrm{is}}^2 e^{-2D/\ell}} = \frac{(1 - g)D/\ell - \frac{(1 - z_{\mathrm{s}}) - \rho_{\mathrm{is}}^2 (1 + z_{\mathrm{s}}) e^{-D/\ell}}{1 - \rho_{\mathrm{is}}^2 e^{-2D/\ell}} (1 - e^{-D/\ell})}{(1 - g) D/\ell + z_{\mathrm{s}} + z_{\mathrm{i}}}$.

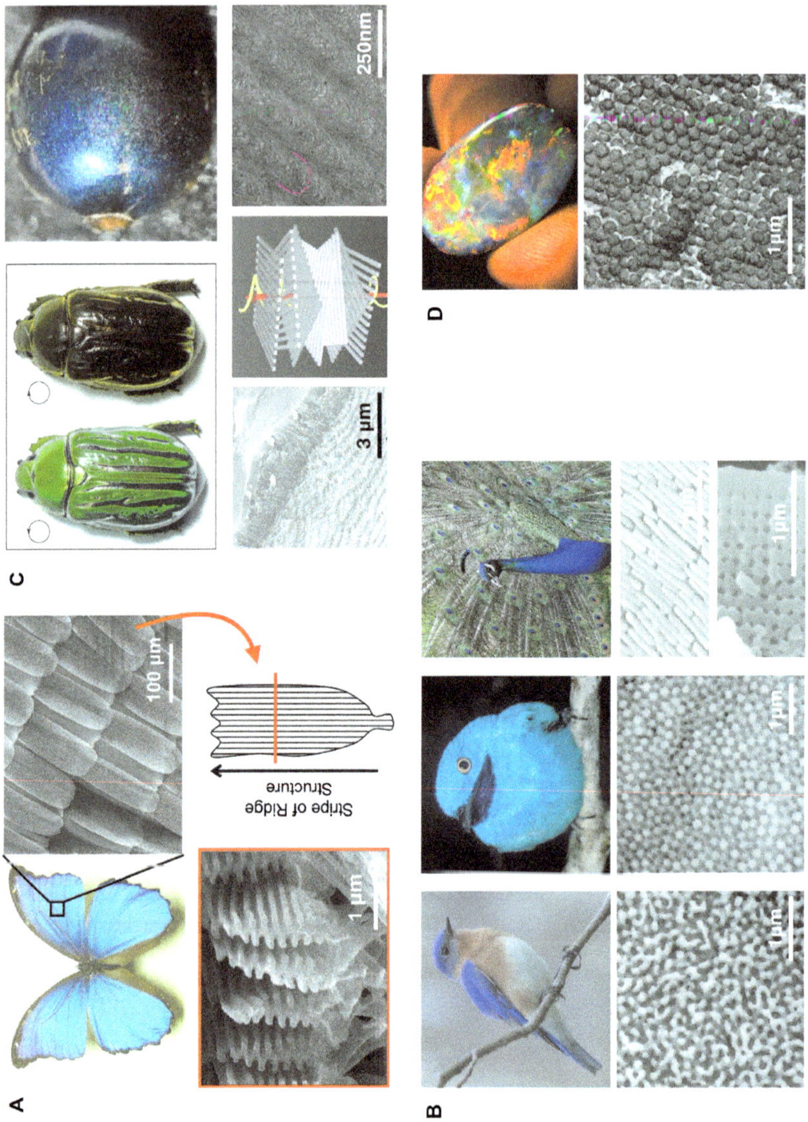
A
100 µm
1 µm
Stripe of Ridge Structure
C
3 µm
250nm
B
1µm
1µm
1µm
D
1µm

Light scattering produces the apparent colours of bird feathers,[73] beetle scales,[72,74] or fish scales,[75] for example.

4.2.2 Model Systems of Natural Structural Colours for Biomimicry

One of the most studied species, in terms of structural colour production and biomimetics, is the *Morpho* butterfly, whose wing scales exhibit a unique metallic blue colour. The *Morpho* is an excellent example of the combined diversity of physical mechanisms in colour production (see Figure 4.5A), and hence all the more challenging to mimic. The iridescent blue colour with high reflectivity is a result of coherent scattering in the periodic arrays of the scales.[76,77] Periodic structures, such as gratings or multilayers, produce colours with high dependency on the angle of reflection. However, for the *Morpho* species, the reflected blue colour has a low angle dependency due to the presence of multilayer surfaces that exhibit a distribution of tilts with respect to the scales' substrate. The strong diffraction is further caused by a second layer of periodic ridges above the layer of highly iridescent ground.[78] As a result, *Morpho* butterfly wings exhibit a complex optical response

Figure 4.5 Structural colours in nature that have attracted significant biomimicry efforts. (A) Typical *Morpho* butterfly (*Morpho didius*) and SEM images of the scale, each covered with ridges whose lateral profile has the typical "Christmas tree" shape (adapted from ref. 82 with permission from The Authors and the Society of Photo Optical Instrumentation Engineers, Copyright 2006). The ridges are supported by a gyroid crystal structure that also produces structural colours. (B) Structural colours of the plumage in eastern bluebird (*Sialia sialis*, left) originate from quasi-order β-keratin tubular nanostructures, while in the plum-throated cotinga (*Cotinga maynana*, centre), the structures are spheres (reproduced from ref. 83 with permission from the Royal Society of Chemistry). The peacock feathers (right) show 2D photonic crystals of melanin rods embedded in keratin (image of blue peacock *Pavo cristatus*: reproduced from https://commons.wikimedia.org/wiki/File:Peacock_front02_-_melbourne_zoo.jpg under the terms of the GFDL v1.2 license, https://www.gnu.org/licenses/old-licenses/fdl-1.2.en.html; micrographs reproduced from ref. 62 with permission of the authors, Copyright 2006 National Academy of Sciences). (C) The jewelled beetles (*Chrysina gloriosa*) and the *Pollia condensata* fruit display circularly polarised iridescence thanks to a structure of chitin (left micrograph) and cellulose (right micrograph) fibrils, respectively, similar to assemblies found in the cholesteric liquid crystal shown in Figure 4.10 (adapted from ref. 50 with permission from AAAS, Copyright 2009). (D) Natural opals display iridescent colours because of the crystal arrangement of silica spheres (reproduced from https://commons.wikimedia.org/wiki/File:16.42cts_Lightning_Ridge_black_Opal.JPG under the terms of the CC BY 3.0 ShareAlike license, https://creativecommons.org/licenses/by-sa/3.0/deed.en; electron micrograph reproduced from ref. 64 with permission from Springer Nature, Copyright 1964.

by combining multilayer interference, diffraction, scattering, and even pigment-induced absorption, to produce its singular, angle-independent brilliant blue colour. From the material design point of view, there are many elements to bring together to attain such complex optical effects: the reflecting elements must be on a subwavelength scale, need to be produced on a large scale, and must be sufficiently ordered to produce the desired colour and reflectivity. But it must also bear some disorder, on the subwavelength scale, in order to eliminate the directionality and sharp reflectance peaks associated with multilayered interference.

The colours in bird feathers are produced in two different ways, either from pigments or from light refraction caused by the structure of the feathers. In some cases, the combination of both effects may create more complex colour arrangements.[79] While structurally coloured barbules are normally composed of well-ordered melanin granules and rods,[80] the spongy keratin structures of weakly localised particles is responsible for structural colours in non-iridescent feathers[81] (see Figure 4.5B and Section 4.2.1).

Another intriguing phenomenon developed in nature is circular dichroism. The metallic green colour of the beetles *Cetonia aurata* and *Chrysina gloriosa* is due to the chitin layers assembling into a highly uniform, helicoidal stack, which is optically analogous to a cholesteric liquid crystal. These circularly polarising multilayers reflect polarised light with a left-handed (anticlockwise) rotation. Some plants, such as *Pollia condensata*[55] and *Margaritaria nobilis*,[84] also use this strategy to produce iridescent colours. Analogous to beetles, their cellulose microfibrils form helicoidal stacks in the cell walls, which selectively reflect circularly polarised light of a specific wavelength (see Figure 4.5C).

The iridescent colour of nacre, on the other hand, is the result of the biomineralisation process, during which amorphous calcium carbonate ($CaCO_3$) and organic material sheets (proteins and chitin) are sequentially deposited in the isolated space between the shell and the epithelium mantle cells. Calcium carbonate at low temperatures leads to formation of lamellar stacks of $CaCO_3$ in aragonite form, separated by organic layers.[85] The resulting ordered multilayer structure of crystalline $CaCO_3$ platelets causes the multilayer interference, displaying the characteristic pale green to pink iridescent coloration (see the spectra shown in Figure 4.2D).[86]

Natural mineralisation processes also engineer one of the most studied structurally coloured systems in biomimicry. Precious opal is an amorphous form of silica ($SiO_2 \cdot nH_2O$) that contains water and that can be found in almost any kind of rock. Under suitable conditions, water percolates through the earth and drags silicates encountered in the soil. When this silicate-rich solution enters a cavity and the water evaporates, the silicates deposit as small spheres (with diameters ranging from 150 nm to 400 nm, in the case of rainbow-coloured opals). When the precipitated silica spheres are monodispersed in size, and stack in an ordered, close-packed fashion, they produce a play of colours by hugely coherent scattering in the visible light range (see Figure 4.5D).[64,87]

4.3 Biomimicry of Natural Structural Colours

These unique characteristics of the structural colours developed in nature, such as iridescence or metallic appearance, cannot be obtained by mere chemical dyes or pigments. It is also worth noting that the micro-architectures producing these optical effects bring further functionalities to the natural materials, such as anti-reflection, specific thermal response, selective vapour response, directional adhesion or superhydrophobicity. Again, such multifunctionalities cannot be achieved *via* chemical coloration alone. Therefore, in order to mimic the biological species, one needs to strategically design material composites with controlled variable nano-structures that can reproduce the biological structure itself.[88] Recent advances in controlling, and fine-tuning, the production and organisation of subwavelength structures have led to a burst of activity in the nanophotonics field. The fabrication methods can generally be classified into two groups.[89]

- The *top-down* strategies involve using microfabrication tools to print computer-generated patterns using lithography techniques or multi-layer deposition onto a larger piece of substrate. Although it is possible to produce a wide range of high-quality photonic nanostructures with these technologies, the cost, efficiency and technical limitations (*i.e.* structures must be larger than 100 nm) of top-down approaches prevent their use in many applications with high-throughput needs.
- The *bottom-up* approaches, on the other hand, use physicochemical interactions for the hierarchical organisation of nanostructures *via* self-assembly of basic building blocks; these approaches are more cost and time effective, and widen the range of applications towards smaller accessible length scales.

4.3.1 Top-down Strategies

4.3.1.1 Lithography

In the last five decades, micro- and nano-lithography techniques have been developed to optimise the manufacture of integrated circuits and microchips. Lithographic patterning can directly control the submicrometre structure on the reflecting surface of a dielectric material, and can thus be used to generate structural colours. Lithography techniques may be divided into two types on the basis of whether they use masks and templates, or not.[90]

Masked lithography transfers patterns from a mould or a mask template over a large substrate area simultaneously, thus enabling high-throughput fabrications. These lithography techniques include photolithography, holographic lithography,[91] nano-imprinting lithography,[92] and soft lithography.[93]

The second type of technique is maskless and fabricates arbitrary patterns by serial writing or etching. Examples of maskless lithography include electron-beam lithography,[94] focused ion beam lithography, and scanning

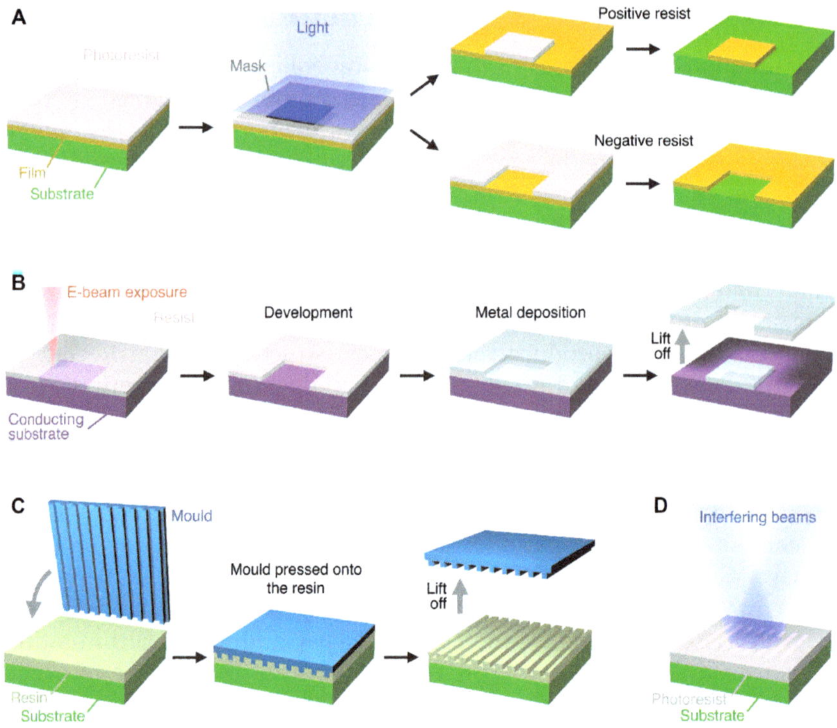

Figure 4.6 Some top-down, micro- and nano-fabrication techniques used for the production of structurally coloured biomimetic materials. (A) In photolithography, a pattern is transferred from a mask to a photoresist by a curing light; further chemical treatment develops the profiled layer. (B) Electron-beam lithography serially registers a resist with a narrow, highly focused e-beam. (C) Nano-imprint lithography press-stamps a mould on a soft resin, which is then processed for its subsequent use. (D) Interference lithography does not use a mask, but exploits interference patterns to print a photoresist.
Reproduced from ref. 25 with permission from the Royal Society of Chemistry.

probe lithography. These techniques allow ultra-high-resolution patterning of arbitrary shapes with a minimum feature size as small as a few nanometres. They are also often used to create the masks and templates employed by masked lithography.

We show in Figure 4.6 the processes behind the lithographic techniques discussed here, and below give a few examples of their recent applications for the production of bioinspired structural colours.

4.3.1.1.1 Photolithography. In photolithography, a photomask with a geometric pattern is transferred onto a light-sensitive chemical substrate (photoresist) by using visible or UV curing light (Figure 4.6A). Lateral periodicities required to produce structural colours in the visible range are usually large enough to be patterned with photolithography.[95] Yet, in many cases, the photolithography is coupled with self-assembled colloidal systems (see Section 4.3.2.1) to further guide and control the shape and surface structure of the assembly.

Lee *et al.*,[93] for example, successfully demonstrated the use of such coupled techniques to develop a 3D photonic crystal for reflection-mode display.[93] They first fabricated colloidal crystals that were embedded in a photoresist matrix on the surface of a substrate. The further photolithographic patterning of the embedded crystal enabled the creation of hybrid, micropatterned structures. Selective removal of the colloids subsequently left a regular array of cavities, thus forming an inverse opal structure (see also Figure 4.9C in Section 4.3.2.1) with a high index contrast between the air and the polymerised photoresist. Such structures produce a high reflectivity at the wavelengths falling within the photonic band gap of the resulting crystal.

Moreover, photolithography can be used to pattern materials that already exhibit structurally coloured surfaces. Hence, a recent work by Song *et al.*[96] focuses on understanding the role of interstructural disorder on the reflectance, which opens the possibility for alternative and easier production of broad-angle reflection of the *Morpho*-blue. They created irregular ridges to mimic the *Morpho*'s metallic blue by first depositing several sequential multilayers of silicon oxide (SiO_2, or silica) and titanium oxide (TiO_2, or titania) on randomly distributed colloidal silica spheres, with diameters ranging from 250 to 440 nm, on a silicon substrate. The multilayers were protected by chromium coating and a photoresist. Subsequent photolithography and etching of this multilayered structure created a dense array of ridges with a periodicity of 700 nm, with great variation between the ridge structures (as opposed to highly regular ridges formed on a silicon wafer without the silica particles). The ridge structure with regular layers demonstrated a strongly angle-dependent colour reflection. The irregular continuous films and the ridged films, however, showed a very similar reflection spectra to the *Morpho*, thus highlighting the crucial role of the interstructural disorder in angle-dependent reflection.

4.3.1.1.2 Interference and Holographic Lithography. Interference and holographic lithography is a versatile technique to fabricate both 2D and 3D periodic photonic structures in a scalable manner. In laser interference lithography, for example, the inhomogeneous distribution of energy, generated by the superposition of electromagnetic plane waves from two beams, is registered in a photoresist material coating a substrate (Figure 4.6D).[91] There, 3D periodic patterns can be printed by adjusting the intensity, geometry, polarisation, and phase of the applied laser beams.[97]

Elaborating on this technique, Siddique *et al.*[98] were recently able to fabricate the complex, hierarchical "Christmas tree"-like morphology of the *Morpho* butterfly shown in Figure 4.5A. Notably, they used a reflective

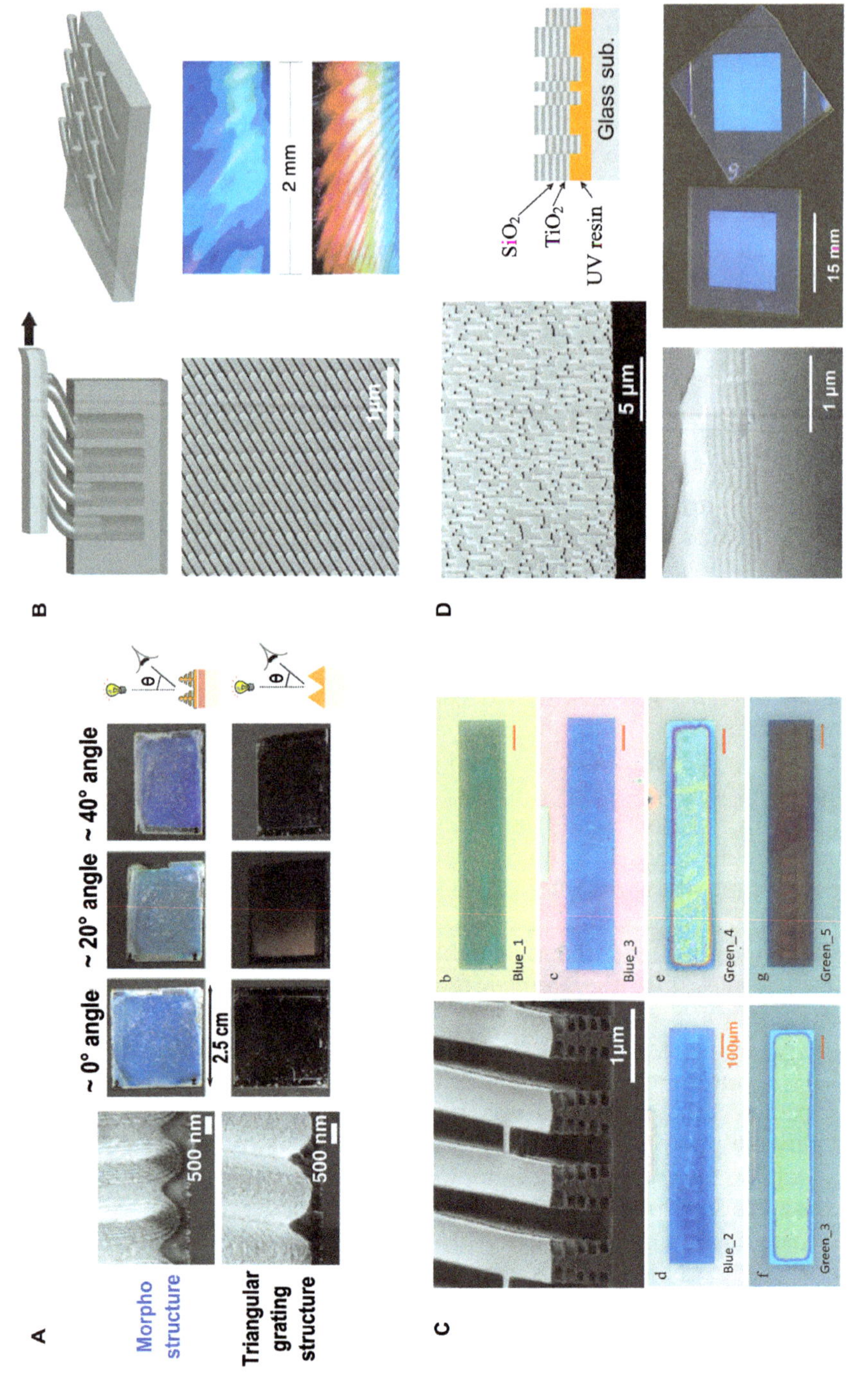
A
Morpho structure
Triangular grating structure
500 nm
500 nm
~ 0° angle
~ 20° angle
~ 40° angle
2.5 cm
θ
B
1μm
2 mm
C
1μm
100μm
Blue_1
Blue_2
Blue_3
Green_3
Green_4
Green_5
D
5 μm
1 μm
SiO2
TiO2
UV resin
Glass sub.
15 mm

coating between the glass and the photoresist to create an additional interference patterning perpendicular to the one obtained from straight dual-beam laser interference.[98] While the plain triangular morphology without lateral lamellae exhibited the usual grating effect at high angles, their *Morpho*-like structure reflected a blue colour by about 30%, over an extended range of reflection angle, between 0° and 40° (see Figure 4.7A).

4.3.1.1.3 Nano-imprint Lithography. The nano-imprint lithography (NIL) process creates patterned, 3D structures with feature sizes as small as 50 nm by imprinting a stamp into a low viscosity resist *via* moulding (see Figure 4.6C).[101] Since the process is based on direct mechanical deformation, the resolution constraints are only due to the structuring of the mould, as opposed to the usual limitations from light diffraction or beam scattering observed in conventional nanolithography methods. As long as the cast is stable, it is possible to produce many replicas from a single, prefabricated mould using this method.[92]

In an attempt to mimic the photonic nanostructures found on butterfly wings, Kustandi *et al.*[99] used NIL to produce tall and dense polycarbonate (PC) nanopillars. The surface adhesive force between those pillars and the intrinsically low modulus of the imprinted structures may cause an irregular collapse of the nanopillars in various directions. Therefore, in order to form an ordered structure, the researchers simultaneously applied a horizontal, shear-patterning force to the polymer during the mould release. This step allowed them to control the lateral collapse on the imprinted structures. The periodic, unidirectional ridge arrangement made the structure work like a diffraction grating, while the multilayer seen from the top, due to the inclination of the pillars, gave the effects of interference and scattering. Consequently, their synthetic butterfly wings had angle-dependent optical properties, and showed both rainbow-type colour patterns and single colour patterns ranging from purple, to blue, to yellow, and to red.[99]

Figure 4.7 Lithographic techniques employed to replicate *Morpho*-blue. (A) Laser interference lithography to produce a 3D surface patterning similar to the wing scales of *Morpho* butterflies; the top is a blue-reflecting grating where each feature is laterally lamellar, the bottom is a simple grating with the same periodicity, but lacking coloration (reproduced from ref. 98, https://doi.org/10.1364/OME.5.000996, under the terms of the CC BY 4.0 license, https://creativecommons.org/licenses/by/4.0/). (B) Nano-imprint lithography, combined with shear patterning, to fabricate uniformly collapsed pillars (adapted from ref. 99 with permission from John Wiley and Sons, © 2009 Wiley-VCH Verlag GmbH & Co. KGaA, Weinheim). (C) Electron-beam lithography can precisely sculpt lamellae structures similar to the ones found on the *Morpho* wing scales (adapted from ref. 100). (D) Nano-imprint or e-beam lithography is combined with layer-by-layer deposition to fabricate a multilayer-coated random grating (adapted from ref. 82 with permission from The Authors and the Society of Photo Optical Instrumentation Engineers, Copyright 2006).

NIL provides a simple, fast and cheap approach for the fabrication of nanofeatures with a high precision. Furthermore, it is possible to develop a high-throughput manufacturing process to imprint the patterns in a continuous manner, *via* roll-to-roll NIL.[102] While NIL exhibits a very promising future, it still has some issues in processing conditions, due notably to the mould's limited durability and the formation of defective structures when air is trapped in the gaps between the resist and mould cavities.

Based on NIL and soft lithography, other nanofabrication techniques, such as nanocasting lithography (NCL), have been developed with similar attributes, yet with some additional advantages. In their work, Saito *et al.*[82,103] produced a *Morpho*-type substrate by using NCL to produce a specific surface pattern. The created pattern was then coated with titanium oxide, TiO_2, and silicon oxide, SiO_2, using successive electron-beam depositions, to form a multilayer with sufficient refractive index contrast. The optical properties of the resulting patterned multilayer showed many attributes similar to the *Morpho*-blue: hot and brilliant coloration over a wide angular range, high reflectance, slight change of the colour tone at a shallow observation angle, speckling, and one-dimensional anisotropy of the brilliance. Using such fabrication methods, one can further tune the colour by patterning the ridges, choosing the combination of inorganic materials, and controlling the multilayer thicknesses.[82]

4.3.1.1.4 Electron-beam Lithography. Electron-beam lithography (EBL) is one of the most important techniques of maskless nanofabrication. It uses a highly focused electron-beam ("e-beam") to modify the solubility of an electron-sensitive resist material, thus allowing selective removal of either the exposed or non-exposed regions upon dissolution in a liquid developer (see Figure 4.6B).[94]

In an another effort to reproduce the properties of butterfly wing coloration, Wong *et al.*[104] used EBL to design groove-based multi-grating structures that generate a specific blue structural colour observable over a wide range of reflection angles (16° to 90°). Exploiting the accuracy of EBL, they could change the geometry of the grooves on multiple hexagonal cells patterning the resist (poly(methyl methacrylate), PMMA, in that case), each endorsing the effect of the grating size, orientation, density, and depth, on the colour formation and on the efficiency of light diffraction.

In a more recent study, Zhang and Chen[100] also mimicked the coloration of the *Morpho* butterfly wing scales by creating well-aligned and alternating lamellae structures of PMMA and lift-off-resist (LOR) using a novel process based on EBL. To create the air layers of the "Christmas tree"-like gratings in *Morpho* butterflies, the researchers selectively dissolved the LOR layers with a diluted alkali. Using this method, the colour reflection can be controlled by fine-tuning the layer thicknesses of both the PMMA and the LOR. Various colours, other than the *Morpho*-blue, can then be obtained.[100]

4.3.1.2 Layer-by-layer Deposition Techniques

Typically, multilayer photonic nanostructure films are fabricated using inorganic dielectric materials or inorganic nanoparticle/polymer composites. To obtain larger, variable stop band gaps, the layers may be designed with porous structures and/or composed of stimuli-responsive polymer materials. Layer-by-layer (LbL) deposition methods, such as atomic layer deposition, electrodeposition, the Langmuir–Blodgett technique, and sputtering, lead to the formation of multilayers that are composed of periodic stacks with alternating high and low refractive indices. Although these methods usually result in well-controlled film growth, they often acquire high cost and suffer from scale limitations. Sol–gel-based depositions and spin-coating can offer more cost-effective routes to produce structural colours, but a high uniformity of deposition on curved substrates is then hard to achieve using these methods.

4.3.1.2.1 Liquid-based Deposition. Silicon oxide and titanium oxide are widely used to fabricate biomimetic multilayered structures. Indeed, they have very different refractive indices ($n_{SiO_2} = 1.45$ and $n_{TiO_2} = 2.44$) and can easily be deposited from a liquid phase by spin- or dip-coating. Choi *et al.*[105] demonstrated the production of Bragg stacks out of porous layers by sequentially spin-casting meso-TiO_2 and meso-SiO_2 into a multilayer whose reflecting wavelengths' bandwidth exceeded 200 nm. In another work, Calvo *et al.*[106] demonstrated all-TiO_2 Bragg mirror production by changing the particle size in the sequential layers.[106] Fuertes *et al.*[107] assembled the same silica/titania nanoparticle pairs using a dipping process followed by humidity (24 hours) and thermal (>48 hours) curing. They were able to achieve refractive index values of 1.35 and 1.82 for the resultant porous SiO_2 and TiO_2 nanoparticle stacks, respectively. Scharf *et al.*[108] constructed an optical system that combined Bragg reflectors and microlenses from poly(vinyl alcohol) (PVA) and poly(*N*-vinyl carbazole) (PVK). To do so, they used spin-coating of organic interference layers and soft replication of microlenses. Although their refractive index contrast between alternating layers was low, they nevertheless achieved a reflection band centred at $\lambda = 570$ nm.[108]

Kolle *et al.*[109] took the LbL technique one step further by combining spin-coating with multilayer-rolling in order to mimic the concentric layers found in the plant *Margaritaria nobilis*.[109] They fabricated a thin glass fibre, on which they rolled bilayers of two elastomeric polymers with sufficient refractive index contrast: polydimethylsiloxane (PDMS) with $n_{PDMS} = 1.41 \pm 0.02$ and polystyrene–polyisoprene triblock copolymer (PS-PI) with $n_{PS-PI} = 1.54 \pm 0.02$. Since the thickness of the layers controls the colour, the team could alter the appearance of these elastic rubber fibres upon the application of a thinning tensile stress. This example shows how structural colours can add functionalities to materials (see Section 4.4), and such colour-morphing fibres could indeed create textile fibres for responsive athletic wear, for example.

Mimicking the multilayered structure of *Margaritaria nobilis* could be achieved using a simple spin-coating of synthetic polymers. However, in some cases, it is necessary to follow the steps taken in biogenic material synthesis in order to achieve the multifunctionality and optimised effects of optical and mechanical properties. This may indeed shed light on the fabrication of novel or low-cost materials. For this reason, Ullrich Steiner's group at the University of Cambridge (now at the Adolphe Merkle Institute) proposed a simple and robust approach to replicate the *natural* LbL process of nacre formation. The researchers created the multilayer by immersing a glass slide into poly(acrylic acid) (PAA) and poly(4-vinyl pyridine) (PVP) solutions sequentially. To introduce porosity into the PAA layers, they immersed the film in a basic solution that dissolved PAA, but also introduced surface functionalisation to the substrate and the PVP layers, then stabilised by UV cross-linking. Mineral films were formed by ammonium carbonate diffusion technique, with a PAA solution that contained Ca^{2+} and Mg^{2+} ions to induce the formation of a modified calcite (cal_{org}) film. The film was further crystallised by exposure to high humidity. They repeated this procedure to create the organic/inorganic lamellar composite of the nacre structure. Through fine control over the layer periodicity, the researchers were finally able to reproduce the iridescence of the nacre.[86]

4.3.1.2.2 Vapour-based Deposition. Chemical vapour deposition (CVD), on the other hand, allows high growth rates, as well as a profitable uniformity over large areas.[110] It is possible to deposit a great variety of inorganic structures through CVD, but the incorporation of organic materials instigates the creation of deformable and tuneable photonic structures. Hence, in their work, Karaman *et al.* reported a Bragg reflector that consists of titania, TiO_2, and an organic material, poly(2-hydroxyethyl methacrylate) (pHEMA). This hybrid material can shift colour very rapidly and reversibly upon exposure to saturated water vapour *via* the rapid swelling of the pHEMA layers.[111]

Plasma-enhanced CVD (PECVD) is also used to produce transparent long-pass filters made of SiO_2/SiN_x layers, inspired by a combination of optical effects such as anti-reflection of the moth-eye, Bragg scattering, and thin-film interference. In order to achieve such a complex effect, Lal *et al.*[112] produced silver nanoparticles on the surface of the SiO_2/SiN_x multilayered structure, *via* inductively coupled plasma reactive-ion etching (ICP-RIE) and chemical wet etching with HNO_3 and HF, thus fabricating conically shaped Bragg long-pass filters, which demonstrates the selectively transparent optical scattering component. The SiO_2/SiN_x multilayers give strong forward scattering at narrow angles of long wavelengths, and strong back scattering at high angles of short wavelengths due to the variation in the refractive index contract and the geometry of the structures.

Another gas-phase deposition technique, atomic layer deposition (ALD), is used to produce thin films with a variety of materials. Based on sequential, self-limiting reactions, ALD offers exceptional conformity on high-aspect ratio structures, thickness control at the Ångstrom level, and tuneable film composition.[113] Kolle *et al.*,[114] for example, used a combination of techniques, including ALD, colloidal self-assembly, and sputtering, to fabricate photonic structures that mimic the colour-mixing effect found on the wings of the Indonesian butterfly *Papilio blumei* (see Figure 4.8).[114]

4.3.1.2.3 Further Examples of Lbl-based Biomimicry. The *Chrysochroa* genus of the metallic wood-boring beetles exhibit glossy iridescent colours. The beetle cuticles are an example of one-dimensional Bragg mirrors with a polarisation- and angle-dependent reflectance spectrum, typical for multilayer structure (see Section 4.2.1). Furthermore, slight variations in layer thickness and/or refractive index cause variations in the coloration.[115] Inspired by the iridescent colour formation of the *Chrysochroa rajah* beetle, Tzeng *et al.*[116] applied the LbL deposition technique to polyethylenimine (PEI) and anionic vermiculite clay bilayers (high refractive index layer component), combined with cationic colloidal silica, SiO_2, and anionic cellulose nanocrystals (CNCs, low refractive index layer) (presented in Section 4.3.2.1). The films produced had an even green colour, resulting from the high consistency in layer thickness at the origin of the relatively narrow reflection band. The iridescence of the films was notably characterised by a blue-shift of the reflected wavelength with increasing incidence angle.[116]

As explained in Section 4.2.2, *Morpho* butterfly wings exhibit a complex optical response by combining several phenomena to produce a unique angle-independent brilliant blue colour. *Morpho* butterflies create these complex structures spontaneously, through self-assembly, without any directed construction. In order to address all the challenges of its biomimicry at the same time, Chung *et al.*[117] first produced a semi-ordered monolayer of silica nanospheres (200–400 nm in diameter) on a silicon wafer by spin-coating, and then further deposited blocks of TiO_2 and SiO_2 on top *via* sputtering.[117] By controlling the thickness of the TiO_2 and SiO_2 layers, they tuned the colour ranging from deep blue, through green, to coppery red. Notably, their deep-blue films are comparable to *Morpho rhetenor*, with a peak reflectance of 55%, and the colour and brightness do not show significant changes across all viewing angles.[117]

4.3.1.3 Iterative Size Reduction

Iterative size reduction (ISR) is an alternative top-down technique to produce a flexible polymer fibre that consists of arrays of millions of ordered, indefinitely long, nanowires and nanotubes that can exhibit structural colours. Inspired by the composite fibre drawing process from polymer reels, Mehmet Bayindir's group at Bilkent University demonstrated the fabrication of ordered arrays of nanowires by thermal ISR.[118,119]

The process comprises of a multi-step drawing of a fibre, starting with a macroscopic polymer rod. Each step starts with structures obtained from the

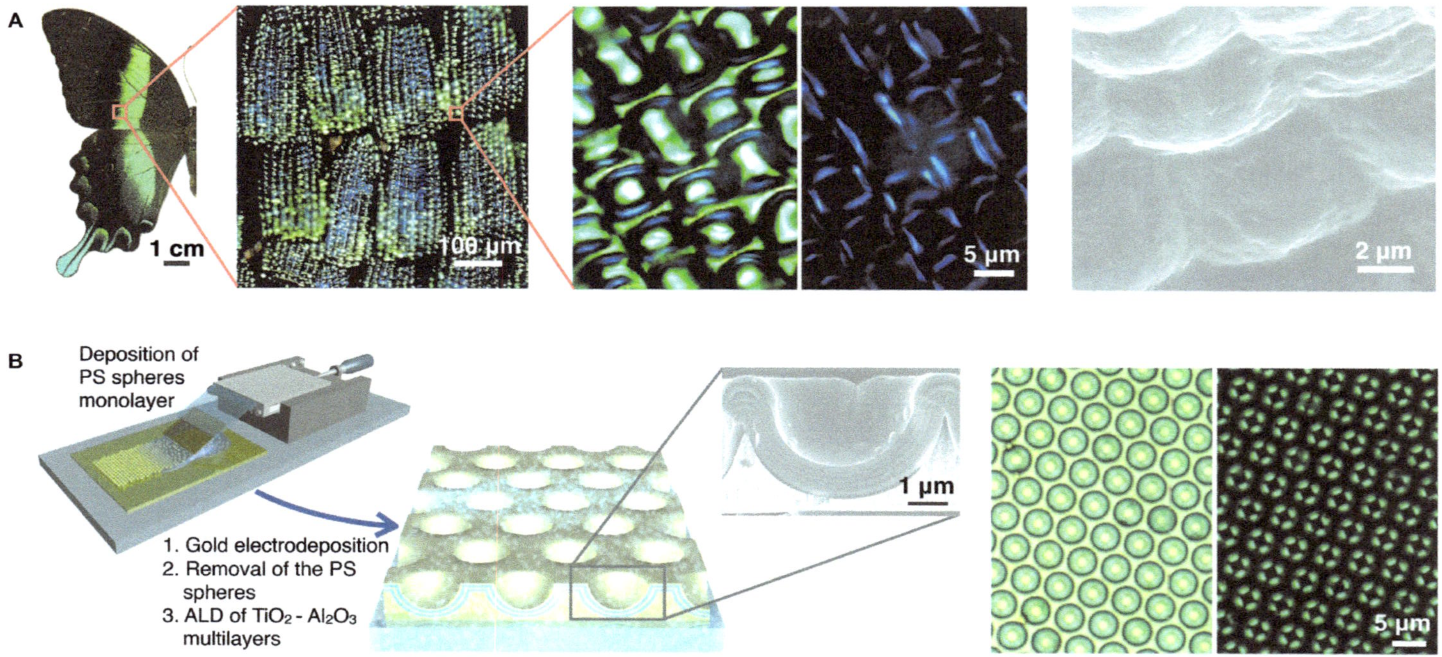

Figure 4.8 *Papilio blumei* butterfly, biomimetic sample fabrication and optical mimicry. (A) The bright green wing of the *Papilio blumei* butterfly is shown at various magnifications (from left to right): a photograph of the wing; a zoomed in image on the wing scales; further magnified optical micrographs of the wing scales observed without (left) and with (right) cross-polarisers; the rightmost image is an SEM of the *Papilio blumei* wing-scale surface showing the concavities. (B) The deposition of polystyrene colloids on a substrate is followed by gold electrodeposition on the interstitial sites, removal of the PS monolayer and ALD deposition of Al_2O_3 and TiO_2 layers, to obtain concavities covered by a conformal multilayer stack of Al_2O_3-TiO_2 alternating layers, as shown in the SEM micrograph inset. The rightmost pair of images are light microscopy of the film that shows that the concavity edges are green and the centres and interstitial regions are yellow (left), while only the green concavity edges are visible under crossed polarisers (right).
Adapted from ref. 114 with permission from Springer Nature, Copyright 2010.

previous one, resulting in geometrical size reduction and increment in wire number and length. At the end of the first thermal size reduction cycle, the cross-section of the fibres reduces from a few millimetres to a few micrometres. These fibres are then cut and arranged into a hexagonal lattice inside a protective jacket, which is vacuum consolidated and further redrawn for the second and third time. The authors showed the possibility of producing a variety of multimaterial nanowire and nanotube structures, including non-linear glass, As_2Se_3, and ordered core–shell nanowires of As_2Se_3-PVDF (polyvinylidene fluoride) in polyethersulfone (PES), which displayed structural colours. The colour formation, in this case, is explained by a combination of thin-film interference from the core–shell nanowire structure, and Mie scattering from the core region (see Section 4.2.1). The refractive index of each layer, as well as the core diameter, determine the scattered resonant modes in the vicinity of the nanowire, although thin-film interference may dominate for appropriate shell thicknesses.

A more recent study from Bayindir's group reported the imitation of the peculiar 2D photonic structure in the neck feathers of mallard drakes. Using the ISR technique, they produced a polymer composite fibre system based on polycarbonate (PC) and PVDF, which have low refractive index contrast ($n_{PC} = 1.58$ and $n_{PVDF} = 1.41$) but are thermally compatible, for successful processing with ISR.[120]

4.3.1.4 *Combining Top-down Strategies*

Control over the shape and spacing of fetaures, when patterning using these sophisticated lithography techniques, is particularly necessary for the production of defect-free photonic structures with a well-defined and complete band gap. By combining multiple lithography techniques together, or with other top-down fabrication processes such as deposition or chemical etching, a high-resolution 3D topography may be ultimately produced. In addition, when further cycles of microfabrication steps are repeated successively several times, even more complex structures may be formed.

Hence, elaborating on their previous work[82] (described above in Section 4.3.1.1) to replicate the delicate structural effects and optical response of *Moprho*-blue, Saito *et al.*[121] subsequently employed EBL and dry-etching to first fabricate a quasi-one-dimensional, patterned quartz structure, whose random discrete step-profiles have optimised widths and gaps sequences. They next used LbL to deposit a multilayer composed of alternating layers of TiO_2 and SiO_2, the patterned quartz substrate. The fabricated film shows the *Morpho*-blue, with some of its fundamental characteristics, such as hot brilliant blue in a wide angular range, and high reflectance. However, they were not able to reproduce the angular dispersion and wavelength dependence of the natural *Morpho*-blue.[121]

Nevertheless, and in spite of the advantage of fabricating nanostructures with precise morphology, a cost-effective manufacturing scheme to generate multicoloured structures over a large area is a challenge with top-down strategies, because of their usual requirement for a vacuum process.

4.3.2 Bottom-up Strategies

4.3.2.1 *Self-assembly*

Self-assembly is the process by which structural order at different scales is achieved without any direct external action.[122] One way to fabricate materials with structural colours is to create periodic nanostructures through self-assembly using various building blocks, such as block copolymers, liquid crystals, and colloids. Nature, of course, uses this phenomenon in numerous cases to produce optimal colours and multifunctional materials. Therefore, it is important to understand the self-organised systems and processes found in nature, in order to replicate them "efficiently" from a technological point of view, with the goal to create defect-free, structurally coloured materials at large scales.

4.3.2.1.1 Colloidal Self-assembly. Dielectric structures of colloidal systems, in the submicrometre length scale, can interact strongly with light. With specific designs, these can produce various and remarkable optical responses, which can be tuned further by altering the types and organisation of these structures. Self-assembled colloidal systems include colloidal crystals, composite and inverse opals, and photonic glasses. Owing to their unique optical properties and their applications as photonic crystals they have been studied for a long time.[93] Colloidal self-assembly has significant advantages and, notably, it offers the most facile and economical way to create 2D and 3D photonic crystals over large areas, and with a wide variety of possible shapes.[123] The optical properties of self-assembled, 3D colloidal beads (artificial opals) were first demonstrated by Astratov *et al.*[124] Since then, a great variety of materials, for both organic and inorganic microspheres, have been used to produce biomimetic structural coloration through colloidal self-assembly. Among these, self-assembly of polymer opals, mainly consisting of polystyrene (PS) and PMMA monodisperse microspheres in a face-centred cubic (fcc) arrangement, have been studied extensively.[122,125–129]

Several methods have been used to achieve long-range 3D arrangements through colloidal self-assembly. These include natural sedimentation of SiO_2 spheres followed by sintering,[130] electrophoresis of SiO_2 monodisperse nanospheres or SiO_2/TiO_2 core–shell particles,[131] or injection of PS beads into a confined environment.[132] Currently, the most commonly used method is the vertical, or convective, deposition method. This technique relies on capillary forces to organise colloid bases during the evaporation of the liquid (ethanol or water), which fabricates colloidal crystal multilayers (see Figure 4.9).[133,134]

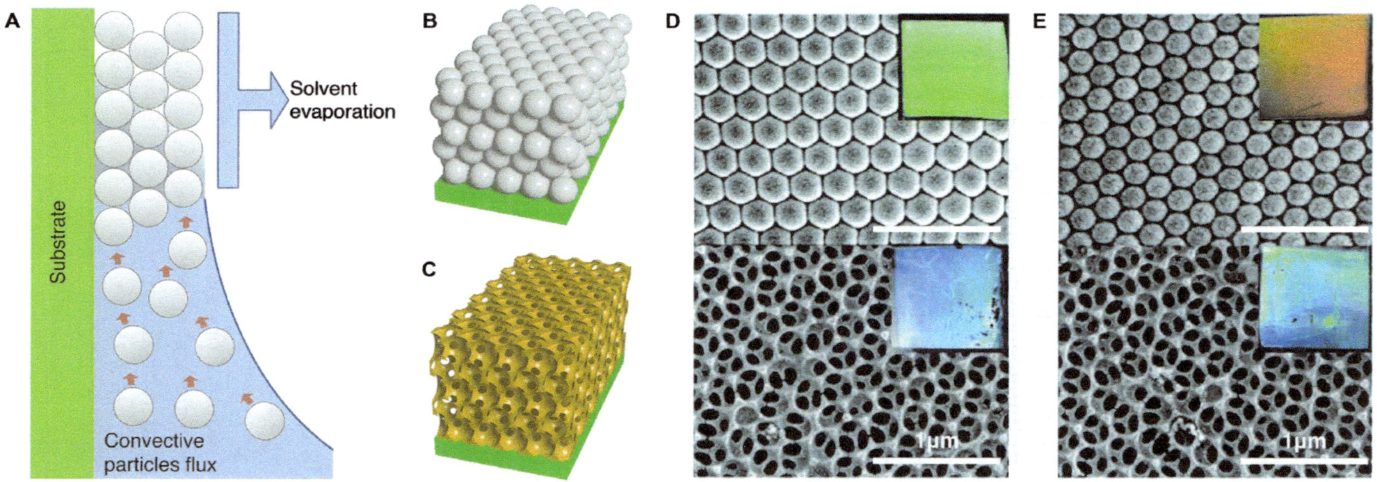

Figure 4.9 (A) Self-assembly of multilayer colloidal arrays through the vertical deposition technique. (B) Colloidal crystal in the opal structure. (C) The inverse opal structure produced through infiltration of the opals, and subsequent selective removal of the colloidal spheres. (A)–(C) Reproduced from ref. 25 with permission from the Royal Society of Chemistry. (D), (E) SEM images of PS colloidal crystal templates (top) and their corresponding inverse opal structures constructed using 250 nm and 263 nm spheres, respectively. (D), (E) Adapted from ref. 135 with permission from the Royal Society of Chemistry.

In order to tailor the photonic band gap to a specific structural colour response, with added functionalities such as chemical sensing through a colour change, core–shell particles with various compositions have been explored over the years. Hence, to control the position and the width of the wavelength band gap, the core to shell compositions can be varied. Alternatively, it is possible to incorporate particles of different sizes into the interstices of crystals of core–shell particles, in order to modify the final colour of the surface.[136] Moreover, one can also target the effect of refractive index contrast in the interparticle medium, as well as composition of the chemical core-interlayer-shell precursor particles, to effectively tune the colour.[137–139]

Natural materials are also often produced by self-assembly processes, and bear optimised structural order over multiple length scales to sustain maximum performance and multifunctionality. Most photonic crystals produced *via* colloidal self-assembly show brilliant structural colouring due to the Bragg diffraction effect, which depends on the angle of observation (Section 4.2.1). Therefore, a number of colloidal photonic crystals were recently developed with a macroscopic, spherical bead appearance.[140] Due to the spherical symmetry, the reflection spectra are unchanged upon rotating these photonic structure constructions under illumination of the surface at a fixed incident angle of the light. This packaging may broaden the range of applications for photonic crystals. Tailoring the colloidal particles into a spherical shape that exhibits structural colour can be achieved with different arrangements within the macroscopic sphere: close-packed colloidal assembly,[141–143] non-close-packed colloidal assembly,[144,145] or inverse opals,[146] for example. Joanna Aizenberg's group at Harvard University also demonstrated colloidal crystallisation in an emulsion droplet, producing micrometre-sized superstructures, which they called "photonic balls".[147] While the layered morphology of the colloidal spheres in the photonic balls gave rise to structural coloration, *via* Bragg diffraction, the natural curvature of the ball created a uniform, angle-independent colour formation. Such systems can have promising applications in optical devices, such as diffraction gratings, colorimetric sensors, or dispersible structural colour pigments.

Melanins are natural pigments found in several organisms, especially in the feathers, hair, or skin of animals. They are thought to absorb UV radiation to protect the living organisms.[148] Birds use melanosomes to form organised structures for producing colours. Although such structures have already inspired the fabrication of photonic crystals, the use of melanins, or melanin-like materials, in structural colour biomimicry is a novel approach that was recently demonstrated by Xiao *et al.*[149] They prepared nanoparticles with an average diameter of 146 ± 15 nm and made of polydopamine (PDA), a type of synthetic melanin with refractive index $n_{PDA} = 1.7$. This particle size is within the diameter range of the rod-like melanosomes found in duck and peacock feathers, and thus replicates their structural features. They used a vertical evaporation-based self-assembly system to produce structurally coloured thin films. By tuning the thickness and/or concentration of the assembled nanoparticles they were able to produce different colours, which all showed a good agreement with their theoretical model based on thin-film interference.[149]

The promise of colloidal self-assembly for high-throughput production of structural colours is indeed remarkable. However, the presence of unavoidable point or extended defects and cracks, accompanied by incomplete photonic band gaps, still leaves colloidal self-assembly far from being routinely used in photonics devices.[150] To overcome this barrier, many new strategies are being developed. For example, the use of a selective chemical "glue", such as DNA or peptides, on particle surfaces has been proposed.[151] Further developments might enable the fabrication of better photonic devices based on colloidal particles. Moreover, recent progress in the synthesis of non-spherical particles is accelerating research into the colloidal assembly of new crystal phases.[152] One promising strategy, with strong potential for producing new photonic properties, is to use directional interparticle interaction *via* chemical patterns on the surface of the colloids to create and design new types of colloidal lattices.

Colloidal self-assembly is a powerful platform for producing 3D-patterned structures, with feature sizes ranging from a few nanometres to several micrometres. However, the variety of patterns that can be made through these bottom-up techniques is very limited, in comparison to top-down techniques. Therefore, recent years have seen several attempts to combine the top-down and bottom-up strategies for producing patterned colloidal photonic crystals. Hence, the production of uniform and crack-free structures, with high colour definition, has been demonstrated by using lithography techniques (Section 4.3.1.1) on various inverse opal structures.[153–155] However, the processes used in the fabrication of such structures are generally complex, time consuming and expensive.

Jeremy Baumberg's group at the University of Cambridge, on the other hand, developed a method to directly pattern colloidal crystals by using micro-imprint lithography, a facile and cost-effective technique. Notably, they assessed the effect of annealing the colloidal particles prior to the imprinting step, which affected the optical behaviour of the films. If the colloids are pre-annealed, micro-imprinting causes a compression of the imprinted region and changes the lattice separation normal to the surface. This effect results in dual-colour patterns. If the colloid crystals are not pre-annealed, micro-imprinting selectively peels the contacted region off the substrate, thus forming negative crystal patterns on the substrate. Such spatial control of colloidal assemblies and complex patterning may enable more advanced applications in, for example, sensing devices, pixelated arrays of structural colour units, or antifouling materials.[156]

In biological species, the growth and assembly of colloidal particles is even more complex, since it often requires phase separation of proteins and inorganic materials within living cells. In order to understand the mechanisms of droplet formation on bird feathers, such as the *Cotinga maynana* bird[157] (Figure 4.5B), Style *et al.*[158] devised an experiment to understand the key processes playing a role in the phase separation of stable, uniform droplets of a solvent mixture inside a cross-linked polymer network. The growth of the droplets was notably affected by the cross-linking density of the polymer network, supersaturation, and the quench rate. This method offers an alternative way of assembling submicrometre spherical particles in a photonic array, with potentially novel optical or mechanical properties. Indeed, the method does not rely on using monodisperse colloidal templates, and hence may facilitate the production of large-scale composites with well-defined microstructures.[158]

4.3.2.1.2 Anisotropic Particle Self-assembly. Self-assembly of nanomaterials is a function of interparticle interactions (*i.e.* ionic charges, capillary forces, *etc.*), particle sizes, their distribution, and particle shapes. Various liquid crystalline order structures, such as nematic, columnar, and smectic phases are observed for rod-like and plate-like particles at high volume fractions[159] (see Figure 4.10). According to Onsager's theory, hard

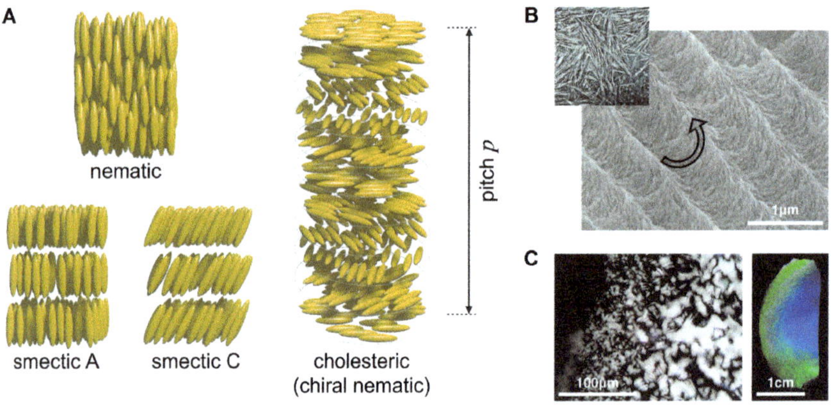

Figure 4.10 (A) Schematic illustration of a variety of liquid crystal mesophases, the nematic phase with only orientational order of the long axis of elongated molecules, the smectic phases (smectic A and smectic C), which exhibit additional one-dimensional positional order, and the chiral nematic (cholesteric) phase where the individual nematic layers stack together with a twist, leading to helicoidal order with a pitch *p*, which refers to the distance over a full 360° twist (reproduced from ref. 25 with permission from the Royal Society of Chemistry). (B) SEM micrographs of the cross-sectional view of chiral nematic mesoporous silica templated from CNCs, showing a left-handed chiral nematic structure; the inset is a transmission electron microscopy image of negatively stained unordered CNCs dropcast from a dilute suspension. (C) Polarisation optical microscopy image of the CNC–silica composite displaying the characteristic fingerprint-like structure showing the formation of a chiral nematic phase (left), and a photograph of a resulting iridescent organosilica film (right). (B), (C) Adapted from ref. 161 with permission from American Chemical Society, Copyright 2012, and from ref. 162, with permission from Springer Nature, Copyright 2010.

slender rods interacting with repulsive forces can exhibit orientational ordering at densities far below the closest packing. The thermodynamic stability of these ordered structures arises from a gain in translational entropy, which overrules the loss of orientational entropy associated with particle alignments.[160] A combination of a wide range of material properties, coupled with the self-assembly, could lead to numerous liquid crystalline phases and new materials with a wide variety of new applications.

As mentioned in Section 4.2.2, nature also uses this strategy to produce iridescent coloration in plants.[55,84] The helicoidal-ordered stacks of cellulose are responsible for the bright colours in fruits and leaves of very different species of plants. Similar photonic structures can be artificially produced using the same constituent material, that is, cellulose nanocrystals (CNCs). Hence, the slow evaporation of a CNC suspension gives rise to their spontaneous assembly into a chiral nematic liquid crystalline phase that can be preserved in the dry state[163] (see Figure 4.10).

Optically, chiral nematic phases have a number of interesting properties. First, they cause a significant optical rotation of light, that is much stronger than that of the individual chiral components. Second, they strongly reflect light of a wavelength that is comparable to the length scale on which the nematic director makes one full turn, that is, the chiral nematic pitch p (Figure 4.10). Interestingly, only circularly polarised light of the same chirality as the substrate is reflected, and the reflected light has the same polarisation (normal reflection on a metallic mirror, on the other hand, would cause the circular polarisation to invert). De Vries[48] explained these phenomena theoretically, and the reflected peak wavelength is given by[48] the Bragg–Snell law, eqn (4.16), where the half-pitch $p/2$ supplants the thickness d, and $\bar{n}$ is the average refractive index of the film ($\bar{n} = 1.55$ for CNCs[164]).

Several methods have been proposed to tune the chiral nematic pitch and consequently the final colour formation in these cholesteric cellulose films.[165–167] Further investigations, notably, revealed that within a single batch of cellulose crystallites there are distinct domain formations of different periodicities, with different colour transitions, which can be correlated with the chiral nematic pitch and layering (see Figure 4.11).[168] Dumanli *et al.*[169] also demonstrated ways to control the CNCs self-assembly process, in order to produce a range of different colours selectively, yet starting from the same suspension.[169] In this work, they also developed an optical microscope set-up to observe the film formation dynamics, while controlling the temperature and relative humidity, during self-assembly, in an environmental chamber. Their findings revealed that the CNCs self-assembly starts from a completely random organisation and, with gel formation, the small domains of chiral nematic stacks progressively appear. Just before the complete drying, they observed that the films formed display colours in the red, and then subsequently shifts colour due to evaporation of water.

Chitin nanocrystals are also found to form chiral nematic phases in concentrated suspensions, in which the chiral nematic phase can be preserved upon drying of the film. However the film does not show structural

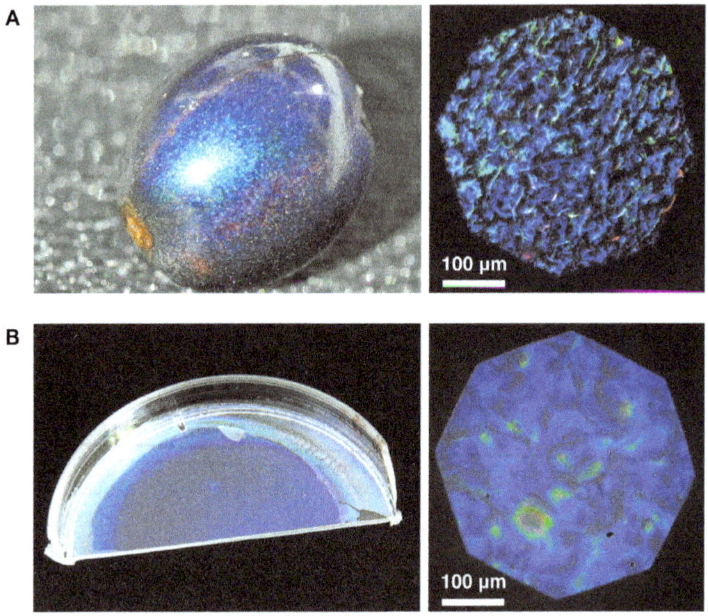

Figure 4.11 (A) Photograph (left) of the *Pollia condensata* fruit and optical microscopy image (right) of its surface in cross-polarisation configuration. (B) Photograph (left) of the chiral nematic cellulose nanocrystal (CNC) film and optical microscopy image (right) of the CNC film in cross-polarisation configuration.

colour formation in the visible range, and the optical activity for such chiral nematic systems was indeed observed in the infrared region.[170]

4.3.2.1.3 Block Copolymer Self-assembly. Polymers made of a single building block can form, *via* phase separation in solution,[171] highly scattering porous films that mimic the network of the *Cyphochilus insulanus* beetle,[72] a well-known example of whiteness found in nature. Nanostructures based on block copolymer (BCP) self-assembly have also been employed in the fabrication of a large number of photonic crystal structures, from one to three dimensions. BCPs are macromolecules that consist of different monomer sequences organised into continuous blocks. Typical BCPs are linear and contain two or three major monomer blocks, say A, B and C, organised such as AB, ABA or ABC.[172]

Chemical incompatibility between the constituent blocks results in phase separation between the two blocks, while the covalent linkage between the blocks restricts the phase separation to the length scale of the polymer molecules (typically, a few tens of nanometres). The spontaneous aggregation of the individual blocks towards an equilibrium condition yields well-defined structures, based on non-covalent interactions. There are four classes of morphologies arising from such phase separation: lamellae (L), hexagonally packed cylinders (H), body-centred spheres (Q^{229}) and the double-gyroid phase (Q^{230}),[173] as shown in Figure 4.12. The main parameters that determine the phase morphology of an AB diblock copolymer are the volume fractions of the blocks f, and the degree of segregation, χN (see Figure 4.12B).

It is evident that multidirectional photonic band gaps exist in these structures, and the reflected wavelength λ_m is proportional to the material periodicity d, and can be again approximately expressed using the Bragg–Snell law, eqn (4.16), for each order of diffraction m. As before, $\bar{n}$ is the average refractive index of the reflecting elements. Most polymers have a refractive index around 1.5. Thus, to obtain a photonic band gap of visible wavelengths, the periodic spatial modulation of the refractive index has to be of the order of 120 nm or larger. The typical unit cell dimensions for self-assembled BCP morphologies are of the order of 5–100 nm, which makes them unsuitable for visible photonic band gaps.[174] Therefore, the main challenge with BCP-based photonic structures is to produce larger structures with periodicity comparable to the wavelength of visible light. Increasing the unit cell size in two- and three-dimensionally continuous morphologies is arduous. Nevertheless, the use of the gyroid phase for optical illusions can actually be seen in nature; for example *Callophrys rubi* butterfly wing scales exhibit a highly ordered, porous, three-dimensional single gyroid structure that extends throughout the scale, beneath the ribs and the cross-ribs. Due to the chiral orientation of the gyroid phase, the butterfly does indeed display circular dichroism.[175]

There are numerous efforts being put into the development of BCP-based photonic crystals that reflect light, as well as change colour in the visible region. A photonic response in the visible range requires reliable phase separation, and a high degree of long-range order in, usually, very high molecular weight BCPs.[176] However, high molecular weight (that is, greater than 1 mg mol^{-1}) materials are extremely viscous in the melt, and do not easily self-assemble into well-ordered structures due to their strongly entangled configurations. Yet, self-assembled BCP photonic crystals are highly responsive to different kinds of external stimuli such as solvent, temperature, or compressive mechanical strain.

One interesting example of solvent stimuli was recently demonstrated by Chiang *et al.*[178] They fabricated a responsive BCP photonic crystal system based on high molecular weight polystyrene–polyisoprene (PS-PI) block copolymers, with a lamellar morphology in toluene. When the polymers were in a relatively low-concentration regime, that is, at a concentration ϕ_p only slightly higher than the order–disorder transition concentration, $\phi_{ODC} \lesssim \phi_p$, a red-shift reflectivity band could be seen. This was due to the thermodynamically controlled swelling of the long period of the BCP by the intensified segregation strength of the BCP. By contrast, in the high-concentration

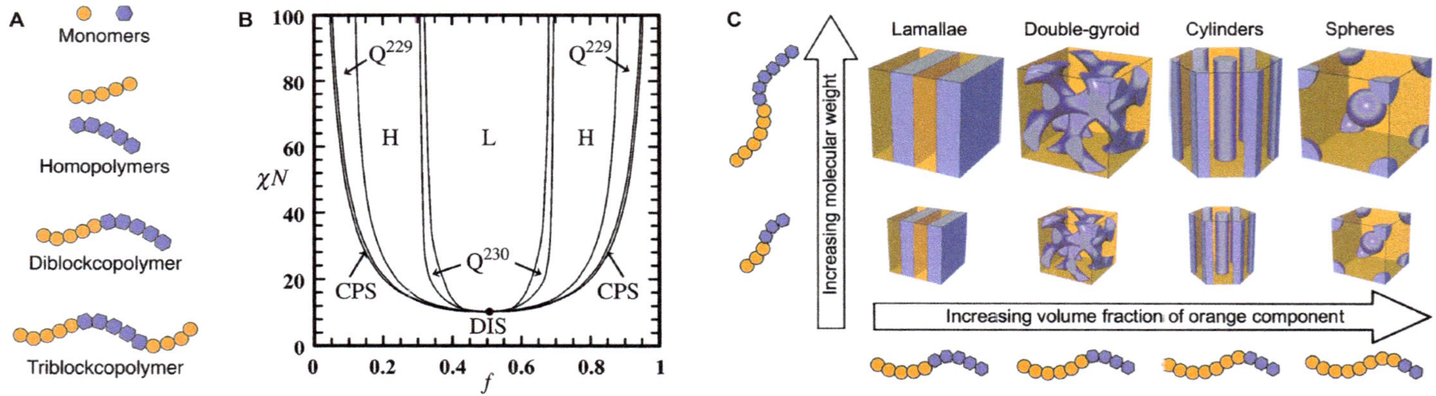

Figure 4.12 (A) Block copolymer system building blocks. (B) Theoretical phase diagram for a linear diblock copolymer (adapted from ref. 173 with permission from American Chemical Society, Copyright 2006) in the degree of segregation χN *versus* the block's volume fraction f space; the various equilibrium morphologies are lamellae (L), hexagonally packed cylinders (H), body-centred spheres (Q^{229}), the double-gyroid phase (Q^{230}) and the close-packed spheres; otherwise the system is disordered (DIS). (C) Block copolymers and self-assembled equilibrium morphologies. Parts (A) and (C) reproduced from ref. 177 with permission from the author.

regime $\phi_{ODC} \ll \phi_P \leq 1$, a blue-shift reflectivity band was seen, and attributed to the kinetically controlled deswelling of the long period of the BCP by the collapse of the block chains.[178] The mixtures of selective and neutral solvents for diblock BCPs are quite successful in fabricating structures with finely tuned photonic properties. Matsushita and Okamoto[179] demonstrated an example of the effect of the solvent, which can selectively swell one or both of the blocks of a symmetric polystyrene-*block*-polyisoprene (PS-*b*-PI) diblock copolymer system.[179]

Alternatively, by blending the BCP system with non-volatile solvents or homopolymers, the periodicity of the structures can be tuned over length scales that enable them to interact with visible light.[180] For example, Urbas *et al.*[181] demonstrated a mix of diblock PS-*b*-PI copolymer with its constitutive homopolymers that formed lamellar structures with reflection peaks in the 330–620 nm range, depending on the blend composition.[181] A reversible switching of an optical band gap is also possible, as demonstrated by Valkama *et al.*,[182] using polystyrene-*block*-poly(4-vinylpyridinium methanesulfonate) and 3-*n*-pentadecylphenol supramolecular diblock systems (PS-*b*-P4VP(MSA)$_{1.0}$ PDP$_{1.5}$). The prepared films had long-ordered periodicity, around 160 nm, and reflected green colour. As the system was heated above 125 °C, the film lost its colour due to disruption of hydrogen bonding. Indeed, this induced a very strong decrease in the long period of the lamellar structure, within a narrow temperature range, because of more compact polymer coiling.[182]

Interaction of the polymer blocks with its own chemical environment and applied fields can also alter the total volume fractions of the lamellae. An example of a pronounced manifestation of this effect was shown by Hwang *et al.*[183] in an electrochemical cell that consists of a BCP system of alternating PS and swellable poly(2-vinylpyridine) (P2VP) gel layers, with pH-dependent photonic band gaps. The optical response of these photonic gels was tuned, and further optimised, by controlling the ion-pairing affinity between the protonated pyridine groups and their counter anions.[183] The photonic band gap of the PS-*b*-P2VP can be tuned by different strategies, such as UV exposure[184] or by adding reactive monomers to the film.[185] It is also possible to process the PS-*b*-P2VP in polyvinyl alcohol to form ellipsoidal nanoparticles. This eliminates the angle-dependent view arising from one-dimensional construction in the films.[186] More recently, Park *et al.*[187] reported an electrically switchable, free-standing film from a self-assembled poly(styrene-*block*-quaternised 2-vinylpyridine) (PS-*b*-QP2VP) copolymer as well.[187]

Alternatively, block copolymer micelles (BCMs) can form micellar photonic crystals. Spherical BCMs have a narrow size dispersity and can act as building blocks for long-range ordered 3D superlattices with body-centred cubic and face-centred cubic crystal structures. Hence, Poutanen *et al.*[188] produced photonic crystals in the visible region, by self-assembling PS-*b*-P2VP block copolymer BCMs into a crystal array. The corona (the outer layer) of the BCMs was expanded by introducing ionic charges, to satisfy the

200–600 nm diameter length scale required for photonic effects. By changing the corona length and ionic strength, they further produced tuneable photonic crystals.

Two recent reviews on photonic structures using self-assembly of BCP rationalise the prospect of these as follows: while it is still a very promising approach for the scalable production of photonic crystals and structurally coloured, functional materials inspired by nature, the working examples are still mostly limited to one-dimensional structures.[174,189] On the other hand, the production of 3D-nanostructured photonic materials of BCP for the visible range remains an important challenge.

4.3.2.2 Biomimetic Templates

Templating is a common technique in materials science that permits control of the morphology of materials at different scales. By using the template strategy, it is even possible to achieve material properties that are different from the host (template) material itself. The infiltration of the secondary components is usually done *via* the deposition of the other compounds (guests) onto the template by means of different techniques, such as sol–gel processes,[162,190–192] ALD[193] and CVD[194] (see Section 4.3.1.2). Replicas are created after the sacrificial scaffold is removed from the system without disrupting the morphology, *via* heat treatment, chemical treatment or subsequent growth.

While transparent polymers are quite successful at mimicking the natural structural colour formation phenomenon through self-assembly, added functionalities often rely on high refractive index contrast or surface plasmon effects that are not yet accessible with these soft materials. A great advantage of the templating method is that it is very general with regard to the types of materials that can be prepared. Hence, it is possible to extend the material properties of soft materials with added compositions, such as conductive polymers, metals, semiconductors or graphitisable carbons.[195]

One can usually reproduce the exact intricate nanostructures seen in nature, by using the actual biological structures as templates, through gas- and solution-based deposition techniques (see also Section 4.3.1.2). Direct replicas, or inverse of the structures, can be generated after the removal of templates. For example, Huang *et al.*[193] directly deposited alumina (Al_2O_3) onto the scales of *Morpho peleides* using ALD.[193] When the biological template was removed, they obtained an inverted Al_2O_3 shell structure, which replicated the morphology of the wing-scale structure, as seen in Figure 4.13. Furthermore, some of the optical properties, such as the photonic band gap, were also inherited by the alumina replica. The reflection spectral analysis of the replicas indicated a shift in the reflection peak to longer wavelength than the typical purple-blue reflection of the *Morpho*, because of the change of periodicity and refraction index. The researchers were also able to tune the shift in the reflection wavelength by changing the thickness of the Al_2O_3 layer (10–50 nm).

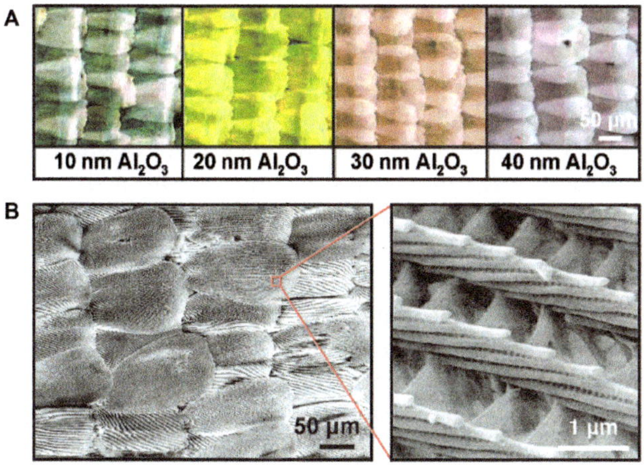

Figure 4.13 Images of the Al_2O_3 replicas of the *Morpho peleides* butterfly wing scales. (A) The optical microscope images of the alumina-coated butterfly wing scales, with different thicknesses of Al_2O_3 deposition. (B) SEM image of the alumina replicas of the butterfly wing scales after the butterfly template was completely removed (left); a higher magnification SEM image of an alumina-replicated scale shows that the replica exhibits the same fine structures (right).
Adapted from ref. 193 with permission from American Chemical Society, Copyright 2006.

Instead of removing the biological template, it is also possible to deposit the inorganic material and leave the biological template as the organic scaffold, in order to retain the mechanical properties. Hence, Liu *et al.*[196] produced such organic–inorganic hybrid material by coating the *Morpho menelaus* wing scales with Al_2O_3 in an ALD chamber. Their work on optical properties of both natural and organic–inorganic hybrids reveals that the iridescence and diffraction characters of the fabricated hybrid structures are homologous to the untreated specimen. Besides, both the uncoated and coated wing scales showed hydrophobic characters, with similar wetting contact angles of 114° and 119.5°, respectively.[196,197] In a similar study, Gaillot *et al.*[198] tried to work with amorphous TiO_2 deposited onto (route I: deposition on the outer surfaces) and into (route II: deposition on both outer surfaces and inner surface through pores and cracks) the scales of *Papilio blumei* butterfly, by using low-temperature ALD.[198] In this work, the researchers demonstrated that the TiO_2 layers form a unique, hybrid organic–inorganic Fabry–Pérot resonator. It was also possible to control the spectral intensity and coloration, by controlling the thickness of the deposited TiO_2.

Solution-based deposition methods involve the chemical treatment of the wing scales to remove pigments, dipping the wing scales in the sol–gel precursor, the heat treatment process for crystallisation of the metal oxide, and, finally, the removal of the biological template. Although the possible wet-chemistry treatment on *Morpho* butterflies was demonstrated for ZnO and TiO_2 in rutile form,[191,199] these replicas did not display the optical properties of the template itself. This was due to disruptions of the order and periodicity in the metal oxide structure. Chen *et al.*,[200] on the other hand, managed to deposit ZrO_2 onto *Euploea mulciber* wing scales using a modified sequential dip-coating, sol–gel method. They started with bleached wing scales, to remove all organic species. Upon removal of the biological template, the inorganic replicas showed angle-dependent iridescent coloration due to the high refractive index of ZrO_2 ($n_{ZrO_2} = 2.16$). Their work indicated that the ZrO_2 precursor was able to deposit onto the intricate structure of the butterfly wing scales without disrupting its morphology.[200]

Biological templates are important to both study and understand the hierarchy and the evolutionary pathways employed to create colour through material structure. However, the biomimetic structures themselves can also be used as templates in a more scalable manner. By using the self-assembled systems as a template, for example, the ability to control the periodicity and structure at nanoscales has a tremendous potential for developing many novel functional materials. The self-assembled systems can be used as direct templates, by infiltrating the second material once the self-assembly process is finished, or they can be used as structure-directing agents that allow the insertion of the second material to follow the specific morphology.

4.3.2.2.1 Direct Templating and Inverse Opals. The development of synthetic opals as 3D photonic crystals naturally pushed researchers to fabricate these systems with material composites, in order to boost their potential. This was done by either increasing their dielectric contrast, or most notably, incorporating active elements (chiefly emitters) to modify their spontaneous emission for low-threshold laser sources. These pursuits have favoured the enormous developments observed in the last few years regarding experimental realisation of opal composites with a large variety of materials. These efforts have undoubtedly pushed and enriched the state of the art in materials science.[201]

The general concept for producing inverse opals is simple. First, one has to form the opal itself, then fill the interstitial spaces with a liquid precursor material (polymer solution, sol–gel precursor or a homogeneous dispersion of nanoparticles), and, finally, once the precursor has solidified within the template, the opal structure is removed through chemical dissolving, radiation or heat treatment. Inverse opals offer a great palette for the production of tuneable structural colour and have been the subject of several review papers.[123,202] Many examples of chemical infiltration in the opal structure, *via* inorganic sol–gel precursors, including TiO_2, SiO_2 and Al_2O_3, have been successfully demonstrated.[203–206] The inverse opals not only show promising applications in thin-film configuration, but can also be processed in a powder form to produce additive structural colour pigments, as demonstrated by Josephson *et al.* in 2014.[207] Gas-phase infiltration methods, such as CVD[208,209] or ALD[210,211] (see Section 4.3.1.2), have also extended the possibility of production of inorganic inverse opal materials in various configurations. These techniques indeed allow very conformal deposition around the spherical units, with controllable thickness, which is essential to control both the homogeneity and the optical properties.

Silica opals, on the other hand, are very useful for the fabrication of inverse opal structures where the guest materials require further thermal and chemical treatment. Silica is thermally stable at elevated temperatures (approximately up to 700 °C) and chemically inert in the presence of most solvents. It is also possible to infiltrate the silica opal structure with polymers to form structurally coloured, free-standing films with long-range order.[212] Even more intriguing systems, where the biological materials are themselves constructed in a photonic organisation by the template infiltration method, may also be explored. Hence, in a recent communication, Kim *et al.*[213] reported the use of silk in an inverse opal structure. Starting from PMMA opal templates, they infiltrated fibroin solutions extracted from *Bombyx mori* silkworm cocoons, and allowed the silk to solidify into an amorphous, free-standing silk–PMMA film. The PMMA templates were then removed by dissolution in acetone, which led to the formation of iridescent silk films.[213]

The combination of templates, materials composition and engineered photonic nanostructures are endless, thus offering vast possibilities for various applications (see Section 4.4). The further incorporation of functional dopants, photoactive or photochemical compounds, or non-linear optical elements are also routes for even more sophisticated properties.

4.3.2.2.2 Co-assembly. As mentioned in Section 4.3.2.1, the use of self-assembly processes to form large-area colloidal crystal films typically results in the formation of cracks, domain boundaries, colloid vacancies, and other defects. The subsequent solution-based infiltration of such templates causes additional cracking of these mechanically fragile templates. In order to avoid further deformations in opal (or inverse opal) structures, and increase the strength of self-assembled templates, partial sintering,[214] deposition onto topologically patterned substrates, or changing the evaporative deposition conditions have been suggested.[215] However, excessive infiltration and deposition may still result in overlayer formation, whereas incomplete, or non-conformal, deposition often leads to structural collapses during removal of the template. Therefore, Aizenberg's group recently made significant efforts to produce crack-free and uniform large-area inverse opal films. In their work, they discovered that letting the colloids and a silicate sol–gel precursor co-assemble in a single step, rather than the classical sequential replication, actually generates highly ordered, crack-free, multilayered inverse opal films on the scale of centimetres. They also reveal the mechanism to achieve such long-range order. Co-assembly indeed eliminates cracking and inhomogeneity, which are associated with liquid infiltration into a pre-assembled opal. Furthermore, this method takes advantage of the interplay between the template and the matrix's assembly, leading to the further correction of incipient defects.[216]

At the other end of the scope, chiral nematic self-assembled systems can also be used as structure-directing agents for self-assembly. Interestingly, for such systems, infiltration with a sol–gel precursor leads to the production of mesoporous metal oxides without replicating the chiral nematic CNC structure.[217,218] Once the CNC self-assembly takes place, the strong interaction between nanocrystals leads to the formation of a compact structure, which does not allow the guest molecules (or particles) to infiltrate. Therefore, Mark MacLachlan's group at the University of British Columbia developed a method where the CNCs self-assemble in the presence of a silicate sol–gel precursor, tetramethylene orthosilicate (TMOS). The researchers were able to demonstrate the first example of using CNCs as a structure-directing agent for the production of free-standing SiO_2 chiral nematic films with structural colour.[162] The helical pitch of the silica films could be tuned across the entire visible spectrum, and into the near-infrared, by adjusting the proportion of the silica precursor TMOS to the CNCs. The same approach was extended to other sol–gel precursors, such as organosilicates, $(RO)_3Si–R'–Si(OR)_3$ where $R' =$ aliphatic/aryl, to produce periodic mesoporous organosilicas.[161] The same team was also able to transfer the chiral nematic structure to water-soluble resins.[219] With regard to further functionalisation of the surface charges of the CNC, they were able to disperse the CNCs in different solvent systems to fabricate chiral nematic composite films of CNCs with PMMA and PS.[220] The group recently reported homogeneous, stretchable CNC/elastomer composites that change colour with mechanical deformation,[221] as well as electrically conductive composites made of CNCs and polypyrrole.[222] These systems all had chiral nematic pitch in the visible region (see Figure 4.10C).

Chemical infiltration into the structurally coloured media entails the transport of chemicals to the reaction site, which takes place in pores of only few hundreds of nanometres (sometimes, even less). Therefore, access to the reaction sites is often limited. Furthermore, when the chemical precursors initiate material growth on the walls of the pores, access to these sites diminishes as the reaction simultaneously proceeds. This effect puts a limit on what thicknesses can be grown. Hence, co-assembly, potentially solving these problems, seems to be a very promising method for the template approach, and more generally, for the synthesis of diverse new porous materials with attractive optical properties.

4.3.3 Scaled-up Production

Synthetic biomimetic materials need to be finely engineered to feature the nano- and micro-scale order required to display structural colours. The top-down and bottom-up techniques explained above are generally limited to a laboratory-scale production: indeed, patterning only small areas, at a high cost, and using sophisticated and uncommon instruments. Bottom-up approaches are intrinsically better suited for larger production scales, but they currently tend to use expensive materials (including nanostructured colloidal materials), require many processing steps and, due to the sensitivity of the self-assembly process to the physical conditions and the nature of the building block that constitutes the process, the resulting materials are fragile, often inhomogeneous, and unstable.[223]

Few recent attempts have been made to offer simple, low-cost, high-throughput, and scalable methods for fabricating structurally coloured, robust and stable materials over large areas. Hence, inspired by the micro-structure of the feather of *Cotinga maynana* (Figure 4.5B), Galinski *et al.*[223] used dealloying of $Pt_{0.14}Y_{0.06}Al_{0.80}$ films to assemble a porous nanoscale metallic network with controllable features. Their method exploits the selective dissolution of the less noble constituent (Al) of the alloy during wet etching by a solution of NaOH. The nanoporous film is subsequently coated with an ultra-thin layer of Al_2O_3 using ALD (see Section 4.3.1.2). The resulting structure is rendered in Figure 4.14A, and the resulting structural colours are shown in Figure 4.14B for various thicknesses of the Al_2O_3 nanolayer. Being based on relatively simple wet chemistry and coating technologies, their method may be suitable for real-world industrial applications.

Similarly, Ruiz-Clavijo *et al.*[224] presented a technique to make periodic nanoporous alumina 3D structures, *via* solution-based anodisation of aluminium to controllably create subwavelength periodic features in nanoporous dielectric materials, thus forming multilayer-based structural colours. Again, their method is well suited for low-cost (less than 0.3 € mm^{-2})

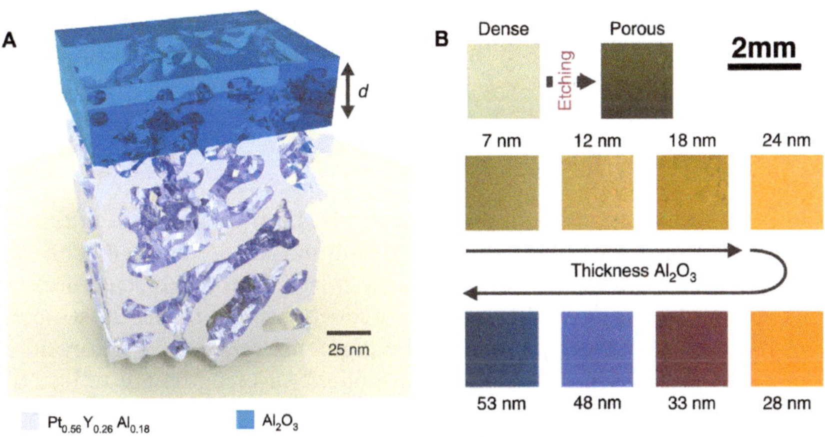

Figure 4.14 (A) Schematic illustration of an Al_2O_3-coated dealloyed PtYAl nanomaterial, based on a 3D reconstruction of the nanoarchitecture obtained *via* thin-film tomography. (B) Photographs of the nanomaterial film, illustrating the formation and continuous change of structural colour with increasing Al_2O_3coat thickness.

and large-scale fabrication of a whole gamut of visible structural colours in a single material.[224]

The rolled multilayered fibres recently developed by Kolle *et al.*[109] show a promising way of producing rather large-scale materials. A recent work by Liang *et al.*[225] demonstrated metre-scale roll-to-roll manufacturing of gel-like hydroxypropyl-cellulose (HPC) laminates in cholesteric liquid crystalline mesophase using continuous coating and encapsulation (see Figure 4.15). Roll-to-roll processing technologies have notably allowed the development of thin-film technologies for flexible electronics,** as well as new production systems†† to allow scaling up the colloidal photonic crystal thin films on flexible substrates. The latter offers a great variety of applications, including anti-reflection coatings, light-trapping layers for photovoltaic devices, smart optical sensors and optical communication systems.

It is thus possible to produce large-scale 3D complex architectures with 1–100 nm features and made of metallic or dielectric materials, which can exhibit advanced optical properties, including structural colours. However, translating academic studies of structural colours that use soft matter-based photonic and plasmonic arrays, into real-world applications still remains a challenge. Although there has been a great deal of investment made in top-down nanofabrication, little commercial return has been produced. Alternative approaches based on self-assembly have already gathered attention, and indeed offer routes to mass-scale production with a cost model that is realistic. Success in this domain depends heavily on industrial collaborations with academia, but the subsequent transformative approaches to manufacturing are still in their infancy. In a recent review, McDougal *et al.*[33] suggest that fabrication processes should also be inspired by the routes chosen by nature, and indeed these remain largely unexplored in industrial settings.

4.4 Applications

As already explained, colours play a vital role, both in the plant and in the animal kingdoms, by providing ways to communicate, to entice or to camouflage.[227] The interaction of light with micro- and nano-structures can produce various optical phenomena, some of which result in structural coloration without using pigments. The pigments rely on the absorption of certain wavelengths of light. As a result, other wavelengths are reflected or scattered, which causes the observer to see the corresponding colours. When chemical pigments absorb light, they trigger a series of chemical interactions within the material, which may cause them to fade over time. Structural colours, on the other hand, potentially endure for much longer timescales. Furthermore, from an environmental standpoint, biomimetic structural coloration has ecological merit since it is not produced through

**https://www.fep.fraunhofer.de/en.html
††https://www.tyndall.ie

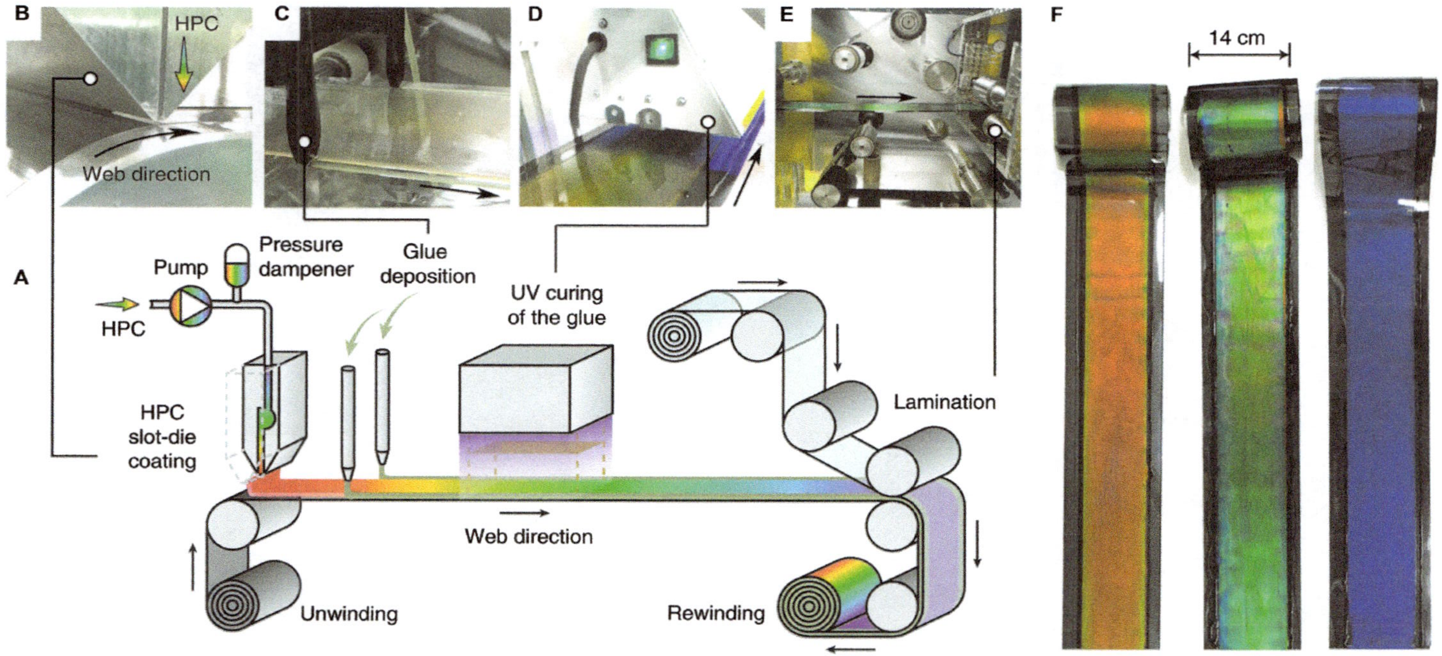

Figure 4.15 (A) Schematic of the roll-to-roll fabrication for HPC-laminated films showing the sequential processes of: (B) slot-die coating of HPC; (C) edge sealing by glue deposition; (D) UV curing of the edge glue with mask shielding the HPC coating; (E) lamination for final packaging and rewinding. (F) Black poly(ethylene terephthalate)-backed product rolls of red, green and blue HPC laminates with HPC concentrations of 63, 66 and 70 wt%, respectively.
Adapted from ref. 225, https://doi.org/10.1038/s41467-018-07048-6, under the terms of the CC BY 4.0 license, http://creativecommons.org/licenses/by/4.0/.

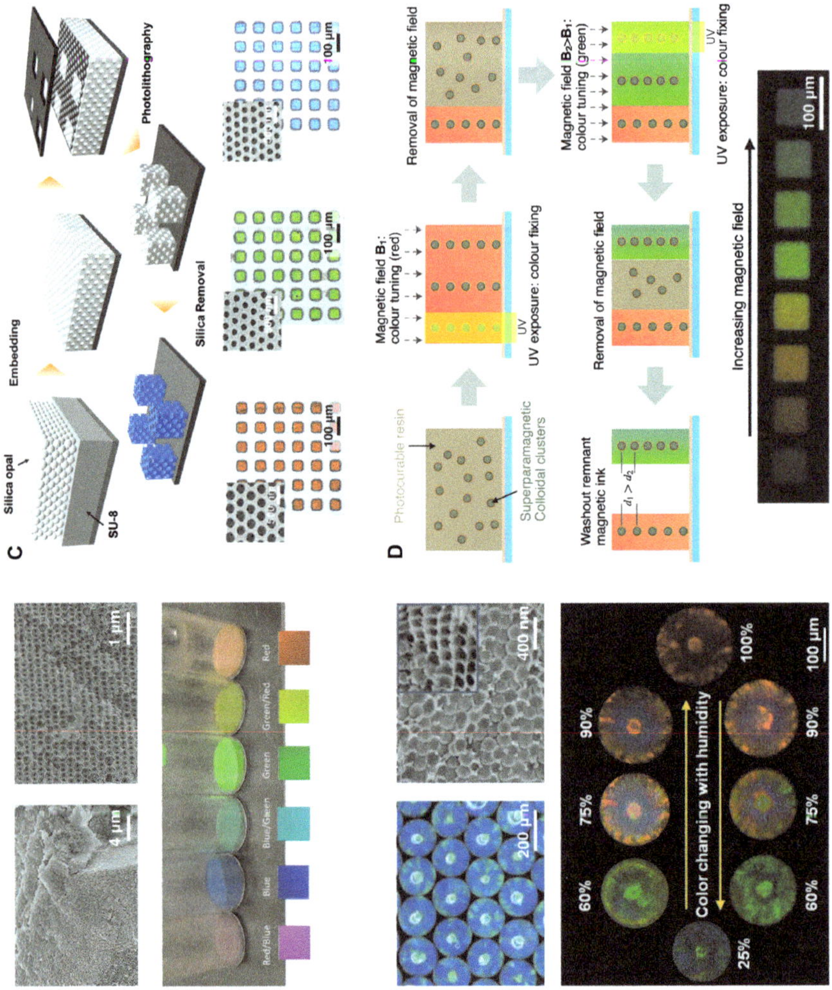
A
4 µm
1 µm
Red/Blue
Blue
Blue/Green
Green
Green/Red
Red
B
200 µm
400 nm
25%
60%
75%
90%
100%
Color changing with humidity
100 µm
C
Silica opal
SU-8
Embedding
Silica Removal
Photolithography
100 µm
D
Photocurable resin
Superparamagnetic Colloidal clusters
Magnetic field B_1: colour tuning (red)
UV
UV exposure: colour fixing
Removal of magnetic field
Magnetic field B_2>B_1: colour tuning (green)
Washout remnant magnetic ink
$d_1 > d_2$
Increasing magnetic field

chemical processes. Understanding the mechanisms behind structurally coloured systems in nature is therefore of high importance. Not only might it shed some light on evolutionary processes, but also it could offer a route to as-yet unknown technological prowess. Nature uses cost-effective materials and simple, finely selected processes in order to develop multifunctional materials. Therefore, mastering nature's concepts for material design opens new doors for today's advanced technologies.

4.4.1 Responsive and Tuneable Structural Colours

One of the most studied areas for structurally coloured materials is that of their response to surrounding environmental stimuli to produce a change in colour. The colours of such stimuli-responsive photonic band gap materials are changeable by varying either the distance of two neighbouring lattice planes or the refractive index contrast between two media. Therefore, various responsive structural coloured materials have great potential for the development of materials in many different industries, from advanced textiles to optical sensing devices and actuation systems (see Figure 4.16).

Mechanochromic solid sheets, or structural colours in hydrogel form, have been widely explored as the basis of smart and functional fabrics. Various methods have been proposed for the fabrication of solid mechanochromic sheets, including cross-linking of monodispersed core–shell microspheres consisting of a rigid core and a soft, elastomeric shell. These offer rapid, reversible and repeatable colour transitions upon application of a stress and its removal. Further methods include the infiltration of an elastomer into assembled hard microspheres, to form non-close-packed

Figure 4.16 Examples of recent applications of bioinspired structural colours. (A) SEM images of inverse opal SiO_2 powders at two magnifications (top) and photography of the mixed primary colour pigments (adapted from ref. 207 with permission from John Wiley and Sons, © 2014 WILEY-VCH Verlag GmbH & Co. KGaA, Weinheim). (B) Optical microscope and SEM images of PS colloids immobilised in polyacrylamide hydrogel (top), forming the photonic supraballs; reversible colour changes of the supraballs with varying relative humidity (adapted from ref. 142 with permission from the Royal Society of Chemistry). (C) Illustration of SU-8 photoresist inverse opal fabrication, and optical microscopy images (corresponding SEM in insets) of the pixelated inverse opals prepared to reflect a single colour: red, green or blue (adapted from ref. 153 with permission from John Wiley and Sons, © 2014 WILEY-VCH Verlag GmbH & Co. KGaA, Weinheim). (D) Illustration of the patterning of the "magnetic ink" made of superparamagnetic colloidal nanocrystal clusters embedded in photocurable resin; the diffraction wavelength (d_1 and d_2) is tuned by varying the strength of the magnetic field. The spatially patterned UV light polymerises the resin and fixes the position of ordered colloidal clusters; the bottom images are optical micrographs of the multicoloured structural colour generated by gradually increasing magnetic fields in reflection mode (adapted from ref. 226 with permission from Springer Nature, Copyright 2009).

colloidal, inverse opal structures.[228–233] It is also possible to produce photonic hydrogels by fixing monodispersed magnetic nanoparticles, *i.e.* Fe_3O_4, within the hydrogels under a magnetic field, and by further *in situ* photopolymerisation in a cross-linked acrylamide (AM) matrix.[234]

It is also possible to process similar biomimetic photonic structures in a fibre form, in order to produce flexible and "smart" photonic textiles with colours that are tuneable over the entire visible spectrum. These can then also act as optical sensors for both strain and pressure. Hence, as mentioned in Section 4.3.1.2, Kolle *et al.*[109] produced a multilayered fibre that is spun around a glass filament, and that essentially forms a planar Bragg stack. Upon removal of the glass core, the resulting flexible fibre showed sensitive colour changes against an applied strain. Such biomimetic fibres could potentially find applications in mechanically tuneable light guides or optical strain sensing.[109]

Similarly, Sun *et al.*[235] demonstrated the fabrication of mechanochromic fibres based on aligned carbon nanotube (CNT) sheets. These were bound to an elastic PDMS fibre, with polymer microspheres deposited on the CNT layer electrophoretically, followed by further embedding in PDMS to fix the microspheres.[235] An alternative method for the production of structurally coloured strain-responsive fibres was suggested by Shang *et al.*,[236] again based on fixing magnetic nanoparticles (Fe_3O_4) in a polyacrylamide glycol gel matrix. Upon the application of an external magnetic field, the randomly dispersed spherical particles are embedded in the matrix as 1D chain-like structures. The distance between each sphere in the matrix can be reversibly changed by elastic deformations of the matrix, which produce a visible colour change.[236] Zhang *et al.*[237] recently developed a different strategy to prepare mechanochromic fibres. They coated, continuously, commercially available black spandex fibres with microspheres made with a hard PS/PMMA core and a soft PEA (poly(ethyl acrylate)) shell. They demonstrated that the microspheres self-assemble into a photonic crystal structure with brilliant structural colours that cover the visible light region. They could also control the colours of these fibres by varying the diameters of the core–shell microspheres.[237]

Novel materials, made *via* the rapid formation of ordered arrays through a change in magnetic or electric fields, and with widely, rapidly, and reversibly tuneable structural colours are desirable for various chromatic applications: colour display, security, camouflage, or information storage, for example. The key to tune the self-assembly, and hence the colour variations of such systems in response to a magnetic field, lies with the formation of a dynamic balance between magnetic dipole–dipole interactions and long-range electrostatic repulsive interactions. This process, however, requires the use of specialised magnetic colloidal particles. Different magnetic colloidal systems, based on superparamagnetic Fe_3O_4 nanoparticles or nano-ellipsoids made of a metallic Fe core inside a SiO_2 shell, were shown to align under magnetic fields spontaneously. These systems are very promising to produce magnetically responsive photonic systems as a new platform for chromatic applications.[238–240] Electric fields, on the other hand, can also assemble colloidal particles into ordered structures that dissipate into a disordered phase on removal of the electric field.[241] It is indeed common to examine the reorganisation of colloidal crystals under an electric field in a liquid medium.[242,243] Further to this, Baumberg's team devised a method to fix the electrically induced particle chains in a polymer matrix permanently, which can potentially be the basis of a new scanning printing process. Since it is difficult to replicate the patterning process and the structural colours, such systems can find a market value where iridescent, non-fading structural colours are desirable, such as printing on banknotes, passports, and certificates.[244]

The colorimetric response of structurally coloured systems to temperature changes,[245–247] as well as the colour changes against chemical interactions in the gas phase or through liquid–solvent interactions (such as solvent vapour infiltration,[142,248,249] ionic media and pH changes,[250,251] and interaction with special molecules[252]) can be used in various sensing applications.[105,156,190] Designing such bioinspired sensors to achieve a sensory response within the visible region relies on the relationship between the properties of the materials, their intrinsic nanostructures and the reflected light. Bioinspired colour sensors within the visible region provide a simple, yet powerful, detection mechanism, which holds great potential for various applications in different industries, including diagnostics,[253,254] environmental monitoring, workplace hazard identification, and threat detection.[255–257] Consequently, there are several review articles focusing on the development of bioinspired colorimetric sensors, their fabrication methods, their operation mechanisms, and the possible industries involved.[95,258–261]

Another system that can also exhibit changes in colour under a wide range of stimuli is chiral nematic photonic structures. Their use in various sensing applications have also been discussed in a recent review.[262] One of the most studied chiral nematic photonic structures is produced through the self-assembly of cellulose nanocrystals and their organic/inorganic composite structures (see Section 4.3.2.1.2). Cellulose is a highly hydrophilic material, and upon exposure to water (in liquid or vapour), the helicoidal pitch of its CNC assembly (hence the colour) shifts reversibly along with the film swelling.[161,263] When the porosity of such compact chiral nematic films is increased through supramolecular methods (using urea formaldehyde, for example), the final mesoporous chiral cellulose material displays dynamic photonic properties, which may make them ideal for optical filters and/or chemical sensors.[264]

4.4.2 Surface Engineering with Structurally Coloured Systems

The production of photonic arrays, using either top-down or bottom-up strategies, not only produces tuneable colours, but also offers a great variety of applications, notably due to the use of advanced surface engineering involved in such processes. It is possible to use a great variety of materials to combine different effects, such as hydrophobicity or electric field-driven colour tuneability.

The iridescent and metallic appearances of the structurally coloured species are, notably, used to attract the attention of potential mates, or to startle predators. An obvious application for these visually attractive, and optically sophisticated, constructions is to create security encoding and anti-counterfeiting features. As we have seen in Section 4.3, replicating these features is highly challenging. Therefore, significant efforts are made to develop security features and encryption applications by using both top-down and bottom-up strategies. The use of self-assembled dielectric materials offers an accessible way to produce security features.[93,265] Additional optical illusions can be achieved by using different material properties: for example, employing a superparamagnetic iron oxide core (Fe_3O_4) inside SiO_2 colloidal spheres, Kim *et al.*[226] demonstrated the patterning of multiple structural colours with a magnetic field. Such systems are also lithographically fixable, which could allow the industrial production of these high-resolution multicoloured patterns.[226] In a similar work, Hu *et al.*[266] used superparamagnetic colloidal particles to produce dual photonic band gap heterostructures, which showed switchable colour changes upon application of a magnetic field. This process adds another security element for anti-counterfeiting applications.[266] Plasmonic materials, on the other hand, exhibit extraordinary enhancements in confining optical fields with well-controlled intensity, phase and polarisation of light, beyond the diffraction limit.[267] Therefore, these structural colours also hold great potential in the development of materials for forgery protection.[36,268]

Interestingly, apart from their structural colour, *Morpho* butterflies have also attracted much interest due to the excellent water repellency of their wings. This effect is caused by the multiple superhydrophobic ridges, giving them a self-cleaning ability.[269] Replication of structural coloration, together with the "lotus effect" (that is, the self-cleaning effect due to the very high water repellence caused by both the nanostructures and the material's chemistry) was studied by Gu *et al.*[203] in inverse opal systems. In their work, they demonstrate the long-order range organisation through a bicomponent nanoparticle (that is, monodispersed polystyrene spheres, in the range of 300–500 nm along with silica nanoparticles of size 6 nm) using vertical convective self-assembly (see Section 4.3.2.1). Upon removal of the polystyrene opal frame, and after surface modification of the silica inverse opal substrates with a fluoroalkylsilane, they achieved superhydrophobicity with a 155° contact angle.[203] In another study, the same silica inverse opal structure was prepared similarly, but the substrate was covered with azobenzene by electrostatic layer-by-layer self-assembly. Such films showed a reversible change in the wettability of the surface depending on the applied irradiation, which caused a change in the azobenzene isomeric conformations.[270] The extensive research on the production of bioinspired photonic crystals with structural colour and tuneable wettability has been summarised in a review.[271] On the side of top-down production, Wang *et al.*[272] fabricated periodic grating structures on a graphene oxide (GO) film using two-beam laser interference. The process seemed to simultaneously remove the oxygen groups on GO. With increasing laser power, the contact angle of the surfaces was increased to a value of 156.7°.[272] The pillar structures on the surface are usually responsible for the coloration and the hydrophobicity. For the glasswing butterfly, *Greta oto*, the small nanopillars covering the transparent regions of its wings actually cause anti-reflection behaviour, which inspires applications for omnidirectional anti-reflective coatings.[273] Moth eyes are also famous examples of anti-reflective structures, and have been the subject of extensive biomimetics studies, with promising applications in display technologies, solar panels, and light-emitting diodes.[23,274,275]

Patterning colloidal crystals, or dynamically tuning their periodicity, could potentially be used as a basis for colour displays in reflection mode.[126,153,156,241,276,277] Furthermore, the deliberate insertion of organised defects into colloidal crystals, or doping such crystals with light emitters, enables the creation of optical waveguides[278,279] and lasers.[280]

Three-dimensional photonic crystals, based on colloidal self-assembly and inverse opal structures, possess a photonic band gap based on their intrinsic periodic structure, which can selectively modify the propagation of light with a specific wavelength. Ideally, it may be desired to maximise the coupling of light into a structure, while minimising the amount of light that escapes. Having such control on the propagation of light, and localisation of photons, is, without doubt, very beneficial for photocatalytic and photovoltaic applications. Researchers have studied various inverse opal systems that implicate metal oxides such as TiO_2, ZnO, $BiVO_4$ or WO_3, for both dye-sensitised solar cell devices (DSSCs) and as photoanodes in photoelectrochemical cells.[210,281–284] One of the most studied metal oxides in such systems is TiO_2, which is usually applied as a bulk film in DSSCs. The bulk TiO_2 layer scatters and reflects the light randomly, whereas the reflectance of TiO_2 in the inverse opal geometry is highly wavelength selective. In that case, the incident light could be managed effectively to enhance light harvesting, by matching the electronic band gap of the metal oxide to its photonic band gap,[210,281] or by matching the photonic band gap to the absorption of the sensitising dye.[285] Organic–inorganic halide perovskite-based solar cells are yet another class of photovoltaic system with a photon-to-electron conversion mechanism that is quite different from that of the classical metal oxide solar cells. Henry Snaith's team at the University of Oxford produced perovskite solar cells on a biomimetic scaffold of alternating layers of TiO_2 and porous SiO_2, and demonstrated that it is possible to tune the hue of the perovskite device. Structural coloration in perovskite solar cells is proven to bring many advantages, such as tuneability of the colours for specific panchromatic photovoltaic absorbers, and the fact that coloration will not bleach or fade with time, which offers great stability for such systems.[286]

Nature offers an extended array of nanostructures, with smart selection of cost-effective materials to sustain numerous physical functions. In addition to mechanical integrity, coloration and hydrophobicity, the nanostructures originally designed to attain structural colours can also help us to produce highly light-absorbent biomimetic materials. For example, in order to maintain a constant body temperature, the *Troides magellanus* butterfly also exhibits an unusual absorption of visible light (98%), while emitting radiation in the infrared region, due to a special organisation of chitin.[287] Possibly, this added functionality is used to dissipate heat for temperature regulation, and is also found in the *Morpho* species.[288] Mimicking the dissipative property of chitin may also be beneficial for photovoltaic applications.

4.4.3 Structural Colours in Art, Cosmetics, Paints and Textiles

The list of potential applications for structurally coloured materials is indeed quite large. Yet, the most obvious way to use these materials is by exploiting the aesthetic aspect of their vivid coloration. Hence, relevant industries, such as art, cosmetics, textiles and decorations, have adopted structural colours to create non-traditional pigments and dyes long ago.

Structurally coloured species, such as opals, nacre and pearls, have been used in jewellery and the fabrication of decorative items (such as pots, boxes, *etc.*) for centuries.[289] Some recent modern art pieces use the brilliant *Morpho*, and other butterflies, as coloured entities, by incorporating entire specimens as themselves (for example, Damien Hirst's "Psalm 6: Domine, ne in furore." 2008). Artists can take advantage of structural colours to create specific coloration and aesthetic expressions. Some artistic paint suppliers, such as Golden Artist Colors[‡‡] or Liquitex,[§§] are producing a large variety of acrylic-based, structurally coloured paints that impart shimmering iridescent or interference reflective qualities. These usually arise from nanoflakes of mica, coated with a thin layer of titanium dioxide, metal or iron oxide pigments.

Some artists, indeed, go beyond what is available on the market, in terms of mimicking iridescence effects, and actually create their own structural colours to achieve unique displays.[290,291] Such artists, like Kate Nichols,[¶¶] Paul Evans,[‖‖] and Franziska Schenk,[***] usually hold artist-in-residence positions in laboratories, or collaborate closely with scientists, to synthesise their own nanoparticles that mimic various optical effects seen in

‡‡http://www.goldenpaints.com
§§http://www.liquitex.com
¶¶http://www.katenicholsstudio.com
‖‖http://www.pkevans.co.uk
***http://www.franziskaschenk.co.uk

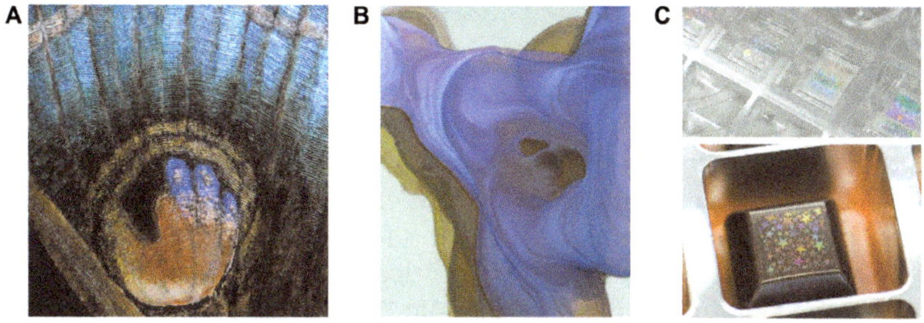

Figure 4.17 (A) Franziska Schenk's "See Change, Painting VI", 2007 uses iridescent paint. (© Franziska Schenk.) (B) Kate Nichols' "Through the Looking Glass", 2011 employs silver nanoparticles (image courtesy of Kate Nichols). (C) Holographic patterns made from chocolate (bottom) *via* direct casting with prefabricated polycarbonate moulds (top); images courtesy of Morphotonix.

Morpho butterflies, beetles and bird feathers (see Figure 4.17A and B). This way, they also construct non-reproducible pieces by making their own base materials.

As we have seen in Section 4.3.2, a way to fabricate structural colours is to produce nanoparticles that self-organise. Yet, colloidal self-assembly has several limitations, due to the sensitive nature of the process: the crystal orientation in large scales is not fully controllable, the process is time consuming, and the success of the self-assembly is heavily dependent on the substrates, where an even coloration can only be achieved on perfectly flat and uniform surfaces. The best way to use the structurally coloured colloidal crystals as pigments is to process them into a powder form.[207] Hence, in order to overcome the shortcomings of solvent-based self-assembly processes, Park *et al.*[292] introduced paints that produce structural colours by rubbing a nanoparticle powder on an elastomer surface. Their technique is a quick, highly reproducible way to fabricate a single crystal monolayer assembly of particles over an unlimited area.[292,293]

Pearlescent pigments, on the other hand, can be produced by coating mica particles with thin layers of metal oxides (TiO_2, ZrO_2 or SiO_2) to create lustrous, iridescent and other angle-dependent optical effects. Such pigments are indeed used extensively in a variety of industrial products.[294,295] Similar pearlescent pigments have notably captured the interest of the cosmetics industry. Again inspired by the structural colours of butterfly wings, researchers at L'Oréal, for example, developed multilayers of mica, silica, and networks of polymers or silica beads with remarkable optical effects. By varying the number of layers, they could achieve different optical effects: with a simple thin layer, they obtained the classical iridescence effect (see Section 4.2.1); but by multiplying the layers of particles or metallic oxides (up to 200 layers), they could produce a mirror effect, where the colour of the light transmitted is complementary to the light reflected.††† Colour-shift, colour-flip or holographic effects on curved surfaces (such as cars) can also be achieved using microflakes and chameleon pearl-based pigments. These can be readily applied onto the body of the car, by mixing them with different paint matrices for coating.‡‡‡§§§

Further to these new generations of powder forms of structurally colouring paints, pre-defined and tuned colloidal suspensions can also be directly used as paints.[296,297] Alternatively, to fabricate a paint composition, photo-curable pre-assembled colloidal photonic crystal films can be ground to microscopic photonic crystal flakes and redispersed into a solvent.[298] Similarly, the aforementioned "photonic balls" (Section 4.3.2.1) may be used directly in paint formulations.[299,300] Additionally, it is possible to enhance the colour effect of these opalescent photonic structures by adding small amounts of highly light-absorbing material, such as carbon, which incorporates in the interstitial space surrounding the high refractive index spheres.[301–303]

In textile-related applications, bioinspired opalescent films can also be produced at industrial scales, by using an edge-shearing method. In this process, colloidal particles are sandwiched as a thin film between two removable polymer sheets, which are then drawn over a sharp edge. This way, the particles rearrange themselves to form a highly periodic structure, greatly strengthening their response to light and making their colour particularly intense.[304] The edge-shearing process notably permits the rapid production of large-scale films. Along the same lines, researchers at the Nissan Motor Company demonstrated that it is possible to produce structurally coloured fibres, and a woven fabric, through a conjugated melt-spinning method.[305] In this process, the non-circular and structurally coloured fibres are prefabricated, prior to melt-spinning, by stacking alternating polymer layers with large refractive index contrast. Their colour can be tuned by changing the thicknesses of the constituent polymer layers.

Finally, it is possible to produce coloration in food without using any artificial colouring agents, by directly using top-down methods. For example, the Swiss nanotechnology company Morphotonix¶¶¶ developed a way to create holographic patterns on the surface of chocolates, without using any chemical additives, by directly casting the chocolate into diffractive holograms (see Figure 4.17C). For this purpose, the company developed a technology to micropattern bulk, free-form metallic templates employed in the fabrication of the polycarbonate moulds that are subsequently used in chocolate production. The diffractive patterns from the metal are thus replicated in the chocolate during its casting process. Even though the process and design are not directly inspired by nature, the use of an edible substance

†††http://www.loreal.com
‡‡‡http://www.pearlsandpigments.com
§§§http://www.sfxc.co.uk
¶¶¶http://www.morphotonix.com

ostensibly shows the variety of potential materials that could be manufactured with structural colours.

4.5 Conclusions

The structural colours of living creatures are designed for various types of biological function. They originate from basic optical phenomena, such as diffraction, interference and scattering, which are often finely combined and balanced to optimise the coloration effects for the particular function sought. From an engineering point of view, natural structural colours exhibit attractive features, such as iridescence, polarisation-dependent response, or long life and high environmental resistance. They also tend to be particularly energy efficient. For these reasons, the structural colours found in nature have attracted intense research and biomimicry efforts in the recent years.

In this review, we have briefly recalled the physical principles behind structural colours in nature, before focusing on the recent achievements in manufacturing these bioinspired structures. We have also reported the current or potential applications of structural colours in industry, arts and sciences. Currently, the most successful approaches for fabricating structural colours usually associate several top-down and bottom-up processes. Still, these techniques are often limited in terms of consistency, object dimensions and throughput.

We have presented strategies using self-assembly in great detail in this review (see Section 4.3.2.1). Aside from being the fabrication process for most materials found in living species, self-assembly also appears to be the best method for building structured materials that exhibit coloration: first, it is both economical (because it typically uses few materials) and environmentally friendly (since the process generally requires little external intervention or additional sources of energy); and second, it allows for industrial-scale production with a high degree of accuracy, in terms of structure conservation and periodicity over large objects.

Controlling self-assembly, however, is an arduous task, which demands further research in chemistry and physics. Additionally, it may be appealing to also mimic the material properties of natural self-assembling building blocks, such as chitin, cellulose, melanin, *etc.* Often, in addition to coloration, the nanostructures that are formed from these constituents serve multiple other functions (self-cleaning, heat dissipation, *etc.*), which are inherited from the building material itself. We have tackled some of these aspects in Section 4.4.2, and shown how multifunctionalities are also an attractive feature for engineers over an even wider range of applications (*e.g.*, solar panels, packaging, *etc.*).

Most strikingly, living creatures employ a very narrow range of materials, when compared with the variety of materials deployed in synthetic functional objects. The optimisation of material selection found in nature should also be a good inspiration for future manufacturing strategies. Natural substances used for structural colours in living species, in particular, are abundant, sustainable and safe for the environment. They may as well be used as the basis for bioinspired structured materials, but they may also form the basis for future custom-designed structures that go beyond those found in nature.

Hence, not only the structural arrangements, but also the fabricating processes and materials may be biologically inspired. A concerted effort, across many disciplines, is needed to master the reproduction of the highly efficient designs and processes found in nature. This is an essential step to undertake in order to invent our own structured materials. Much is left to investigate, and indeed this field of research is growing rapidly, as nature's evolutionary engineering is progressively elucidated.

References

1. A. R. Parker, *J. Opt. A: Pure Appl. Opt.*, 2000, **2**, R15–R28.
2. S. Kinoshita, *Structural Colors in the Realm of Nature*, World Scientific, Singapore, 2008.
3. J. Sun, B. Bhushan and J. Tong, *RSC Adv.*, 2013, **3**, 14862–14889.
4. S. Kinoshita, S. Yoshioka and J. Miyazaki, *Rep. Prog. Phys.*, 2008, **71**, 076401.
5. T. F. Anderson and A. G. Richards Jr., *J. Appl. Phys.*, 1942, **13**, 748–758.
6. E. Yablonovitch, *Phys. Rev. Lett.*, 1987, **58**, 2059–2062.
7. J. D. Joannopoulos, S. G. Johnson, J. N. Winn and R. D. Meade, *Photonic Crystals: Molding the Flow of Light*, Princeton University Press, Princeton, 2008.
8. F. P. Barrows and M. H. Bartl, *Nanomater. Nanotechnol.*, 2014, **4**, 1.
9. M. J. Madou, *Fundamentals of Microfabrication and Nanotechnology*, CRC Press, Boca Raton, 2012.
10. M. F. Land, *Prog. Biophys. Mol. Biol.*, 1972, **24**, 75–106.
11. M. Srinivasarao, *Chem. Rev.*, 1999, **99**, 1935–1961.
12. A. R. Parker, *Mater. Today*, 2002, **5**, 26–31.
13. P. Vukusic and J. R. Sambles, *Nature*, 2003, **424**, 852–855.
14. A. R. Parker, *Philos. Trans. R. Soc., A*, 2004, **362**, 2709–2720.
15. L. P. Lee and R. Szema, *Science*, 2005, **310**, 1148–1150.
16. A. R. Parker and H. E. Townley, *Nat. Nanotechnol.*, 2007, **2**, 347–353.
17. S. M. Doucet and M. G. Meadows, *J. R. Soc., Interface*, 2009, **6**, S115–S132.
18. A. R. Parker, *Philos. Trans. R. Soc., A*, 2009, **367**, 1759–1782.
19. J.-P. Vigneron and P. Simonis, *Adv. Insect Physiol.*, 2010, **38**, 181–218.
20. P. V. Braun, *Nature*, 2011, **472**, 423–424.
21. H. Fudouzi, *Sci. Technol. Adv. Mater.*, 2011, **12**, 064704.
22. Y. Zhao, Z. Xie, H. Gu, C. Zhu and Z. Gu, *Chem. Soc. Rev.*, 2012, **41**, 3297–3317.
23. A. Saito, *Sci. Technol. Adv. Mater.*, 2011, **12**, 064709.
24. J. Xu and Z. Guo, *J. Colloid Interface Sci.*, 2013, **406**, 1–17.
25. A. G. Dumanli and T. Savin, *Chem. Soc. Rev.*, 2016, **45**, 6698–6724.

26. S. Tadepalli, J. M. Slocik, M. K. Gupta, R. R. Naik and S. Singamaneni, *Chem. Rev.*, 2017, **117**, 12705–12763.
27. M. Kolle and S. Lee, *Adv. Mater.*, 2018, **30**, 1702669.
28. M. Kolle, PhD thesis, University of Cambridge, 2010.
29. *Biomimetics in Photonics*, ed. O. Karthaus, CRC Press, Boca Raton, 2012.
30. *Photobiology: The Science of Light and Life*, ed. L. O. Björn, Springer, New York, 2015.
31. D. Zhang, W. Zhang, J. Gu, T. Fan, Q. Liu, H. Su and S. Zhu, *Prog. Mater. Sci.*, 2015, **68**, 67–96.
32. Y. Fu, C. A. Tippets, E. U. Donev and R. Lopez, *WIREs Nanomed. Nanobiotechnol.*, 2016, DOI: 10.1002-wnan.1396.
33. A. McDougal, B. Miller, M. Singh and M. Kolle, *J. Opt.*, 2019, **21**, 073001.
34. M. Born and E. Wolf, *Principles of Optics*, Pergamon, Oxford, 1980.
35. H. A. Macleod, *Thin-Film Optical Filters*, CRC Press, Boca Raton, 2010.
36. K. Kumar, H. Duan, R. S. Hegde, S. C. W. Koh, J. N. Wei and J. K. W. Yang, *Nat. Nanotechnol.*, 2012, 7, 557–561.
37. T. Smith and J. Guild, *Trans. Opt. Soc.*, 1931, **33**, 73–134.
38. V. E. Johansen, J. Andkjær and O. Sigmund, *J. Opt. Soc. Am. B*, 2014, **31**, 207–217.
39. H. M. Whitney, M. Kolle, P. Andrew, L. Chittka, U. Steiner and B. J. Glover, *Science*, 2009, **323**, 130–133.
40. A. L. Ingram, V. Lousse, A. R. Parker and J.-P. Vigneron, *J. R .Soc., Interface*, 2008, **5**, 1387–1390.
41. T. L. Tan, D. Wong and P. Lee, *Opt. Express*, 2004, **12**, 4847–4854.
42. Z. Knittl, *Optics of Thin Films*, Wiley, London, 1976.
43. B. E. Perilloux, *Thin-Film Design: Modulated Thickness and Other Stop-band Design Methods*, SPIE Press, Bellingham, WA, 2012, vol. TT57.
44. D. G. Stavenga, *Mater. Today Proc.*, 2014, **1**, 109–121.
45. E. Shevtsova, C. Hansson, D. H. Janzen and J. Kjærandsen, *Proc. Natl. Acad. Sci. U. S. A.*, 2011, **108**, 668–673.
46. S. Yoshioka, E. Nakamura and S. Kinoshita, *J. Phys. Soc. Jpn.*, 2007, **76**, 013801.
47. R. Maia, J. A. V. O. Caetano, S. N. Báo and R. H. Macedo, *J. R. Soc., Interface*, 2009, **6**, S203–S211.
48. H. de Vries, *Acta Crystallogr.*, 1951, **4**, 219–226.
49. A. E. Seago, P. Brady, J.-P. Vigneron and T. D. Schultz, *J. R. Soc., Interface*, 2009, **6**, S165–S184.
50. V. Sharma, M. Crne, J. O. Park and M. Srinivasarao, *Science*, 2009, **325**, 449–451.
51. D. G. Stavenga, H. L. Leertouwer, N. J. Marshall and D. Osorio, *Proc. R. Soc. B*, 2011, **278**, 2098–2104.
52. S. Yoshioka, B. Matsuhana, S. Tanaka, Y. Inouye, N. Oshima and S. Kinoshita, *J. R. Soc., Interface*, 2011, **8**, 56–66.
53. D. J. Brink, N. G. van der Berg and A. J. Botha, *Appl. Opt.*, 2002, **41**, 717–722.

54. L. M. Mäthger, E. J. Denton, N. J. Marshall and R. T. Hanlon, *J. R. Soc., Interface*, 2009, **6**, S149–S163.
55. S. Vignolini, P. J. Rudall, A. V. Rowland, A. Reed, E. Moyroud, R. B. Faden, J. J. Baumberg, B. J. Glover and U. Steiner, *Proc. Natl. Acad. Sci. U. S. A.*, 2012, **109**, 15712–15715.
56. K. Sakoda, *Optical Properties of Photonic Crystals*, Springer, Berlin, 2005, vol. 80.
57. J.-M. Lourtioz, H. Benisty, V. Berger, J.-M. Gérard, D. Maystre, A. Tchelnokov and D. Pagnoux, *Photonic Crystals: Towards Nanoscale Photonic Devices*, Springer, Berlin, 2008.
58. I. A. Sukhoivanov and I. V. Guryev, *Photonic Crystals: Physics and Practical Modeling*, Springer, Heidelberg, 2009, vol. 152.
59. D. M. Whittaker and I. S. Culshaw, *Phys. Rev. B*, 1999, **60**, 2610–2618.
60. E. Pavarini, L. C. Andreani, C. Soci, M. Galli, F. Marabelli and D. Comoretto, *Phys. Rev. B*, 2005, **72**, 045102.
61. T. M. Trzeciak and P. Vukusic, *Phys. Rev. E*, 2009, **80**, 061908.
62. J. Zi, X. Yu, Y. Li, X. Hu, C. Xu, X. Wang, X. Liu and R. Fu, *Proc. Natl. Acad. Sci. U. S. A.*, 2003, **100**, 12576–12578.
63. E. Lee, J. Miyazaki, S. Yoshioka, H. Lee and S. Sugita, *Ornithol. Sci.*, 2012, **11**, 59–64.
64. J. V. Sanders, *Nature*, 1964, **204**, 1151–1153.
65. J. W. Galusha, L. R. Richey, J. S. Gardner, J. N. Cha and M. H. Bartl, *Phys. Rev. E*, 2008, **77**, 050904.
66. V. Saranathan, C. O. Osuji, S. G. J. Mochrie, H. Noh, S. Narayanan, A. Sandy, E. R. Dufresne and R. O. Prum, *Proc. Natl. Acad. Sci. U. S. A.*, 2010, **107**, 11676–11681.
67. J.-P. Vigneron and P. Simonis, *Physica B*, 2012, **407**, 4032–4036.
68. B. D. Wilts, K. Michielsen, H. De Raedt and D. G. Stavenga, *J. R. Soc., Interface*, 2012, **9**, 1609–1614.
69. H. Noh, S. F. Liew, V. Saranathan, S. G. J. Mochrie, R. O. Prum, E. R. Dufresne and H. Cao, *Adv. Mater.*, 2010, **22**, 2871–2880.
70. C. F. Bohren and D. R. Huffman, *Absorption and Scattering of Light by Small Particles*, Wiley, Weinheim, 1998.
71. R. A. Sturm and T. N. Frudakis, *Trends Genet.*, 2004, **20**, 327–332.
72. M. Burresi, L. Cortese, L. Pattelli, M. Kolle, P. Vukusic, D. S. Wiersma, U. Steiner and S. Vignolini, *Sci. Rep.*, 2014, **4**, 6075.
73. R. O. Prum, E. R. Dufresne, T. Quinn and K. Waters, *J. R. Soc., Interface*, 2009, **6**, S253–S265.
74. P. Vukusic, B. Hallam and J. Noyes, *Science*, 2007, **315**, 348.
75. J. T. Bagnara, P. J. Fernandez and R. Fujii, *Pigm. Cell Res.*, 2007, **20**, 14–26.
76. S. Kinoshita, S. Yoshioka and K. Kawagoe, *Proc. R. Soc. B*, 2002, **269**, 1417–1421.
77. R. O. Prum, T. Quinn and R. H. Torres, *J. Exp. Biol.*, 2006, **209**, 748–765.
78. P. Vukusic, J. R. Sambles, C. R. Lawrence and R. J. Wootton, *Proc. R. Soc. B*, 1999, **266**, 1403–1411.

79. C. Gower, *Auk*, 1936, **53**, 178–185.
80. D. G. Stavenga, H. L. Leertouwer, D. C. Osorio and B. D. Wilts, *Light: Sci. Appl.*, 2015, **4**, e243.
81. A. J. Parnell, A. L. Washington, O. O. Mykhaylyk, C. J. Hill, A. Bianco, S. L. Burg, A. J. C. Dennison, M. Snape, A. J. Cadby, A. Smith, S. Prevost, D. M. Whittaker, R. A. L. Jones, J. P. A. Fairclough and A. R. Parker, *Sci. Rep.*, 2015, **5**, 18317.
82. A. Saito, M. Nakajima, Y. Miyamura, K. Sogo, Y. Ishikawa and Y. Hirai, *Proc. SPIE*, 2006, 63270Z.
83. E. R. Dufresne, H. Noh, V. Saranathan, S. G. J. Mochrie, H. Cao and R. O. Prum, *Soft Matter*, 2009, **5**, 1792–1795.
84. E. Cazetta, L. S. Zumstein, T. A. Melo-Júnior and M. Galetti, *Revista Brasil. Bot.*, 2008, **31**, 303–308.
85. J. Sun and B. Bhushan, *RSC Adv.*, 2012, **2**, 7617.
86. A. Finnemore, P. Cunha, T. Shean, S. Vignolini, S. Guldin, M. Oyen and U. Steiner, *Nat. Commun.*, 2012, **3**, 966.
87. M. Liesegang and R. Milke, *Am. Mineral.*, 2014, **99**, 1488–1499.
88. K. Yu, T. Fan, S. Lou and D. Zhang, *Prog. Mater. Sci.*, 2013, **58**, 825–873.
89. B. P. Isaacoff and K. A. Brown, *Nano Lett.*, 2017, **17**, 6508–6510.
90. A. Pimpin and W. Srituravanich, *Eng. J.*, 2012, **16**, 37–55.
91. M. Maldovan and E. L. Thomas, *Periodic Materials and Interference Lithography*, Wiley, Weinheim, 2009, pp. 63–96.
92. N. Kooy, K. Mohamed, L. T. Pin and O. S. Guan, *Nanoscale Res. Lett.*, 2014, **9**, 320.
93. H. S. Lee, T. S. Shim, H. Hwang, S.-M. Yang and S.-H. Kim, *Chem. Mater.*, 2013, **25**, 2684–2690.
94. M. A. Mohammad, M. Muhammad, S. K. Dew and M. Stepanova, *Nanofabrication*, Springer, Vienna, 2012, pp. 11–41.
95. I. B. Burgess, M. Lončar and J. Aizenberg, *J. Mater. Chem. B*, 2013, **1**, 6075–6086.
96. B. Song, V. E. Johansen, O. Sigmund and J. H. Shin, *Sci. Rep.*, 2017, 7, 46023.
97. J.-H. Jang, C. K. Ullal, M. Maldovan, T. Gorishnyy, S. Kooi, C. Y. Koh and E. L. Thomas, *Adv. Funct. Mater.*, 2007, **17**, 3027–3041.
98. R. H. Siddique, R. Hünig, A. Faisal, U. Lemmer and H. Hölscher, *Opt. Mater. Express*, 2015, **5**, 996–1005.
99. T. S. Kustandi, H. Y. Low, J. H. Teng, I. Rodriguez and R. Yin, *Small*, 2009, **5**, 574–578.
100. S. Zhang and Y. Chen, *Sci. Rep.*, 2015, **5**, 16637.
101. T. Balla, S. M. Spearing and A. Monk, *J. Phys. D: Appl. Phys.*, 2008, **41**, 174001.
102. S. H. Ahn and L. J. Guo, *Adv. Mater.*, 2008, **20**, 2044–2049.
103. A. Saito, K. Ishibashi, J. Ohga, Y. Hirai and Y. Kuwahara, *Proc. SPIE*, 2018, 105930C.
104. T.-H. Wong, M. C. Gupta, B. Robins and T. L. Levendusky, *Opt. Lett.*, 2003, **28**, 2342–2344.

105. S. Y. Choi, M. Mamak, G. von Freymann, N. Chopra and G. A. Ozin, *Nano Lett.*, 2006, **6**, 2456–2461.
106. M. E. Calvo, S. Colodrero, T. C. Rojas, J. A. Anta, M. Ocaña and H. Míguez, *Adv. Funct. Mater.*, 2008, **18**, 2708–2715.
107. M. C. Fuertes, F. J. López-Alcaraz, M. C. Marchi, H. E. Troiani, V. Luca, H. Míguez and G. J. A. A. Soler-Illia, *Adv. Funct. Mater.*, 2007, **17**, 1247–1254.
108. T. Scharf, S. Jaquet, P. Ruffieux and H. P. Herzig, *Jpn. J. Appl. Phys.*, 2008, **47**, 6699–6705.
109. M. Kolle, A. Lethbridge, M. Kreysing, J. J. Baumberg, J. Aizenberg and P. Vukusic, *Adv. Mater.*, 2013, **25**, 2239–2245.
110. J. D. Rancourt, *Optical Thin Films: User Handbook*, SPIE Press, Bellingham, 1996.
111. M. Karaman, S. E. Kooi and K. K. Gleason, *Chem. Mater.*, 2008, **20**, 2262–2267.
112. N. N. Lal, K. N. Le, A. F. Thomson, M. Brauers, T. P. White and K. R. Catchpole, *ACS Photonics*, 2017, **4**, 741–745.
113. R. W. Johnson, A. Hultqvist and S. F. Bent, *Mater. Today*, 2014, **17**, 236–246.
114. M. Kolle, P. M. Salgard-Cunha, M. R. J. Scherer, F. Huang, P. Vukusic, S. Mahajan, J. J. Baumberg and U. Steiner, *Nat. Nanotechnol.*, 2010, **5**, 511–515.
115. D. G. Stavenga, B. D. Wilts, H. L. Leertouwer and T. Hariyama, *Philos. Trans. R. Soc., B*, 2011, **366**, 709–723.
116. P. Tzeng, D. J. Hewson, P. Vukusic, S. J. Eichhorn and J. C. Grunlan, *J. Mater. Chem. C*, 2015, **3**, 4260–4264.
117. K. Chung, S. Yu, C.-J. Heo, J. W. Shim, S.-M. Yang, M. G. Han, H.-S. Lee, Y. Jin, S. Y. Lee, N. Park and J. H. Shin, *Adv. Mater.*, 2012, **24**, 2375–2379.
118. M. Yaman, T. Khudiyev, E. Ozgur, M. Kanik, O. Aktas, E. O. Ozgur, H. Deniz, E. Korkut and M. Bayindir, *Nat. Mater.*, 2011, **10**, 494–501.
119. T. Khudiyev, E. Ozgur, M. Yaman and M. Bayindir, *Nano Lett.*, 2011, **11**, 4661–4665.
120. T. Khudiyev, T. Dogan and M. Bayindir, *Sci. Rep.*, 2014, **4**, 4718.
121. A. Saito, Y. Ishikawa, Y. Miyamura, M. Akai-kasaya and Y. Kuwahara, *Proc. SPIE*, 2007, 676706.
122. J. F. Galisteo-López, M. Ibisate, R. Sapienza, L. S. Froufe-Pérez, A. Blanco and C. López, *Adv. Mater.*, 2011, **23**, 30–69.
123. S.-H. Kim, S. Y. Lee, S.-M. Yang and G.-R. Yi, *NPG Asia Mater.*, 2011, **3**, 25–33.
124. V. N. Astratov, V. N. Bogomolov, A. A. Kaplyanskii, A. V. Prokofiev, L. A. Samoilovich, S. M. Samoilovich and Y. A. Vlasov, *Nuovo Cimento D*, 1995, **17**, 1349–1354.
125. J. Sussman, D. Snoswell, A. Kontogeorgos, J. J. Baumberg and P. Spahn, *Appl. Phys. Lett.*, 2009, **95**, 173116.
126. L. Mishchenko, B. Hatton, M. Kolle and J. Aizenberg, *Small*, 2012, **8**, 1904–1911.

127. C. E. Finlayson and J. J. Baumberg, *Polym. Int.*, 2013, **62**, 1403–1407.
128. Y. Takeoka, S. Yoshioka, A. Takano, S. Arai, K. Nueangnoraj, H. Nishihara, M. Teshima, Y. Ohtsuka and T. Seki, *Angew. Chem., Int. Ed.*, 2013, **52**, 7261–7265.
129. Y. Takeoka, *J. Mater. Chem. C*, 2013, **1**, 6059–6074.
130. R. Mayoral, J. Requena, J. S. Moya, C. López, A. Cintas, H. Míguez, F. Meseguer, L. Vázquez, M. Holgado and A. Blanco, *Adv. Mater.*, 1997, **9**, 257–260.
131. M. Holgado, F. García-Santamaría, A. Blanco, M. Ibisate, A. Cintas, H. Míguez, C. J. Serna, C. Molpeceres, J. Requena, A. Mifsud, F. Meseguer and C. López, *Langmuir*, 1999, **15**, 4701–4704.
132. S. H. Park, D. Qin and Y. Xia, *Adv. Mater.*, 1998, **10**, 1028–1032.
133. A. S. Dimitrov and K. Nagayama, *Langmuir*, 1996, **12**, 1303–1311.
134. P. Jiang, J. F. Bertone, K. S. Hwang and V. L. Colvin, *Chem. Mater.*, 1999, **11**, 2132–2140.
135. M.-L. Zhang, F. Jin, M.-L. Zheng and X.-M. Duan, *RSC Adv.*, 2014, **4**, 20567.
136. O. L. Pursiainen, J. J. Baumberg, H. Winkler, B. Viel, P. Spahn and T. Ruhl, *Opt. Express*, 2007, **15**, 9553–9561.
137. P. Spahn, C. E. Finlayson, W. Mbi Etah, D. R. E. Snoswell, J. J. Baumberg and G. P. Hellmann, *J. Mater. Chem.*, 2011, **21**, 8893–8897.
138. S. Magkiriadou, J.-G. Park, Y.-S. Kim and V. N. Manoharan, *Opt. Mater. Express*, 2012, **2**, 1343–1352.
139. L. Zulian, E. Emilitri, G. Scavia, C. Botta, M. Colombo and S. Destri, *ACS Appl. Mater. Interfaces*, 2012, **4**, 6071–6079.
140. Y. Zhao, L. Shang, Y. Cheng and Z. Gu, *Acc. Chem. Res.*, 2014, **47**, 3632–3642.
141. K. Xu, J.-H. Xu, Y.-C. Lu and G.-S. Luo, *Cryst. Growth Des.*, 2013, **13**, 926–935.
142. S.-N. Yin, C.-F. Wang, S.-S. Liu and S. Chen, *J. Mater. Chem. C*, 2013, **1**, 4685–4690.
143. H. Gu, Y. Zhao, Y. Cheng, Z. Xie, F. Rong, J. Li, B. Wang, D. Fu and Z. Gu, *Small*, 2013, **9**, 2266–2271.
144. J. Hu, X.-W. Zhao, Y.-J. Zhao, J. Li, W.-Y. Xu, Z.-Y. Wen, M. Xu and Z.-Z. Gu, *J. Mater. Chem.*, 2009, **19**, 5730–5736.
145. T. Kanai, D. Lee, H. C. Shum and D. A. Weitz, *Small*, 2010, **6**, 807–810.
146. Y. Zhao, Y. Cheng, L. Shang, J. Wang, Z. Xie and Z. Gu, *Small*, 2014, **11**, 151–174.
147. N. Vogel, S. Utech, G. T. England, T. Shirman, K. R. Phillips, N. Koay, I. B. Burgess, M. Kolle, D. A. Weitz and J. Aizenberg, *Proc. Natl. Acad. Sci. U. S. A.*, 2015, **112**, 10845–10850.
148. J. D. Simon and D. N. Peles, *Acc. Chem. Res.*, 2010, **43**, 1452–1460.
149. M. Xiao, Y. Li, M. C. Allen, D. D. Deheyn, X. Yue, J. Zhao, N. C. Gianneschi, M. D. Shawkey and A. Dhinojwala, *ACS Nano*, 2015, **9**, 5454–5460.

150. J. Wang, L. Yang, D. Lin, Y. Luo, D. Li and Q. Meng, *J. Chem. Phys.*, 2012, **137**, 234111.
151. A. J. Kim, R. Scarlett, P. L. Biancaniello, T. Sinno and J. C. Crocker, *Nat. Mater.*, 2009, **8**, 52–55.
152. S.-M. Yang, S.-H. Kim, J.-M. Lim and G.-R. Yi, *J. Mater. Chem.*, 2008, **18**, 2177–2190.
153. S. Y. Lee, S.-H. Kim, H. Hwang, J. Y. Sim and S.-M. Yang, *Adv. Mater.*, 2014, **26**, 2391–2397.
154. M. Schaffner, G. England, M. Kolle, J. Aizenberg and N. Vogel, *Small*, 2015, **11**, 4334–4340.
155. I. B. Burgess, J. Aizenberg and M. Lončar, *Bioinspir. Biomim.*, 2013, **8**, 045004.
156. T. Ding, S. K. Smoukov and J. J. Baumberg, *Nanoscale*, 2015, **7**, 1857–1863.
157. S. L. Burg and A. J. Parnell, *J. Phys.: Condens. Matter*, 2018, **30**, 413001.
158. R. W. Style, T. Sai, N. Fanelli, M. Ijavi, K. Smith-Mannschott, Q. Xu, L. A. Wilen and E. R. Dufresne, *Phys. Rev. X*, 2018, **8**, 011028.
159. F. M. van der Kooij, K. Kassapidou and H. N. W. Lekkerkerker, *Nature*, 2000, **406**, 868–871.
160. L. Onsager, *Ann. N. Y. Acad. Sci.*, 1949, **51**, 627–659.
161. K. E. Shopsowitz, W. Y. Hamad and M. J. MacLachlan, *J. Am. Chem. Soc.*, 2012, **134**, 867–870.
162. K. E. Shopsowitz, H. Qi, W. Y. Hamad and M. J. MacLachlan, *Nature*, 2010, **468**, 422–425.
163. J.-F. Revol, L. Godbout and D. G. Gray, *J. Pulp Paper Sci.*, 1998, **24**, 146–149.
164. M. Hasani, E. D. Cranston, G. Westman and D. G. Gray, *Soft Matter*, 2008, **4**, 2238–2244.
165. X. M. Dong, T. Kimura, J.-F. Revol and D. G. Gray, *Langmuir*, 1996, **12**, 2076–2082.
166. J. Pan, W. Hamad and S. K. Straus, *Macromolecules*, 2010, **43**, 3851–3858.
167. S. Beck, J. Bouchard and R. Berry, *Biomacromolecules*, 2011, **12**, 167–172.
168. A. G. Dumanli, H. M. van der Kooij, G. Kamita, E. Reisner, J. J. Baumberg, U. Steiner and S. Vignolini, *ACS Appl. Mater. Interfaces*, 2014, **6**, 12302–12306.
169. A. G. Dumanli, G. Kamita, J. Landman, H. van der Kooij, B. J. Glover, J. J. Baumberg, U. Steiner and S. Vignolini, *Adv. Opt. Mater.*, 2014, **2**, 646–650.
170. J.-F. Revol and R. H. Marchessault, *Int. J. Biol. Macromol.*, 1993, **15**, 329–335.
171. J. Syurik, G. Jacucci, O. D. Onelli, H. Hölscher and S. Vignolini, *Adv. Funct. Mater.*, 2018, **28**, 1706901.
172. T. P. Lodge, *Macromol. Chem. Phys.*, 2003, **204**, 265–273.
173. E. W. Cochran, C. J. Garcia-Cervera and G. H. Fredrickson, *Macromolecules*, 2006, **39**, 2449–2451.

174. M. Stefik, S. Guldin, S. Vignolini, U. Wiesner and U. Steiner, *Chem. Soc. Rev.*, 2015, **44**, 5076–5091.
175. C. Mille, E. C. Tyrode and R. W. Corkery, *Chem. Commun.*, 2011, **47**, 9873–9875.
176. T. Deng, C. Chen, C. Honeker and E. L. Thomas, *Polymer*, 2003, **44**, 6549–6553.
177. M. R. J. Scherer, *PhD thesis*, University of Cambridge, 2014.
178. Y.-W. Chiang, C.-Y. Chou, C.-S. Wu, E.-L. Lin, J. Yoon and E. L. Thomas, *Macromolecules*, 2015, **48**, 4004–4011.
179. A. Matsushita and S. Okamoto, *Macromolecules*, 2014, **47**, 7169–7177.
180. A. C. Edrington, A. M. Urbas, P. DeRege, C. X. Chen, T. M. Swager, N. Hadjichristidis, M. Xenidou, L. J. Fetters, J. D. Joannopoulos, Y. Fink and E. L. Thomas, *Adv. Mater.*, 2001, **13**, 421–425.
181. A. M. Urbas, R. Sharp, Y. Fink, E. L. Thomas, M. Xenidou and L. J. Fetters, *Adv. Mater.*, 2000, **12**, 812–814.
182. S. Valkama, H. Kosonen, J. Ruokolainen, T. Haatainen, M. Torkkeli, R. Serimaa, G. ten Brinke and O. Ikkala, *Nat. Mater.*, 2004, **3**, 872–876.
183. K. Hwang, D. Kwak, C. Kang, D. Kim, Y. Ahn and Y. Kang, *Angew. Chem., Int. Ed.*, 2011, **50**, 6311–6314.
184. S. E. Shin, S. Y. Kim and D. M. Shin, *Proc. SPIE*, 2010, 75991K.
185. H. J. Kim and D. M. Shin, *J. Nanosci. Nanotechnol.*, 2014, **14**, 6235–6237.
186. J. Y. Lim, J. H. Yun, D. bin Jang and D. M. Shin, *Proc. SPIE*, 2015, 958611.
187. T. J. Park, S. K. Hwang, S. Park, S. H. Cho, T. H. Park, B. Jeong, H. S. Kang, D. Y. Ryu, J. Huh, E. L. Thomas and C. Park, *ACS Nano*, 2015, **9**, 12158–12167.
188. M. Poutanen, G. Guidetti, T. I. Gröschel, O. V. Borisov, S. Vignolini, O. Ikkala and A. H. Gröschel, *ACS Nano*, 2018, **12**, 3149–3158.
189. J.-H. Lee, C. Y. Koh, J. P. Singer, S.-J. Jeon, M. Maldovan, O. Stein and E. L. Thomas, *Adv. Mater.*, 2014, **26**, 532–569.
190. S. Zhu, X. Liu, Z. Chen, C. Liu, C. Feng, J. Gu, Q. Liu and D. Zhang, *J. Mater. Chem.*, 2010, **20**, 9126–9132.
191. M. R. Weatherspoon, Y. Cai, M. Crne, M. Srinivasarao and K. H. Sandhage, *Angew. Chem., Int. Ed.*, 2008, **47**, 7921–7923.
192. Y. Vasquez, M. Kolle, L. Mishchenko, B. D. Hatton and J. Aizenberg, *ACS Photonics*, 2014, **1**, 53–60.
193. J. Huang, X. Wang and Z. L. Wang, *Nano Lett.*, 2006, **6**, 2325–2331.
194. G. Cook, P. L. Timms and C. Göltner-Spickermann, *Angew. Chem., Int. Ed.*, 2003, **42**, 557–559.
195. J. C. Hulteen and C. R. Martin, *J. Mater. Chem.*, 1997, **7**, 1075–1087.
196. F. Liu, Y. Liu, L. Huang, X. Hu, B. Dong, W. Shi, Y. Xie and X. Ye, *Opt. Commun.*, 2011, **284**, 2376–2381.
197. Y. Liu, L. Huang and W. Z. Shi, *Adv. Mater. Res.*, 2011, **391–392**, 409–417.
198. D. P. Gaillot, O. Deparis, V. Welch, B. K. Wagner, J.-P. Vigneron and C. J. Summers, *Phys. Rev. E*, 2008, **78**, 031922.

199. W. Zhang, D. Zhang, T. Fan, J. Ding, Q. Guo and H. Ogawa, *Microporous Mesoporous Mater.*, 2006, **92**, 227–233.
200. Y. Chen, J. Gu, S. Zhu, T. Fan, D. Zhang and Q. Guo, *Appl. Phys. Lett.*, 2009, **94**, 053901.
201. A. Stein and R. C. Schroden, *Curr. Opin. Solid State Mater. Sci.*, 2001, **5**, 553–564.
202. C. I. Aguirre, E. Reguera and A. Stein, *Adv. Funct. Mater.*, 2010, **20**, 2565–2578.
203. Z.-Z. Gu, H. Uetsuka, K. Takahashi, R. Nakajima, H. Onishi, A. Fujishima and O. Sato, *Angew. Chem., Int. Ed.*, 2003, **42**, 894–897.
204. K. R. Phillips, N. Vogel, Y. Hu, M. Kolle, C. C. Perry and J. Aizenberg, *Chem. Mater.*, 2014, **26**, 1622–1628.
205. O. Sato, S. Kubo and Z.-Z. Gu, *Acc. Chem. Res.*, 2009, **42**, 1–10.
206. G. I. N. Waterhouse, W.-T. Chen, A. Chan, H. Jin, D. Sun-Waterhouse and B. C. C. Cowie, *J. Phys. Chem. C*, 2015, **119**, 6647–6659.
207. D. P. Josephson, M. Miller and A. Stein, *Z. Anorg. Allg. Chem.*, 2014, **640**, 655–662.
208. B. H. Juárez, P. D. García, D. Golmayo, A. Blanco and C. López, *Adv. Mater.*, 2005, **17**, 2761–2765.
209. H. Míguez, E. Chomski, F. García-Santamaría, M. Ibisate, S. John, C. López, F. Meseguer, J. P. Mondia, G. A. Ozin, O. Toader and H. M. van Driel, *Adv. Mater.*, 2001, **13**, 1634–1637.
210. L. Liu, S. K. Karuturi, L. T. Su and A. I. Y. Tok, *Energy Environ. Sci.*, 2011, **4**, 209–215.
211. M. Liu, X. Li, S. K. Karuturi, A. I. Y. Tok and H. J. Fan, *Nanoscale*, 2012, **4**, 1522–1528.
212. H. Míguez, F. Meseguer, C. López, F. López-Tejeira and J. Sánchez-Dehesa, *Adv. Mater.*, 2001, **13**, 393–396.
213. S. Kim, A. N. Mitropoulos, J. D. Spitzberg, H. Tao, D. L. Kaplan and F. G. Omenetto, *Nat. Photonics*, 2012, **6**, 818–823.
214. A. A. Chabanov, Y. Jun and D. J. Norris, *Appl. Phys. Lett.*, 2004, **84**, 3573–3575.
215. S. Wong, V. Kitaev and G. A. Ozin, *J. Am. Chem. Soc.*, 2003, **125**, 15589–15598.
216. B. Hatton, L. Mishchenko, S. Davis, K. H. Sandhage and J. Aizenberg, *Proc. Natl. Acad. Sci. U. S. A.*, 2010, **107**, 10354–10359.
217. E. Dujardin, M. Blaseby and S. Mann, *J. Mater. Chem.*, 2003, **13**, 696–699.
218. A. Ivanova, D. Fattakhova-Rohlfing, B. E. Kayaalp, J. Rathouský and T. Bein, *J. Am. Chem. Soc.*, 2014, **136**, 5930–5937.
219. M. K. Khan, W. Y. Hamad and M. J. MacLachlan, *Adv. Mater.*, 2014, **26**, 2323–2328.
220. C. C. Y. Cheung, M. Giese, J. A. Kelly, W. Y. Hamad and M. J. MacLachlan, *ACS Macro Lett.*, 2013, **2**, 1016–1020.
221. O. Kose, A. Tran, L. Lewis, W. Y. Hamad and M. J. MacLachlan, *Nat. Commun.*, 2019, **10**, 510.

222. E. Lizundia, T.-D. Nguyen, J. L. Vilas, W. Y. Hamad and M. J. MacLachlan, *J. Mater. Chem. A*, 2017, **5**, 19184–19194.
223. H. Galinski, G. Favraud, H. Dong, J. S. T. Gongora, G. Favaro, M. Döbeli, R. Spolenak, A. Fratalocchi and F. Capasso, *Light: Sci. Appl.*, 2016, **6**, e16233.
224. A. Ruiz-Clavijo, Y. Tsurimaki, O. Caballero-Calero, G. Ni, G. Chen, S. V. Boriskina and M. Martín-González, *ACS Photonics*, 2018, **5**, 2120–2128.
225. H.-L. Liang, M. M. Bay, R. Vadrucci, C. H. Barty-King, J. Peng, J. J. Baumberg, M. F. L. De Volder and S. Vignolini, *Nat. Commun.*, 2018, **9**, 4632.
226. H. Kim, J. Ge, J. Kim, S.-E. Choi, H. Lee, H. Lee, W. Park, Y. Yin and S. Kwon, *Nat. Photonics*, 2009, **3**, 534–540.
227. T. Caro, *BioScience*, 2005, **55**, 125–136.
228. Y. Cho, S. Y. Lee, L. Ellerthorpe, G. Feng, G. Lin, G. Wu, J. Yin and S. Yang, *Adv. Funct. Mater.*, 2015, **25**, 6041–6049.
229. C. G. Schäfer, B. Viel, G. P. Hellmann, M. Rehahn and M. Gallei, *ACS Appl. Mater. Interfaces*, 2013, **5**, 10623–10632.
230. C. G. Schäfer, T. Winter, S. Heidt, C. Dietz, T. Ding, J. J. Baumberg and M. Gallei, *J. Mater. Chem. C*, 2015, **3**, 2204–2214.
231. B. Viel, T. Ruhl and G. P. Hellmann, *Chem. Mater.*, 2007, **19**, 5673–5679.
232. L. Duan, B. You, L. Wu and M. Chen, *J. Colloid Interface Sci.*, 2011, **353**, 163–168.
233. E. P. Chan, J. J. Walish, A. M. Urbas and E. L. Thomas, *Adv. Mater.*, 2013, **25**, 3934–3947.
234. X. Jia, J. Wang, K. Wang and J. Zhu, *Langmuir*, 2015, **31**, 8732–8737.
235. X. Sun, J. Zhang, X. Lu, X. Fang and H. Peng, *Angew. Chem., Int. Ed.*, 2015, **54**, 3630–3634.
236. S. Shang, Z. Liu, Q. Zhang, H. Wang and Y. Li, *J. Mater. Chem. A*, 2015, **3**, 11093–11097.
237. J. Zhang, S. He, L. Liu, G. Guan, X. Lu, X. Sun and H. Peng, *J. Mater. Chem. C*, 2016, **4**, 2127–2133.
238. Y. Hu, L. He and Y. Yin, *Small*, 2012, **8**, 3795–3799.
239. L. He, M. Wang, J. Ge and Y. Yin, *Acc. Chem. Res.*, 2012, **45**, 1431–1440.
240. M. Wang, L. He, W. Xu, X. Wang and Y. Yin, *Angew. Chem., Int. Ed.*, 2015, **54**, 7077–7081.
241. A. C. Arsenault, D. P. Puzzo, I. Manners and G. A. Ozin, *Nat. Photonics*, 2007, **1**, 468–472.
242. T. S. Shim, S.-H. Kim, J. Y. Sim, J.-M. Lim and S.-M. Yang, *Adv. Mater.*, 2010, **22**, 4494–4498.
243. A. A. Shah, M. Ganesan, J. Jocz and M. J. Solomon, *ACS Nano*, 2014, **8**, 8095–8103.
244. B. Michaelis, D. R. E. Snoswell, N. A. W. Bell, P. Spahn, G. P. Hellmann, C. E. Finlayson and J. J. Baumberg, *Adv. Eng. Mater.*, 2013, **15**, 948–953.
245. J. J. Walish, Y. Fan, A. Centrone and E. L. Thomas, *Macromol. Rapid Commun.*, 2012, **33**, 1504–1509.

246. H. Xing, J. Li, J. Guo and J. Wei, *J. Mater. Chem. C*, 2015, **3**, 4424–4430.
247. D. Xu, H. Yu, Q. Xu, G. Xu and K. Wang, *ACS Appl. Mater. Interfaces*, 2015, 7, 8750–8756.
248. T. Endo and H. Kajita, *IEEJ Trans. Sens. Micromach.*, 2016, **136**, 115–119.
249. Z. Wang, J. Zhang, J. Xie, C. Li, Y. Li, S. Liang, Z. Tian, T. Wang, H. Zhang, H. Li, W. Xu and B. Yang, *Adv. Funct. Mater.*, 2010, **20**, 3784–3790.
250. H. S. Lim, J.-H. Lee, J. J. Walish and E. L. Thomas, *ACS Nano*, 2012, **6**, 8933–8939.
251. Z. Yang, D. Shi, X. Zhang, H. Liu, M. Chen and S. Liu, *RSC Adv.*, 2015, **5**, 69263–69267.
252. X. Hong, Y. Peng, J. Bai, B. Ning, Y. Liu, Z. Zhou and Z. Gao, *Small*, 2013, **10**, 1308–1313.
253. J. Li, Z. Zhang, S. Xu, L. Chen, N. Zhou, H. Xiong and H. Peng, *J. Mater. Chem.*, 2011, **21**, 19267–19274.
254. Z. Yang, D. Shi, M. Chen and S. Liu, *Anal. Methods*, 2015, 7, 8352–8359.
255. A. M. Ruminski, B. H. King, J. Salonen, J. L. Snyder and M. J. Sailor, *Adv. Funct. Mater.*, 2010, **20**, 2874–2883.
256. Y. Zhang, J. Qiu, R. Hu, P. Li, L. Gao, L. Heng, B. Z. Tang and L. Jiang, *Phys. Chem. Chem. Phys.*, 2015, **17**, 9651–9658.
257. F. Wang, Z. Meng, F. Xue, M. Xue, W. Lu, W. Chen, Q. Wang and Y. Wang, *Trends Environ. Anal. Chem.*, 2014, **3–4**, 1–6.
258. H. Wang and K.-Q. Zhang, *Sensors*, 2013, **13**, 4192–4213.
259. C. Fenzl, T. Hirsch and O. S. Wolfbeis, *Angew. Chem., Int. Ed.*, 2014, **53**, 3318–3335.
260. Q. Li, Q. Zeng, L. Shi, X. Zhang and K.-Q. Zhang, *J. Mater. Chem. C*, 2016, **4**, 1752–1763.
261. K. R. Phillips, G. T. England, S. Sunny, E. Shirman, T. Shirman, N. Vogel and J. Aizenberg, *Chem. Soc. Rev.*, 2016, **45**, 281–322.
262. D. J. Mulder, A. P. H. J. Schenning and C. W. M. Bastiaansen, *J. Mater. Chem. C*, 2014, **2**, 6695–6705.
263. Y. P. Zhang, V. P. Chodavarapu, A. G. Kirk and M. P. Andrews, *Sens. Actuators, B*, 2013, **176**, 692–697.
264. M. Giese, L. K. Blusch, M. K. Khan, W. Y. Hamad and M. J. MacLachlan, *Angew. Chem., Int. Ed.*, 2014, **53**, 8880–8884.
265. Y. P. Zhang, V. P. Chodavarapu, A. G. Kirk and M. P. Andrews, *Proc. SPIE*, 2012, 825808.
266. H. Hu, Q.-W. Chen, J. Tang, X.-Y. Hu and X.-H. Zhou, *J. Mater. Chem.*, 2012, **22**, 11048–11053.
267. J. A. Schuller, E. S. Barnard, W. Cai, Y. C. Jun, J. S. White and M. L. Brongersma, *Nat. Mater.*, 2010, **9**, 193–204.
268. F. Cheng, J. Gao, T. S. Luk and X. Yang, *Sci. Rep.*, 2015, **5**, 11045.
269. F. Xia and L. Jiang, *Adv. Mater.*, 2008, **20**, 2842–2858.
270. H. Ge, G. Wang, Y. He, X. Wang, Y. Song, L. Jiang and D. Zhu, *ChemPhysChem*, 2006, 7, 575–578.

271. J. Wang, Y. Zhang, S. Wang, Y. Song and L. Jiang, *Acc. Chem. Res.*, 2011, **44**, 405–415.
272. J.-N. Wang, R.-Q. Shao, Y.-L. Zhang, L. Guo, H.-B. Jiang, D.-X. Lu and H.-B. Sun, *Chem. Asian J.*, 2012, 7, 301–304.
273. R. H. Siddique, G. Gomard and H. Hölscher, *Nat. Commun.*, 2015, **6**, 6909.
274. H. K. Raut, V. A. Ganesh, A. S. Nair and S. Ramakrishna, *Energy Environ. Sci.*, 2011, **4**, 3779–3804.
275. H. Sai, T. Matsui, K. Saito, M. Kondo and I. Yoshida, *Prog. Photovolt: Res. Appl.*, 2015, **23**, 1572–1580.
276. L. Cui, Y. Li, J. Wang, E. Tian, X. Zhang, Y. Zhang, Y. Song and L. Jiang, *J. Mater. Chem.*, 2009, **19**, 5499–5502.
277. J. Walia, N. Dhindsa, M. Khorasaninejad and S. S. Saini, *Small*, 2013, **10**, 144–151.
278. S. A. Rinne, F. García-Santamaría and P. V. Braun, *Nat. Photonics*, 2007, **2**, 52–56.
279. K. Zhong, P.-J. Demeyer, X. Zhou, O. Kruglova, N. Verellen, V. V. Moshchalkov, K. Song and K. Clays, *J. Mater. Chem. C*, 2014, **2**, 8829–8836.
280. S. Furumi, H. Fudouzi, H. T. Miyazaki and Y. Sakka, *Adv. Mater.*, 2007, **19**, 2067–2072.
281. G. I. N. Waterhouse, A. K. Wahab, M. Al-Oufi, V. Jovic, D. H. Anjum, D. Sun-Waterhouse, J. Llorca and H. Idriss, *Sci. Rep.*, 2013, **3**, 2849.
282. L. Zhang, E. Reisner and J. J. Baumberg, *Energy Environ. Sci.*, 2014, 7, 1402–1408.
283. L. Zhang, C.-Y. Lin, V. K. Valev, E. Reisner, U. Steiner and J. J. Baumberg, *Small*, 2014, **10**, 3970–3978.
284. J. K. Kim, J. H. Moon, T.-W. Lee and J. H. Park, *Chem. Commun.*, 2012, **48**, 11939–11941.
285. X. Zheng, S. Meng, J. Chen, J. Wang, J. Xian, Y. Shao, X. Fu and D. Li, *J. Phys. Chem. C*, 2013, **117**, 21263–21273.
286. W. Zhang, M. Anaya, G. Lozano, M. E. Calvo, M. B. Johnston, H. Míguez and H. J. Snaith, *Nano Lett.*, 2015, **15**, 1698–1702.
287. A. Herman, C. Vandenbem, O. Deparis, P. Simonis and J.-P. Vigneron, *Proc. SPIE*, 2011, 80940H.
288. S. Berthier, *Photonique Des Morphos*, Springer, Paris, 2010.
289. N. Dushkina and A. Lakhtakia, *Proc. SPIE*, 2009, 740106.
290. F. Schenk, *Int. J. Des. Nat. Ecodyn.*, 2009, **4**, 274–284.
291. F. Schenk and A. R. Parker, *Leonardo*, 2011, **44**, 108–115.
292. C. Park, K. Koh and U. Jeong, *Sci. Rep.*, 2015, **5**, 8340.
293. C. Park, T. Lee, Y. Xia, T. J. Shin, J. Myoung and U. Jeong, *Adv. Mater.*, 2014, **26**, 4633–4638.
294. M. Knoth, G. Miehe and H. Fuess, *Adv. Eng. Mater.*, 2005, 7, 928–931.
295. Y. C. Ryu, T. G. Kim, G.-S. Seo, J. H. Park, C. S. Suh, S.-S. Park, S.-S. Hong and G. D. Lee, *J. Ind. Eng. Chem.*, 2008, **14**, 213–218.

296. J. G. McGrath, R. D. Bock, J. M. Cathcart and L. A. Lyon, *Chem. Mater.*, 2007, **19**, 1584–1591.
297. M. S. Vlad-Cristea, V. Landry, P. Blanchet and C. Ouellet-Plamondon, *ISRN Nanomater.*, 2013, **2013**, 930236.
298. S.-M. Yang, S.-H. Kim, J. W. Shim and G.-R. Yi, EP. Pat., 2 371 908, 2011.
299. S.-H. Kim, S. Y. Lee, G.-R. Yi, D. J. Pine and S.-M. Yang, *J. Am. Chem. Soc.*, 2006, **128**, 10897–10904.
300. N. Koay, I. B. Burgess, T. M. Kay, B. A. Nerger, M. Miles-Rossouw, T. Shirman, T. L. Vu, G. England, K. R. Phillips, S. Utech, N. Vogel, M. Kolle and J. Aizenberg, *Opt. Express*, 2014, **22**, 27750–27768.
301. C. H. Lim, H. Kang and S.-H. Kim, *Langmuir*, 2014, **30**, 8350–8356.
302. C. I. Aguirre, E. Reguera and A. Stein, *ACS Appl. Mater. Interfaces*, 2010, **2**, 3257–3262.
303. O. L. J. Pursiainen, J. J. Baumberg, K. Ryan, J. Bauer, H. Winkler, B. Viel and T. Ruhl, *Appl. Phys. Lett.*, 2005, **87**, 101902.
304. C. E. Finlayson, P. Spahn, D. R. E. Snoswell, G. Yates, A. Kontogeorgos, A. I. Haines, G. P. Hellmann and J. J. Baumberg, *Adv. Mater.*, 2011, **23**, 1540–1544.
305. M. Asano, T. Kuroda, S. Shimizu, A. Sakihara, K. Kumazawa and H. Tabata, US Pat., 6 326 094, 2001.

CHAPTER 5

Bioinspired Approaches to Bone

F. NUDELMAN,*[a] S. DILLON[a,b] AND D. ELDOSOKY[a]

[a] School of Chemistry, University of Edinburgh, Joseph Black Building, The King's Buildings, David Brewster Road, Edinburgh EH9 3FJ, UK;
[b] The Roslin Institute and Royal (Dick) School of Veterinary Studies, University of Edinburgh, Easter Bush, Midlothian EH25 9RG, UK
*Email: fabio.nudelman@ed.ac.uk

5.1 Introduction

The development of biologically inspired materials for applications in regenerative medicine and biomedical engineering is a field of research that has been expanding over recent decades. One tissue that has attracted great interest is bone. As part of the vertebrate skeleton, bone is a complex organ that performs several different functions, including: providing a biomechanical and protective scaffold in conjunction with the musculature; regulating calcium and phosphate ion homeostasis; and as an endocrine organ involved with energy homeostasis. Importantly, the ability of bone to self-repair is limited to small defects, decreases with age and is affected by diseases. Therefore, there is a necessity to develop bone-replacement materials that mimic the properties and restore the function of the native tissue in cases where the damage is beyond the ability of bone to self-repair. In this chapter, we will discuss recent approaches aimed at developing and producing artificial bone-replacement materials. Because the properties of bone – which ideally need to be replicated by synthetic materials – arise

Inorganic Materials Series No. 4
Bioinspired Inorganic Materials: Structure and Function
Edited by Simon R. Hall

Published by the Royal Society of Chemistry, www.rsc.org

from its hierarchical structure, the first part of the chapter will be devoted to describing its composition and structure.

5.1.1 The Composition and Structure of Bone

Bone is a biocomposite material containing approximately 65 wt% mineral in the form of hydroxyapatite crystals, 25 wt% of organic material, which is predominantly type I collagen fibrils, and 10 wt% water. Electron microscopy and X-ray diffraction and scattering studies have shown that the mineral phase takes the form of elongated platelets approximately 2.5–4 nm in thickness and 20–50 nm in width, arranged parallel to each other.[1,2] The basic building blocks of type I collagen fibrils are the collagen molecules, which are triple helices of two α1 and one α2 chains stabilised by hydrogen bonds and hydrophobic interactions. Each polypeptide chain is approximately 1000 amino acids and is characterised by repeats of $(\text{Gly-X-Y})_n$, where X and Y are often proline and 4-hydroxyproline, respectively.[3–5] Each molecule is *ca.* 300 nm in length and 1.5 nm in diameter. The molecules are organised into fibrils in a quarter-staggered arrangement that gives rise to a D-band periodicity of 67 nm along the fibril axis, with a 40 nm gap separating two longitudinally adjacent molecules (Figure 5.1a, top). This organisation gives rise to a characteristic banding pattern, with a 27 nm overlap zone appearing as dark bands in transmission electron microscopy (TEM) and a 40 nm gap zone appearing as light bands in TEM (Figure 5.1a, bottom).[4,6,7]

The basic structural unit of bone is the mineralised collagen fibril, which integrates both the mineral and the collagen and serves as the building block of the higher architectural levels[8,9] (Figure 5.1b). The conventional model of collagen mineralisation has been that nucleation occurs within the gap zones of the collagen scaffold and, as the mineral grows, it spreads into the overlap zones[1,2,10] (Figure 5.1c). This results in mineral platelets distributed throughout the internal structure of the fibril, as visualised by Landis *et al.* in calcified avian tendon using electron tomography.[1] More recently, studies using focused ion beam milling were combined with TEM and electron tomography of human cortical bone to visualise mineral lamellae approximately 60 nm wide, 5 nm thick and hundreds of nm long. The authors interpreted electron-transparent holes surrounded by aligned mineral lamellae in their transverse sections as collagen fibrils in cross-section. They therefore proposed a new model wherein mineral is located mainly extrafibrillar, between collagen fibrils in the overall nanoarchitecture.[11–14] Many of these studies use conflicting *in vitro* and *in vivo* models, including reconstituted collagen, naturally mineralising avian tendon, and mouse and human cortical bone, thereby potentially confounding interpretation should multiple mechanisms be at play. Indeed, Reznikov and colleagues found a distinct disordered phase of mineralised collagen fibrils situated continuously between ordered lamellar structures in rat and human cortical bone comprised of both inter- and intra-fibrillar mineral, potentially explaining disparate observations.[15–17] The authors noted that the canalicular network

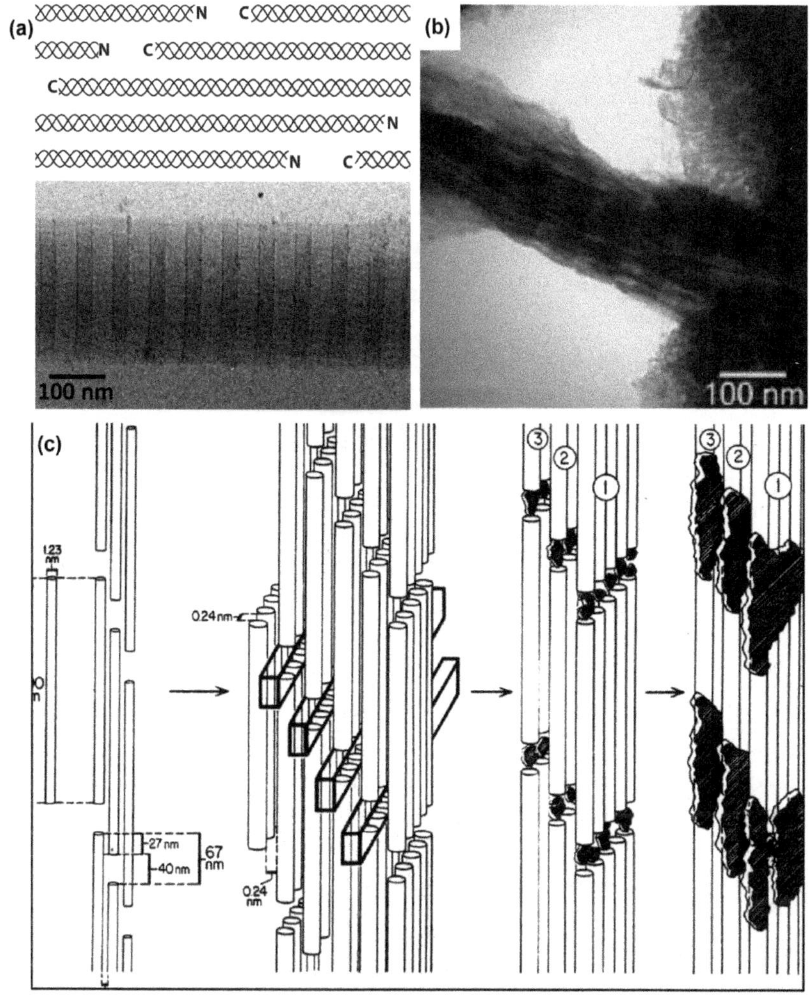

Figure 5.1 (a) Top: schematic representation of the organisation of the collagen molecules into fibrils. Bottom: cryo-TEM image of a collagen fibril, where the dark bands correspond to the overlap region and the light bands correspond to the gap region. (b) TEM image of a mineralised collagen fibril from bone. Reproduced from ref. 3 with permission from Elsevier, Copyright 2007. (c) Schematic representation of the collagen mineralisation model, in which hydroxyapatite crystals nucleate within the gap regions of collagen. Reproduced from ref. 1 with permission from Elsevier, Copyright 1993.

of the osteocytes resided exclusively within the disordered phase, which exhibits more extensive interfibrillar mineral. These conclusions would seem to correlate well with previous observations that mineral is present both inter- and intra-fibrillarly in human fetal woven bone; a prime example of a disorganised bone tissue.[18]

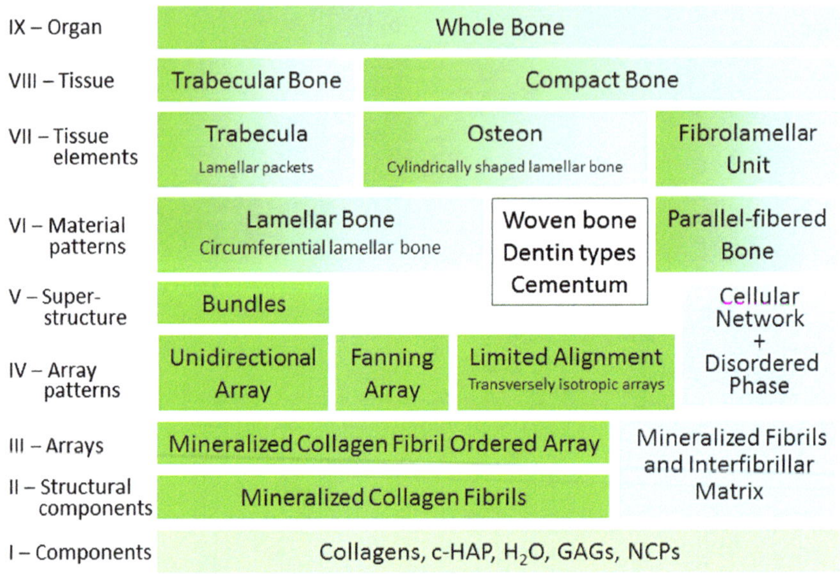

Figure 5.2 Scheme showing the hierarchical organisation of bone. Reproduced from ref. 17 with permission from Elsevier, Copyright 2014.

The mineralised collagen fibrils are then hierarchically structured, first forming fibre bundles at the micrometre level. The bundles are then organised into lamellae, giving rise to lamellae, osteons and trabeculae, forming the overall organ on the centimetre scale[16,19] (Figure 5.2).

5.1.2 Mechanical Properties of Bone

The mechanical properties of mineralised tissues are a function of both the relative proportions of the elementary constituents, in particular when combining organic and inorganic components, and their organisation into hierarchical structures.[20] Bone is a material that exhibits high stiffness and toughness, which are a result of its structure, composition and structural organisation.[19,21] The high stiffness arises from the mineral phase[21] and the exceptional fracture resistance comes from the nanometre size of the apatite crystals, which makes them insensitive to defects and maintains their strength despite pre-existing flaws or cracks.[22] Thus, by being only 3–5 nm in thickness, the apatite crystals in bone are optimised with respect to fracture strength and tolerance of flaws. Additionally, it has also been shown that the smaller the crystals, the larger the optimal aspect ratio, and the larger the aspect ratio, the larger the stiffening effect. Thus, the size of the apatite crystals in bone is optimised with respect to the fracture resistance of bone. The hierarchical structure of bone contributes to the high toughness of the material by creating a large number of interfaces that helps to avoid crack propagation.[23,24] An important aspect of the hierarchical structure of bone is that it allows an additional level of constructional control, where the basic building blocks can be assembled into different structural motifs, resulting in different mechanical properties that will be optimised to particular functions.[19,20] Ossified tendon, for instance, is composed of uniaxially orientated collagen I fibres and can be found in many bird species.[25] The mineral crystals in mineralised turkey tendon have also been found to be generally uniformly orientated with respect to the collagen scaffold.[26] Mechanical testing has shown the impressive stiffness of mineralised tendon when loaded in tension parallel to the collagen fibre orientation, with studies reporting values for Young's modulus up to 15 GPa,[27] compared with that of 1–2 GPa in native tendon.[28] In contrast, the tissue performs very poorly when loaded transversely, perpendicular to the fibre axis.[25] In dentine, on the other hand, the fibrils are organised into radial arrays in the plane parallel to the surface at which dentine formation takes place in the pulp cavity. This type of organisation makes this tissue exceptional at withstanding compressive forces. Finally, the plywood-like structure found in lamellar bone makes the tissue able to withstand stresses in several directions.[29]

The analysis of the structure–function relationship in bone reveals the challenges facing material scientists when trying to design bone-replacement materials: the ideal material should be able to replicate the material properties of native bone. Developing such materials is challenging because the dependence of biomechanics on fine structure is complicated by the varied and diverse architectures that can be found, sometimes across many hierarchical length scales. Furthermore, the properties of a given type of bone are highly optimised to its specific biological function in the degree of mineralisation present and in how mineralised fibrils are arranged in three dimensions to produce higher order structures.[19] In general, the biomechanical optimisation of bone and other mineralised tissues requires stiffness (*i.e.* a high Young's modulus of elasticity, or resistance to changes in geometry under stress) and also a high toughness (*i.e.* a resistance to fracture under stress).[19,30] However, tissues in disparate biological contexts often demonstrate calibration to function in the sacrifice of one for the other. Replicating such calibration synthetically is no mean feat.

In this chapter, we will discuss the desirable features that scaffolds should have in order to promote bone healing, and we will describe two different types of approaches for the development of bone implant materials: materials based on the basic constituent of bone (collagen and hydroxyapatite) and scaffolds based on synthetic materials. We will then discuss some important considerations on cell–scaffold interactions and finally emerging fabrication technologies to synthesise materials with highly controlled shapes, internal and external architectures, and sizes.

5.2 Scaffold Properties

The paramount feature of any implant material is that it is biocompatible and non-toxic. It must promote cell adhesion, migration, proliferation and differentiation, leading to the deposition of new extracellular matrix and the integration of the implant with the surrounding tissue.[31–33] Furthermore, it must be immune-inert, meaning that it elicits a negligible response from the immune system to avoid an inflammatory response that can prevent healing or cause rejection by the body.[31] A second important feature is the biodegradability of the scaffold: the body's own cells should be able to slowly degrade the implant and replace it with newly deposited native tissue.[34] The by-products of this biodegradation must also be non-toxic. Clearly, scaffolds that have similar chemical compositions to native bone will be more likely to combine biocompatibility, non-toxicity, biodegradability and ability to stimulate cellular activity towards healing.

The three-dimensional (3D) architecture of scaffolds is another critical factor. They must contain a network of interconnected pores in order to allow cell penetration, vascular in-growth, adequate nutrient diffusion and the elimination of waste products.[35,36] Indeed, the lack of vascularisation and waste elimination from the centre of implant materials is one of the main concerns in biomedical engineering. Scaffolds should also display ligands on their surface that bind to receptors on the cells in order to promote cell adhesion, proliferation and differentiation. This is less of a concern for materials based on natural extracellular matrix proteins, such as collagen, but scaffolds made from purely synthetic materials may require the incorporation of specific ligands.[31] Finally, there is a pore size range between 200 and 500 μm that is considered optimal for cell interactions and vascularisation.[37]

The ideal scaffold should have mechanical properties consistent with those of the site where they will be placed. The 3D architecture of the scaffold, along with its composition, will also determine the mechanical properties, such as elastic modulus, tensile strength and fracture toughness.[19,20] Therefore, an ideal implant material will have its structure and composition tailored so that the mechanical properties match those of the implantation site.[32,33,38,39] This aspect is perhaps the most challenging to mimic and control since, as mentioned above, different types of bone have different mechanical properties, tailored to their particular function. This means that each fracture site demands implant materials with different properties.

The final consideration is regarding the manufacturing technology: it must be possible to translate the synthesis and manufacture of scaffolds from a laboratory scale to a large scale in a cost-effective manner, otherwise the material is not clinically and commercially viable.[31] Moreover, off-the-shelf products offer an obvious advantage to having to harvest cells for *in vitro* tissue engineering, which would require additional surgical procedures and take several weeks. The implication is that the shelf-life of the implants must be considered.

Ultimately, the choice of bioinspired materials to be used as implants will depend on their capability to combine all the aspects described above.

5.3 Materials Based on Collagen and Hydroxyapatite

Given that the basic constituents of bone are collagen and hydroxyapatite, it is perhaps intuitive to use these same components as the basis for synthetic bone-replacement materials. One approach that has been used over the years is to precipitate hydroxyapatite during collagen fibrillogenesis. Bradt *et al.* mixed a solution of calcium-containing collagen dissolved in hydrochloric acid with a neutralisation buffer containing phosphate, triggering both assembly of the fibrils and precipitation of the mineral at the same time.[40] As a result, they obtained a 3D network of collagen fibrils that were covered with hydroxyapatite crystals. The attachment of the crystals to the surface of the fibrils was further improved by adding polyaspartate to the crystallisation mixture (Figure 5.3). It must be noted that while this type of scaffold mimics the composition of bone, it does not reflect the structure and organisation of the native tissue. Therefore, to what extent it can support cell attachment and growth and lead to bone regeneration still needs to be determined.

A similar approach was employed by Tampieri *et al.*, following two different methods.[41] In the first, they dispersed synthetic hydroxyapatite crystals in a solution where collagen fibrils were undergoing assembly.

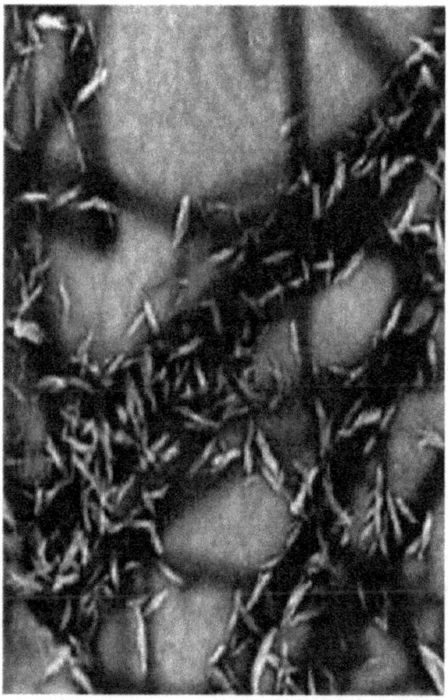

Figure 5.3 Scanning electron microscopy image of collagen fibrils mineralised by promoting collagen assembly together with the nucleation of hydroxyapatite in the presence of polyaspartic acid. The reaction was carried out by mixing a solution of calcium-containing collagen dissolved in hydrochloric acid with a neutralisation buffer containing phosphate. Reproduced from ref. 40 with permission from American Chemical Society, Copyright 1999.

This method resulted in collagen fibrils covered with hydroxyapatite crystals. With the second method, collagen dissolved in phosphoric acid was directly mixed with a $Ca(OH)_2$ solution, raising the pH to 9–10 and starting both fibrillogenesis and hydroxyapatite precipitation. Although the early stages of the process seemed to indicate that the crystals were intrafibrillar, images from the later stages clearly show that the hydroxyapatite crystals are covering the surface of the fibrils, in random morphological and crystallographic orientations. Using this method, a three-layered scaffold was created. The first layer was composed of mineralised collagen, which mimicked subchondral bone; the second layer was also composed of mineralised collagen, but with lower mineral content and it simulated the tidemark layer that separates hyaline cartilage from subchondral bone; and the third layer was composed of hyaluronic acid–collagen, simulating cartilage[42] (Figure 5.4a). The biological activity of these scaffolds was evaluated in tissue culture, where it was demonstrated that chondrocytes produced cartilaginous tissues specifically in the cartilage-mimicking layer. *In vivo*

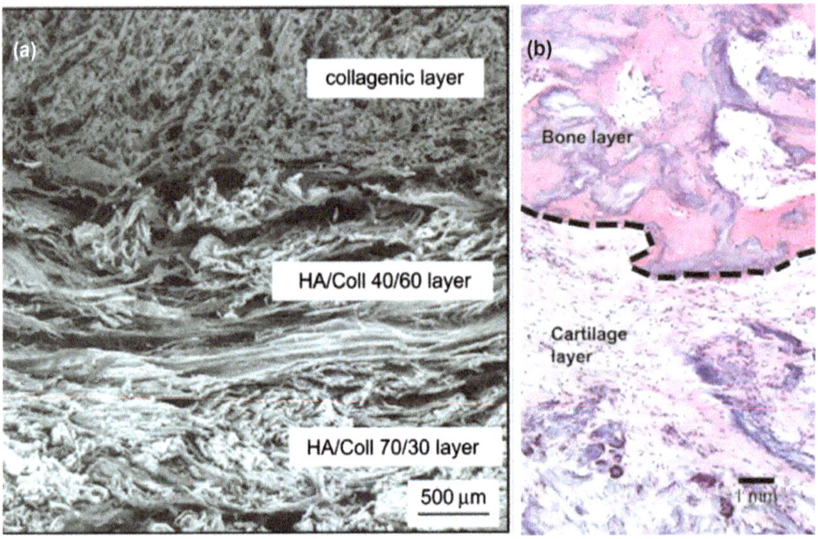

Figure 5.4 (a) Scanning electron microscopy images of the osteochondral scaffold morphology showing the three different layers with different mineral contents. The upper layer, composed of non-mineralised collagen simulating cartilage; the middle layer composed of hydroxyapatite : collagen 40 : 60 mimicking the tidemark layer that separates hyaline cartilage from subchondral bone; and the bottom layer composed of hydroxyapatite : collagen 70 : 30 mimicking subchondral bone. (b) Histological analysis showing newly formed bone tissue in the construct bone layer (top part of the panel) and connective tissue in the construct cartilaginous layer (bottom part of the panel), after 8 weeks of *in vivo* implantation. The dotted line shows the transition between the bone and cartilage layers of the scaffold.
Reproduced from ref. 42 with permission from Elsevier, Copyright 2008.

tests showed that scaffolds seeded with stromal cells and implanted in mice promoted bone formation within the mineralised layer, and connective tissue formation in the cartilage-like layer[43] (Figure 5.4b). These scaffolds were further tested in osteochondral defects in sheep, where it was demonstrated that they promoted the formation of new hyaline-like tissue and were completely resorbed by the body. In contrast, control groups where the osteochondral defect was left void displayed no spontaneous healing. These results are interesting in that they show that an implant material does not actually need to replicate the nano- and microstructure of bone in order to restore the function of the tissue. If the material is able to induce a cellular response and promote regeneration, it will end up being resorbed and replaced by natural tissue, which will, in turn, regenerate the lesion.

Up until 2007, most methods to mineralise collagen *in vitro* produced fibrils with hydroxyapatite crystals on their surface, with little resemblance to the mineralised collagen found in bone. A significant advance in this area was achieved by Olszta *et al.*, who obtained, for the first time, intrafibrillar mineralisation of collagen *in vitro*.[3] This was done by incubating preassembled collagen fibrils in a buffer supersaturated with respect to calcium and phosphate and in the presence of polyaspartic acid, which is an inhibitor of calcium phosphate precipitation in solution. This work was followed by others,[44–46] and it was demonstrated by Nudelman *et al.* that under the experimental conditions used the polymer forms negatively charged complexes with calcium phosphate that interact with positively charged sites on the surface of the collagen fibrils.[45] Given that many of the non-collagenous proteins that control bone formation *in vivo* are negatively charged due to the presence of phosphate or carboxylate groups,[47] it is postulated that polyaspartic acid mimics, at least to some extent, their properties. The polymer–mineral complexes then infiltrate into the collagen as amorphous calcium phosphate and then crystallise into hydroxyapatite[45,48] (Figure 5.5). One main difference between this type of biomimetic mineralisation and *in vivo* collagen mineralisation is that in bone the hydroxyapatite crystals nucleate on the gap zones of collagen. Using the *in vitro* method, crystal nucleation occurs both on the gap and overlap zones. It was later demonstrated that the mechanism of mineral infiltration into the collagen fibrils is not dependent on only the charge interactions between negatively charged polymer–mineral complexes and positively charged sites on collagen. In a polyelectrolyte-directed mineralisation system, there is a need to maintain osmotic equilibrium and electroneutrality. Thus, long-range interactions through the establishment of Gibbs–Donnan equilibrium act together with short-range electrostatic interactions to provide the necessary driving forces to promote the infiltration of the mineral inside the collagen fibril.[49] Finally, it was also demonstrated that intrafibrillar mineralisation of dense collagen matrices can also be obtained in the absence of polymers by using a biomimetic extracellular fluid to promote mineralisation.[50]

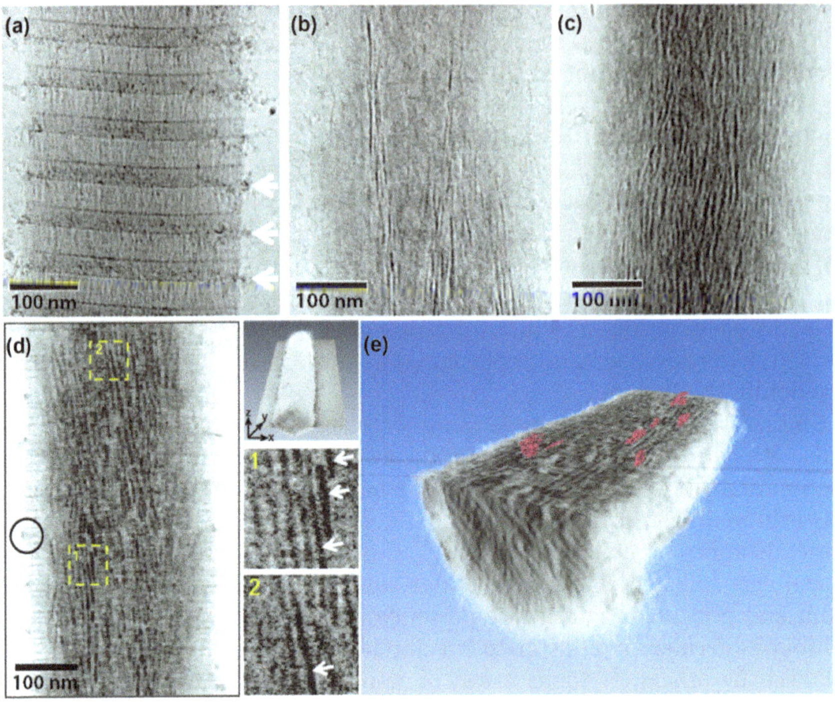

Figure 5.5 (a)–(c) Cryo-TEM images of collagen at different stages of mineralisation in the presence of polyaspartic acid. (a) Mineralisation for 24 h. White arrows show calcium phosphate–polyaspartic acid aggregates infiltrating into the collagen fibril. (b) Mineralisation for 48 h. Several needle-like hydroxyapatite crystals are present within the amorphous calcium phosphate in the fibril. (c) Mineralisation for 72 h. All the amorphous calcium phosphate converted to hydroxyapatite. (d), (e) Cryo-electron tomography of a collagen fibril mineralised for 72 h in the presence of polyaspartic acid. (d) Slice from a section of the 3D reconstruction along the *xy* plane (top-most inset), showing the crystals edge-on (insets 1 and 2, white arrows). Black circle: amorphous calcium phosphate infiltrating into the collagen. (e) Computer-generated 3D visualisation of the mineralised collagen fibril. The fibril is sectioned through the *xy* plane, showing plate-shaped hydroxyapatite crystals (in pink) inside the collagen fibril.
Reproduced from ref. 45 with permission from Springer Nature, Copyright 2010.

In terms of mechanical properties, the intrafibrillar mineralisation resulted in higher Young's modulus and toughness when compared to fibrils containing only extrafibrillar hydroxyapatite. Although this is no surprise, given that the mechanical properties depend on the localisation of the mineral within the fibril, the spacing between the platelets, their organisation and size,[20,51] it is nevertheless important. If collagen/mineral-based scaffolds are to be used as bone-replacement materials, it is necessary that they exhibit good load-bearing properties.

The development of new methods to promote intrafibrillar mineralisation of collagen opens new possibilities for implants; as such materials will mimic much more closely the nanostructure of the building blocks of bone. It must be noted, however, that the studies discussed above focused mainly on individual collagen fibrils, while bone actually consists of a densely packed collagen matrix. Hence, the first challenge consists of extending the methodology for obtaining intrafibrillar mineralisation of collagen to denser substrates. The first steps in this direction were carried out on porous collagen sponges with the aid of polyelectrolytes[52] (Figure 5.6a). Similar to the experiments on single fibres, the sponges were placed in Tris-buffer saline containing $CaCl_2$ and K_2HPO_4 and polyaspartic acid. This resulted in intrafibrillar mineralisation as before, but, interestingly, the extent of mineralisation was dependent on the molecular weight of polyaspartic acid used.[52] Using a molecular weight of 10 kDa resulted in a mineral content of 34 wt% after 8 days of mineralisation, which increased to 60 wt% after 16 days. Polyaspartic acid with a molecular weight of 32 kDa resulted in even higher mineral content, reaching 58 wt% after 8 days and 74 wt% after 16 days of mineralisation. These observations are interesting because they mean that the mineral content of collagen can actually be controlled by altering the length of the polymer. This provides another important tool to control the composition and mineral content of the scaffold and, consequently, its mechanical properties such as stiffness and toughness. It is not known, however, to what extent the mineralisation was homogeneous

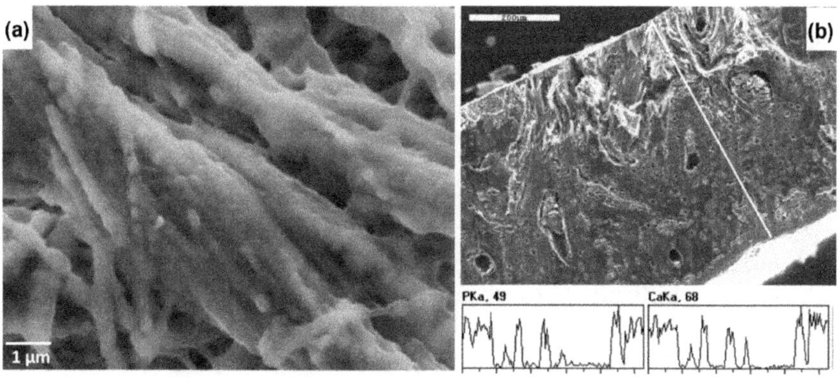

Figure 5.6 (a) Scanning electron microscopy (SEM) image of a collagen sponge mineralised *in vitro* with hydroxyapatite in the presence of polyaspartic acid for 16 days, where the mineral is predominantly inside the collagen fibrils. Reproduced from ref. 52 with permission from Elsevier, Copyright 2010. (b) Top: SEM image of a cross-section of a manatee bone sample, demineralised and remineralised *in vitro* in the presence of polyaspartic acid for 7 days. Bottom: energy-dispersive X-ray spectroscopy line scan showing the presence of Ca and P along the cross-section of the bone. The line scans shows a mineral penetration depth of up to 100 µm during the process. Reproduced from ref. 54 with permission from Elsevier, Copyright 2011.

throughout the scaffold: whether it was more concentrated on the surface rather than deep inside the sponge; whether the majority fibrils were fully or only partially mineralised. An interesting follow-up study was to use an alkaline phosphatase to provide a slow release of inorganic phosphate from a phosphate ester to promote mineralisation.[53] In some ways, this procedure more closely resembles the physiological process, and may provide an additional way to control the kinetics of mineralisation. Importantly, in this study the penetration depth of the mineral into the collagen scaffold was measured and found to reach 500 μm. This is significant, since a potential challenge in the use of 3D collagen scaffolds is to ensure that the mineral infiltrates through the sponge, and does not remain restricted to the fibrils close to the surface. The ability of mineralised sponges to support cell growth and differentiation was also analysed *in vitro* on osteoblast-like MG 63 cells. It was shown that not only was cell proliferation supported, but also that the scaffolds also promoted differentiation into osteoblasts, which was visible by the increase in alkaline phosphatase activity. Additionally, the mineralised sponges facilitated the formation of focal adhesion points, showing good contact between the cells and the substrate. These results demonstrate that these scaffolds have very good potential to integrate well into a lesion and to promote osteoblast activity. To what extent they can promote healing and be completely resorbed and replaced by newly deposited bone still remains to be determined.

While the work on the collagen sponges demonstrated that denser substrates could be mineralised, it could still be argued that their density and organisation do not reflect that of native bone. Therefore, Thula *et al.* applied the same biomimetic approach to remineralise demineralised bone specimens, which are composed of collagen in multiple hierarchical levels or organisations – fibrils, lamellae and osteons[54] (Figure 5.6b). It is interesting to note that in this case remineralisation occurred preferentially on osteonal structures. Additionally, a degree of mineralisation of up to 45 wt% was obtained and the penetration depth was much lower than with sponges – only up to 100 μm (Figure 5.6b). It is very likely that the high density of the bone made the diffusion of the mineral-polyaspartic acid throughout the material more difficult. Altogether, it is clear that using negatively charged polyelectrolytes, such as polyaspartic acid or polyacrylic acid, is an effective way to promote the biomimetic mineralisation of collagen, producing a material that is similar to bone at the nanoscale.

Scaffolds combining both intrafibrillar and extrafibrillar mineralisation are also under development. Ye *et al.*[55] demonstrated that such scaffolds exhibited a compressive modulus of 101.49 kPa, compared to 9.36 kPa for non-mineralised collagen and 26.71 kPa for collagen mineralised without additives. The compressive strength was reported to be 64 kPa, compared to 4.12 kPa and 15.02 for non-mineralised collagen and collagen mineralised without additives, respectively. It is clear that, not only the combination of organic and inorganic phases, but also the biomimetic association between collagen and mineral significantly enhanced the mechanical properties of the material. Importantly, the biological activity was evaluated both *in vivo* and *in vitro*. These scaffolds were shown to promote the adhesion, proliferation and differentiation of human umbilical cord mesenchymal stem cells into osteoblasts *in vitro*. When implanted into bone defects, the scaffolds promoted bone healing and were completely absorbed and replaced by lamellar bone after 12 weeks.

Focusing more on periodontal regeneration, Lausch *et al.* developed a new multiphasic collagen scaffold combining both mineralised and non-mineralised layers.[56] This was done by casting acid-soluble collagen into moulds to create porous scaffolds. These were then mineralised using polyaspartic acid as an additive, to promote intrafibrillar mineralisation. The diffusion of the mineralising solution through the scaffold was controlled, producing a hypermineralised surface layer and a bulk mineralised layer. Acid-soluble collagen was then gelled on the premineralised scaffold to create a biphasic material, which could be mineralised again and attached to another layer of non-mineralised collagen (Figure 5.7). This approach allows the synthesis of stratified collagen scaffolds that would be suitable for the regeneration of hard–soft tissue interfaces, such as the periodontium, which is responsible for tooth attachment.

Scaffolds based on collagen–apatite are very promising for bone regeneration: they replicate the natural composition of bone, providing osteoblasts with substrates that are similar to that of the native tissue. The similar composition also means that osteoblasts can resorb the implant material and replace it with newly deposited bone. In this respect, although it is desirable to use collagen scaffolds that mimic the nanostructure and

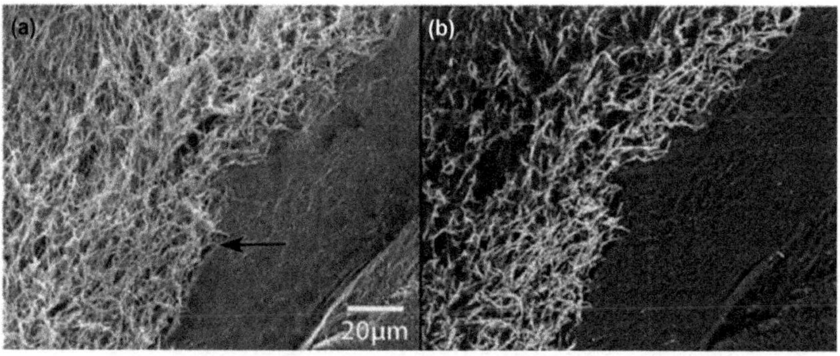

Figure 5.7 SEM images of a biphasic collagen scaffold combining a mineralised and an unmineralised layer. (a) Secondary electrons detector image where contrast is given by topography. The arrow shows the interface between the mineralised and unmineralised layers. (b) Backscattered electrons detector image, where the contrast is given by differences in elemental composition. The mineralised layer appears brighter than the non-mineralised layer.
Reproduced from ref. 56 with permission from John Wiley and Sons, © 2018 WILEY-VCH Verlag GmbH & Co. KGaA, Weinheim.

materials properties of native mineralised collagen, this might not be necessary. As long as the implant can stimulate osteoblast activity and induce the healing of a fracture or defect that would not otherwise regenerate, the cells can resorb the tissue and replace it with bone. Moreover, since the sequence and structure of collagen are highly conserved across species, there is little danger of rejection of the implant by the body.

Collagen–hydroxyapatite scaffolds have been used as coating materials on metal alloys such as titanium used as bone-graft substitutes. Given that such metals and metal alloys are commonly used as bone-graft substitutes in orthopaedics and dentistry due to their excellent mechanical properties and corrosion resistance,[57–60] it is important to optimise their integration into the native bone. Such integration depends largely on the porosity of the scaffold, the mineralisation degree of the coating and the amount of collagen fibrils available for the cells to attach to.[61,62] Therefore, controlling these factors is an important step when designing and optimising coating materials for implants. Xia *et al.*[63] sought to exert such control in a collagen–hydroxyapatite coating onto a Ti–6Al–4V surface by varying the amount of collagen present during the synthesis of the scaffold. The composite coating was formed by coprecipitating collagen and hydroxyapatite (HA) in a collagen-containing modified simulated body fluid (m-SBF), which was deposited onto UV light-treated Ti–6Al–4V substrates. Xia *et al.* showed that modifying collagen concentration in the m-SBF solution altered the composition and therefore the morphology of the collagen–HA microstructure. A higher collagen content in m-SBF allowed for subsequent nucleation and crystal growth of HA onto fibrous collagen and inhibited direct HA growth onto the Ti–6Al–4V substrate. In contrast, the higher collagen concentration affected the thickness of the composite coating, as well as the level of coating adhesion onto the Ti-alloy surface compared with a purely HA coating. The strength of coating bonded to the surface was determined using a tensile test. The hydroxyapatite coating had a bonding strength of 14.66 MPa as opposed to the highest collagen-containing composite, which had a bonding strength of 4.86 MPa, less than half the strength of hydroxyapatite alone. On a final note, the biocompatibility of the collagen–hydroxyapatite coating was evaluated in cell culture, where it was shown that increasing the amount of collagen in the m-SBF resulted in higher cell proliferation and higher expression levels of alkaline phosphatase, an indicator of cell differentiation, at the early stages of the culture when compared to hydroxyapatite alone. This makes this approach a relatively promising alternative for bone substitutes to promote osseointegraton in native bone; however, the composition of the coating needs to be optimised with respect to the osteoinductive properties and the desired mechanical properties of the implantation site.

Finally, electrochemical deposition has also been used to control the formation of collagen–hydroxyapatite coatings on the surface of implant materials. The advantage of this method is that parameters such as deposition time, deposition potential and the presence of additives in the solution can be modulated to control the thickness, porosity and mineralisation extent of the coating.[64–67] Using this technique to deposit collagen–hydroxyapatite on a Ti electrode, Ling *et al.* showed that, depending on these parameters, two types of coatings can be produced: dense coatings with low degrees of mineralisation and porous coatings with high degrees of mineralisation.[64] They proposed that the increase in the local pH in the cathode triggers the nucleation of hydroxyapatite and the self-assembly of collagen molecules, which are drawn to the electrode due to their net positive charge. The collagen fibrils then serve as substrates onto which hydroxyapatite crystals nucleate and deposit, leading to a composite material. It is emphasised that two inherent factors control the degree of mineralisation and porosity of the coatings: the pH gradient at the vicinity of the electrode as well as the low isoelectric point of the collagen fibrils. *In vitro* cell culture experiments showed that the collagen–hydroxyapatite coatings supported cell adhesion and proliferation, which were more pronounced on the denser scaffolds. Further controlling parameters, such as composition of the solution, pH and deposition potential, can be effective in optimising the overall properties of the coating. Providing that these scaffolds are effective materials in promoting bone healing, an additional step will be to scale up their production for clinical use.

5.4 Implants Based on Synthetic Materials

Since bone is a hybrid of collagen and hydroxyapatite, it is only instinctive to develop synthetic alternatives that mimic the combination of this natural polymer with a mineral component. A few types of bone substitutes made from artificial biomaterials include polymer- and protein-based scaffold composites that mimic grafts, injectables, as well as CaP cements. The presence of hydroxyapatite has proven to be a fundamental part in the construction of synthetic alternatives, and this is no surprise considering the percentage composition of this mineral in natural bone is higher than that of its organic counterpart. Hence, having briefly mentioned the types of alternatives, these will be discussed further, outlining the work of researchers and how, ultimately, these bioinspired versions could be innovative materials for clinical use to replace or regenerate bone.

A considerable amount of research has been dedicated to understanding the benefits of using synthetic polymer-based scaffolds, especially ones that are able to intrinsically mimic the highly ordered and porous structure that is found in bone. Such scaffolds are fabricated based on polymers that are able to display good biocompatibility and good osteoconductive properties.[68] In addition to these essential properties, the criteria for durable synthetic polymer replacement materials needs to include the crucial mechanical properties that bone exhibits, and this is determined by its composition and ordered assembly.[69] One polymer that has been at the centre of these types of scaffolds is the fully biodegradable polymer, poly(lactic) acid (PLA). Its favoured use stems from the bioresorbable nature of this polymer, promoting healing without leaving residual foreign bodies.[70] In addition, it is also one that has been approved for use *in vivo* for medical devices by the Food and Drug Administration (FDA), further highlighting its suitability as part of a synthetic scaffold for total bone replacement.[71,72]

The composite of poly(lactic) acid–hydroxyapatite (PLA–HA) is one that has been explored by many researchers and so the following discussion is aimed at probing the technique used to produce this organic–inorganic hybrid scaffold. A notable early study, from the works of Kim *et al.*,[73] was the production of PLA–HA fibrous nanocomposites *via* an electrospinning method. The production of these nanocomposites was achieved by electrospinning a mixture containing hydroxyapatite, PLA and an amphiphilic surfactant, hydroxysteric acid (HSA) (Figure 5.8a). The use of the surfactant was critical to get an even distribution of the mineral within the polymer matrix (Figure 5.8b). The nanocomposite was shown to enhance MG 63 osteoblastic cells in comparison with a PLA scaffold alone, suggesting the necessity of including hydroxyapatite for good cellular response (Figure 5.8c). Alternatively, Peng *et al.*[74] were able to integrate HA into PLA nanofibres *via* electrospinning needle-shaped HA particles that were nanometres or micrometres in size. The ratio between the polymer and the mineral was 80 : 20. The scaffolds obtained displayed homogeneous dispersion of HA in PLA nanofibres, such that the scaffolds exhibited inorganic assemblies that could be either random or aligned. They were able to demonstrate stable compatibility of the components through constant stretching of the fibres during the electrospinning process. They term this as "physical trapping" and also attribute this stability to chemical non-bonding interactions that could be due to Ca^{2+} ions either interacting with the carbonyl groups on the PLA chains, or with the carboxyl/hydroxyl terminal groups also found on the polymer chains. In addition, the authors also demonstrated that the PLA–HA scaffolds showed good biocompatibility and enhanced cell proliferation and expression of alkaline phosphatase, using rat osteosarcoma cells. Interestingly, they observed that electrospun fibres containing micrometre-sized hydroxyapatite particles supported cell proliferation and differentiation better than fibres containing nanometre-sized hydroxyapatite. Additionally, but not surprisingly, the alignment of the fibres also had an impact on the morphology of the cells. These exemplary studies show that scaffolds based on hydroxyapatite and PLA are promising for bone regeneration. More studies still need to be conducted to characterise their mechanical properties, investigate the effect of different polymer:mineral ratios on the cellular activity, and test their regenerative capabilities *in vivo*.

Purely organic polymeric scaffolds have also been studied for their bone-regenerating properties. Notably, Khan *et al.* identified a polymer blend composed of poly(L-lactic acid)/poly(ε-caprolactone) 20 : 80 as a promising scaffold capable of promoting bone healing.[75] This polymer blend was

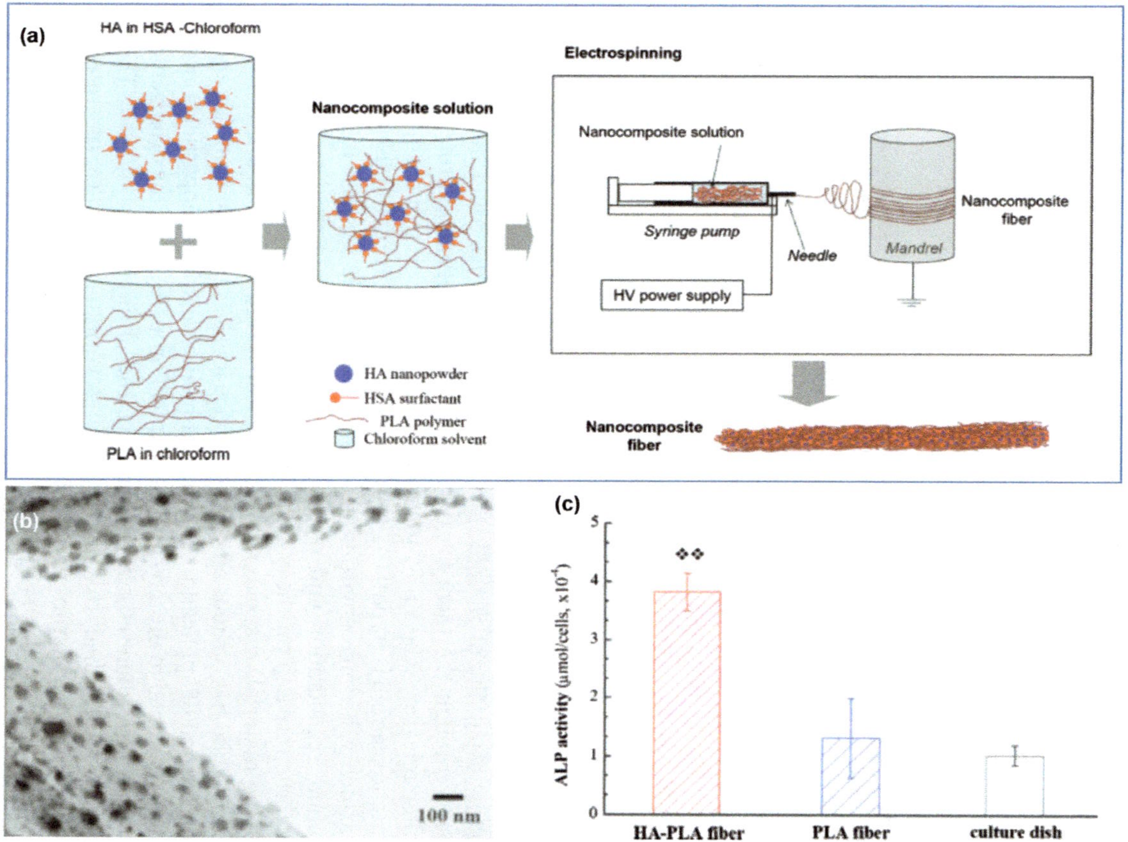

Figure 5.8 (a) Schematic representation of the electrospinning methodology to obtain hydroxyapatite–poly(lactic acid) composites with hydroxysteric acid as a surfactant. (b) TEM image of the nanocomposite fibres, with the hydroxyapatite particles embedded in the polymer fibres. (c) Alkaline phosphatase activity expressed by MG 63 cells cultured on the substrate for 7 days. Reproduced from ref. 73 with permission from John Wiley and Sons, © 2006 Wiley Periodicals, Inc.

shown to display a microporous structure with pores in the range 10–25 μm, comparable with the porosity of bone. This scaffold promoted the adhesion and proliferation of STRO-1$^+$ skeletal stem cells. The cells could also differentiate into mature osteoblasts when placed into osteogenic differentiation medium. Importantly, this scaffold was also able to promote bone healing *in vivo* in tests using femoral defect mouse model, and further work on sheep models is ongoing.[76] Ternary polymer blends combining chitosan, poly(ε-caprolactone), poly(L-lactic acid), poly(vinyl acetate), poly(2-hydroxyethyl methacrylate), poly(ethylamine) and poly(ethylene oxide) also demonstrated potential bone-healing abilities in tissue culture and mouse femur fracture models, suggesting that they could be used for bone bioengineering.[77]

A natural biomaterial that has been emerging as a candidate to synthesise scaffolds for bone regeneration is silk fibroin-based composites. Silk fibroin has the advantages that it is biocompatible, has suitable mechanical properties, has low immunogenicity and is biodegradable.[78–80] Additionally, it can be loaded with growth factors that induce cellular activity, therefore facilitating osteointegration, cell migration, *etc.* The major disadvantage of silk fibroin is that it is not osteoconductive.[81] To circumvent this limitation, silk fibroin scaffolds have been coated with hydroxyapatite by performing several cycles of soaking knitted silk in $CaCl_2$ solution, followed by immersion in Na_2HPO_4. These scaffolds were able to promote the osteogenic differentiation of bone marrow mesenchymal stem cells, and supported osteoblast growth.[82] When implanted in calvarial bone defect in rats, the scaffold promoted the formation of bone-like tissue 16 days post-implantation.[83] The osteogenic capability of the silk–hydroxyapatite construct could be further enhanced through the incorporation of bone morphogenetic protein-2 for controlled release.[79] The ability to incorporate different growth factors into a biomimetic scaffold is important, as it gives the possibility of mimicking with greater fidelity the microenvironment in the native tissue, which is important to promote regeneration and restore the functionality of the tissue. More recently, the research on these silk–hydroxyapatite scaffolds has focused on developing materials that can be directly injected into the lesion site.[83] This delivery method has obvious advantages, since it is less invasive than surgery. For this method of application, the ratio of hydroxyapatite nanoparticles and silk had to be tuned to preserve the mechanical properties and the osteogenic capability of the material, while retaining the ability to be injected. The researchers approached this by combining various quantities of water-dispersible hydroxyapatite nanoparticles with silk nanofibre hydrogels, producing composite materials containing between 20 and 60% of hydroxyapatite. *In vivo* tests in rat calvaria defects showed that the hydrogel with the highest mineral content (60%) of hydroxyapatite was the most effective in promoting bone healing.

Calcium phosphate cements (CPC) are synthetic bone substitutes that were invented by Chow and Brown and have since been extensively studied and used in the clinic.[70] They provide several advantages over other materials: they can be readily prepared during surgery and made to fit the lesion site; the hardening of the cement results in hydroxyapatite formation at physiological pH so that there is no damage to the surrounding tissue during the setting reaction.[70] Furthermore, as the cement is made of hydroxyapatite, it is biocompatible and osteoconductive. The disadvantages of this material, on the other hand, are that it has poor mechanical properties, which limits its use to non-load-bearing tissues; it has poor osteoinductivity and osseointegration; and, in many cases, it does not allow the repair of critical-size defects, resulting in the lesion being filled with connective tissue after the cement is resorbed.[84–86] Aiming to improve the osteogenic capacity of CPC, Shi *et al.*[87] recently investigated the inclusion of chondroitin sulfate (CS) into the cements. They demonstrated that the functionalisation of CPC with CS significantly improved the proliferation and osteogenic differentiation of bone mesenchymal stem cells. Furthermore, the CS can be functionalised with receptors or growth factors to further promote osteogenic stimulation.

An innovative bioinspired approach that is currently under development is exploiting the properties of wood for the production of porous organic–inorganic scaffolds displaying a 3D architecture and hierarchical structure.[88] This process involves the chemical transformation of wood through a series of thermal and hydrothermal processes in five steps, consisting of: (1) pyrolysis of wood to decompose mainly the cellulose, hemicelluloses and lignin and produce a carbon template; (2) vapour or liquid calcium infiltration to transform carbon into calcium carbide; (3) oxidation of the calcium carbide template to produce calcium oxide; (4) carbonation by hydrothermal autoclave treatments to convert the calcium oxide into calcium carbonate; and (5) phosphatisation to transform the calcium carbonate into hydroxyapatite. After this multi-step process the biomaterial retained the structure and morphology of the original wood template, resulting in a porous scaffold hierarchically structured in parallel hollow microtubules. This porous structure is important because it allows cell infiltration and reorganisation, and provides the necessary space for vascularisation. While the first studies resulted in a material with poor mechanical properties and of limited sizes, further optimisation was achieved by controlling the transformation of CaO into $CaCO_3$ through heterogeneous reactions between CO_2 gas in the supercritical state and the solid biomorphic CaO template.[89] During the synthesis process, the scaffold could be doped with Mg^{2+} and Sr^{2+} ions in amounts close to those found in native bone. The porosity of the constructs was found to be approximately 60% of their volume with unidirectional channels ~300 μm in diameter accompanied by smaller pores with multiscale interconnectivity (Figures 5.9a–c). Mechanical testing demonstrated good compressive strength of the construct in the 20 MPa range, with an initial elastic deformation. Moreover, these substituted hydroxyapatite scaffolds were able to bind and support mouse mesenchymal (bone marrow-derived) stem cells (mMSCs) in culture, while stimulating upregulation of osteogenic gene expression. *In vivo* osteoconductivity was also demonstrated

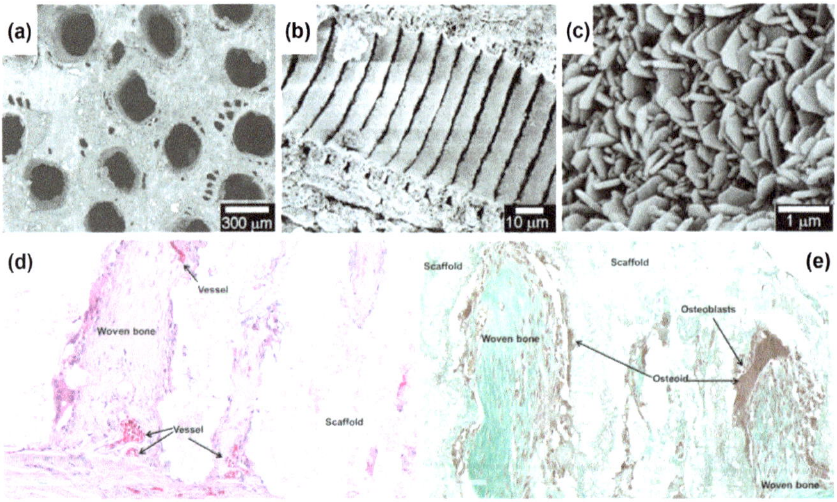

Figure 5.9 (a)–(c) SEM images of the wood-templated hydroxyapatite scaffolds viewed from the top showing macroscopic longitudinal sections (a), transversal section showing interconnected smaller channels (b), and the lamellar structure of the composite (c). (d), (e) Histological analysis after 12 weeks of subcutaneous implantation of the scaffold. (d) Haematoxilin-eosin stain. (e) Goldner's Trichrome stain.
Reproduced from ref. 89 with permission from the Royal Society of Chemistry.

by subcutaneous implantation of scaffolds in a rabbit model for 12 weeks. Upon extraction and histological examination, scaffolds were shown to have mediated bone cell recruitment and exhibited osteoblastic activity (Figures 5.9d and e). One interesting aspect of this scaffold is that the process to convert wood to a biomorphic scaffold is highly versatile and can be adapted to a variety of different structures, allowing the production of scaffolds with different 3D morphologies.

In addition to structural hierarchy, biomaterials that are intended for clinical use in regenerative medicine must also mimic the multiphasic nature of biological tissues at tissue interfaces, to accurately induce appropriate cell phenotypes.[43] Sprio *et al.* integrated biomineralisation with tape-casting and electrospinning to engineer a multiphasic scaffold for periodontal regeneration, imitating the three layers of tissue that occur in this region; namely, alveolar bone, the periodontal ligament and tooth cementum.[90] Furthermore, scaffolds seeded with mMSCs were able to support the cells, with maintenance of cell viability after 24 hours. This construct provides a more biologically relevant morphological architecture, mimicking the actual structures found *in vivo*. In this way, constructs such as these provide chemical and physical stimuli to seeded cells in an attempt to regulate their differentiation and activity to promote regeneration.[43]

5.5 Cell–Scaffold Interactions

As we have discussed, many studies have demonstrated a diversity of approaches in generating constructs for bone regeneration that are engineered to mimic the extracellular matrix and structural organisation of osteochondral tissues, taking a variety of forms, including solid polymer scaffolds, ceramics and layered polymer- and protein-based hydrogels. Much research has demonstrated the power of these scaffolds to maintain osteoblast and/or mesenchymal stem cell (MSC) viability *in vitro* and in animal studies; however, an ever-increasing body of work implicates several properties of the constructs themselves in influencing cell behaviour.

5.5.1 Architecture and Surface Topography

The architecture of engineered scaffolds has long been known to influence cellular activity in bone cells and in those from other tissues. In polymeric scaffolds, several basic design principles have held sway for over a decade. In general, engineered constructs must be resorbable *in situ* and demonstrate mechanical properties akin to native bone, while maintaining a relatively high porosity.[91–94] Scaffold parameters describing overall architecture at the microscopic scale such as porosity (≥90% in some cases), pore size and pore connectivity have been shown to have a profound influence on the success of a particular construct, regulating cell penetration, diffusion of nutrients and metabolites, and vascularisation.[35,92,93] Roy and colleagues used a poly(lactic-*co*-glycolic acid) (PLGA) polymer with 20% β-tricalcium phosphate (β-TCP) scaffold with a controlled porosity gradient (80–88%) to study these phenomena in a calvarial defect model in rabbits.[95] The authors reported a higher degree of bone in-growth in those scaffolds with increased porosity.[95] This result has been replicated using a range of scaffolds, and in a range of *in vivo* models, including in tibial, femoral and mandibular implants.[96,97] Interestingly, hydroxyapatite-based constructs with 70% macroporosity and as seeded with bone marrow stromal cells, rather than differentiated osteoblasts, also demonstrated increased osteogenesis compared with a lower porosity construct when implanted paraspinally in a goat model.[98]

Pore size has also been found to regulate bone cell behaviour independently from porosity. Initial studies found that a minimum pore size of 100 μm is required to induce osteogenesis in implanted calcium aluminate pellets with a 65% porosity.[99] Pores of <100 μm resulted in accumulation of unmineralised osteoid tissue upon histological examination.[99] Other studies have investigated the effect of pore size in hydroxyapatite ceramics coated with recombinant human bone morphogenic protein 2 (BMP-2) implanted subcutaneously, and found a peak of osteocalcin and alkaline phosphatase expression in those with pores of 300–400 μm, which decreased below and above these values.[100] This finding was replicated later by Kuboki *et al.* using both histological and biochemical characterisation methods.[101] Interestingly, some studies have reported an effect of not only pore size, but also of the internal geometry of a scaffold on the tissue type produced. In the same study Kuboki *et al.* found that scaffolds composed of porous particles or hydroxyapatite induced direct bone formation, whereas fibrous glass membranes brought about chondrogenesis.[101] A series of subsequent studies using the same range of scaffolds with a BMP-2 coating demonstrated that those containing tunnels with a diameter less than 100 μm induced the formation of cartilage, whereas those with a larger diameter (~350 μm) induced direct formation of bone.[102,103] In the porous hydroxyapatite scaffold with a honeycomb structure, cartilage was observed to later mineralise, in a process reminiscent of endochondral ossification.[101,102] The authors implicate delayed vascular invasion into the smaller structures as playing a key role in regulating this process; a hypothesis that may integrate well with research providing evidence for enhanced chondrogenic differentiation of MSCs under hypoxic conditions.[104–106]

Along with aspects of scaffold microarchitecture, the nanotopography of the material surface has been shown to influence how both seeded and native cells respond in terms of cell adhesion, proliferation and, potentially, differentiation.[107–109] An elegant early study by Dalby *et al.* engineered poly(methyl methacrylate) (PMMA) surfaces with elevated islands in the nanometre range to investigate the response of osteoprogenitors[110] (Figure 5.10). MSCs were shown to extend membrane protrusions towards these features in culture, spread significantly more compared to controls and also exhibited cytoskeletal rearrangement.[110] These effects have further been observed in MSCs across other materials with nanotopographies including mica and titanium.[111,112] Specific topographies may induce specific changes in stem cell populations with effects such as regulation of cell alignment, which has been shown in MSCs cultured on nanogratings 350 nm wide.[113] These effects appear to be mediated by several properties of the topographies themselves, including the scale of features and the extent of the disorder in their layout.[110,112,114,115] Pertinent to bone tissue engineering, several studies have reported nanotopographical parameters that drive osteoblast differentiation from MSC progenitors. Dalby and colleagues demonstrated increased osteoblast differentiation when MSCs were cultured on disorganised nanopits (120 nm in diameter and 100 nm deep) compared with a regular pattern of pits, as assessed through upregulation of osteoblastic gene and protein expression along with matrix mineralisation[110] (Figure 5.10). An intriguing study furthermore constructed synthetic nanoscale fibrils with a periodicity corresponding to that of collagen I, finding that matrices composed of these fibrils increased osteoblastic differentiation of MSCs in culture compared to fibrils with a higher periodicity.[116]

5.5.2 Matrix Stiffness and Mechanical Stimulation

It has been well established in the scientific literature that a huge variety of cells are sensitive to the mechanical properties of their matrix, and also to direct mechanical stimulation themselves.[117,118] These stimuli can be

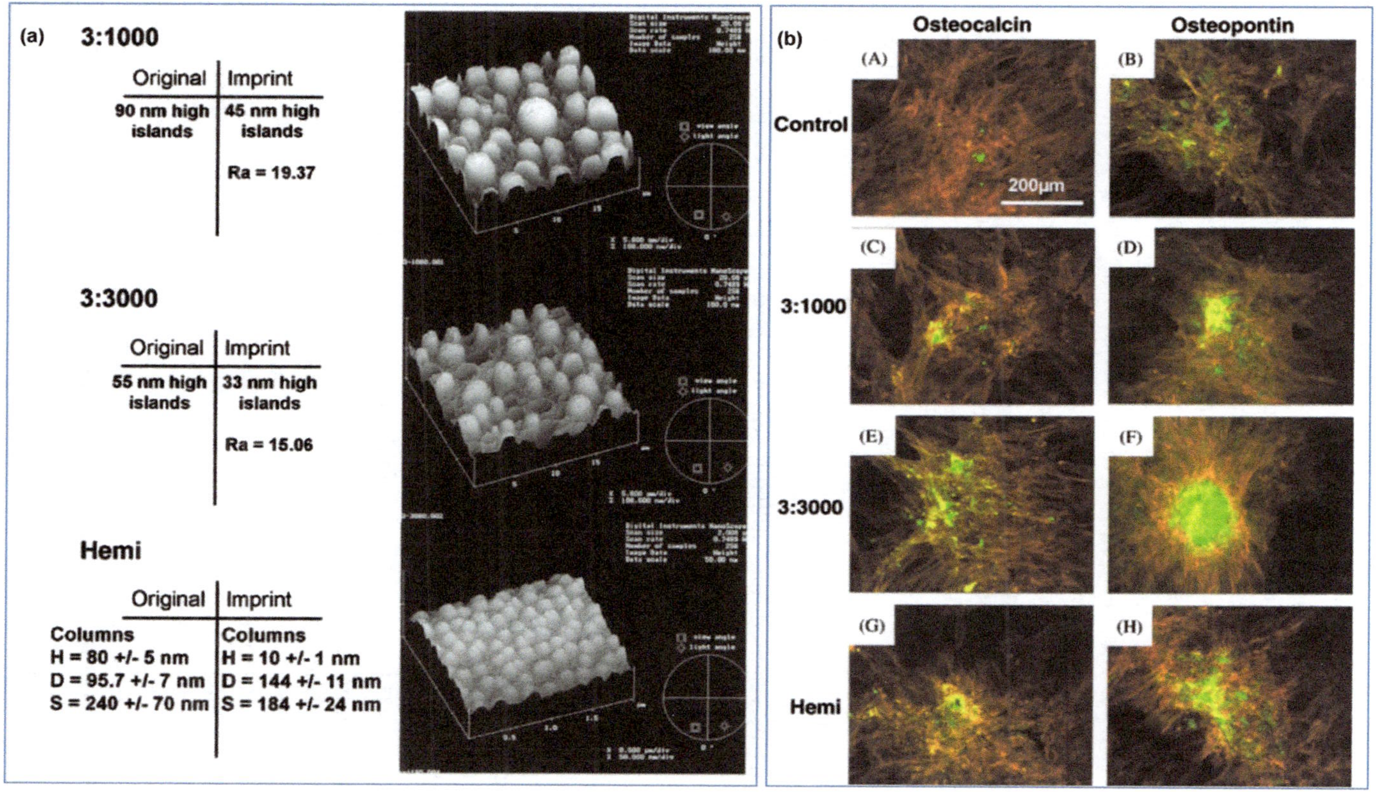

Figure 5.10 (a) Three-dimensional atomic force microscopy image of poly(methyl methacrylate) (PMMA) substrates designed with different nanotopographies. (b) Fluorescent images showing the expression of osteocalcin and osteopontin by human mesenchymal stem cells cultured on the different substrates shown in (a). In particular, bone nodule formation by cells grown on the 3 : 3000 material was observed.
Reproduced from ref. 110 with permission from Elsevier, Copyright 2006.

transduced through several mechanisms, including the extracellular matrix (ECM)-binding integrin proteins in conjunction with protein complexes at the cell membrane and the cytoskeleton, and are critically important during tissue development. Bone, in particular, is known to be a mechanically sensitive tissue, with an increase in habitual loading *in vivo* leading to increased bone formation and geometric change, and *vice versa*.[119,120] Much research has therefore investigated MSC and osteoblast mechanosensitivity, and whether this can be exploited to improve tissue engineering scaffolds for bone regeneration.

Matrix stiffness has been shown to be a critical regulator of MSC fate. Engler and colleagues studied the effect of ECM stiffness using polyacrylamide (PAM) gels with a collagen I coating, with varying bis-acrylamide concentrations allowing modulation of stiffness.[121] The authors reported that MSCs cultured on these gels were driven towards neuronal, myoblastic or osteoblastic differentiation with increasing matrix stiffness, as mediated through non-muscle myosin II (NMII).[121] Furthermore, MSCs were also shown to migrate to the stiffest regions across a PAM gel.[122] This effect has also been observed in protein-based hydrogels, such as collagen–glycosaminoglycan gels with 1-ethyl-3-3-dimethyl aminopropyl carbodiimide (EDAC) cross-linking.[123] Recent studies have begun to design scaffolds for hard tissue regeneration utilising stiffness to drive stem cell differentiation, with much success across a variety of substrates and *in vivo* models.[124–126]

In life the established anabolic response of mature bone to direct mechanical loading is mediated by complex interactions between subsets of bone cells, and their relationship with the surrounding tissues.[127–129] Germane to bone tissue engineering, however, is a large body of work that has investigated the effects of direct mechanical stimulation on osteoblast activity and osteoprogenitor commitment. As comprehensively reviewed elsewhere, MSCs under tension, compression (hydrostatic pressure) and fluid shear stress *in vitro* show upregulation of osteoblastic markers such as the transcription factor Runx2, along with others including alkaline phosphatase and collagen I.[129,130] Biomaterials scientists have applied these findings to bone tissue engineering and have reported improved bone formation and osseointegration across a wide range of scaffolds seeded with MSCs or primary osteoblasts and subjected to loading conditions such as cyclic compression and four-point bending.[131–134] A handful of studies have furthermore replicated these findings in rodent models using *in vivo* loading in conjunction with implanted scaffolds.[135,136]

Low magnitude high frequency (LMHF) vibration has also demonstrated anabolic effects on the skeletal structure in murine models.[137,138] In a series of papers, Curtis *et al.* engineered a novel bioreactor capable of delivering piezo-driven nanodisplacements (subsequently known as nanokicking) in order to examine the effects of LMHF vibration on MSCs.[139–141] The authors reported enhanced osteoblast differentiation in 2D culture, along with changes in cell morphology, adhesion and cytoskeletal remodelling.

Interestingly, some evidence of distinct effects of nanokicking and nanotopography were also found.[142] These analyses were furthermore extended to 3D culture of MSCs in a collagen gel, where nanokicking was found to stimulate osteoblast lineage commitment and matrix mineralisation, despite unfavourable matrix properties.[143]

5.6 Technologies for the Fabrication of Bone-replacement Scaffolds

Much of the research on bone-replacement scaffolds, such as the examples discussed in the previous sections, has focused mainly on optimising the biocompatibility, osteoconductivity, osteoinductivity and mechanical properties of the materials – in other words, in optimising their ability to restore the function of the damaged tissue, either as permanent replacement or as a material to be resorbed and replaced by newly formed bone. The critical-size bone defects in which regenerative scaffolds are most effectively employed are caused by a variety of disorders, including fracture, degenerative changes, trauma and tumours, and often result in lesions with complex 3D morphology. One barrier to clinical translation of many scaffolds is the difficulty in adapting scaffold design to lesion morphology, which is likely to be highly variable between patients. To this end, a number of rapid-prototyping techniques such as stereolithography, fused-deposition modelling, selective laser sintering and 3D printing have been under development.[144] These techniques use computer-aided design (CAD) software with input from computer tomography (CT) scans of the lesions to produce scaffolds that have highly controlled and reproducible structures, such as pore size, shape and interconnectivity, while at the same time being low cost and requiring minimal human input when compared to conventional techniques.

Three-dimensional printing is a technique that was initially developed in the 1990s, and is based on printing a liquid binder onto loose powders in a powder bed. Printing is done layer-by-layer, with each layer following the 2D image of the cross-section of the material, until the 3D structure has been built. The parameters that must be optimised prior to printing are: powder packing density, which is the density of the powder bed after uniform spreading; powder flowability, which refers to the ability of the powder to spread and is determined by the particle size, size distribution, roughness and shape; and the layer thickness, where thin layers can cause the penetration of the binder material and spreading of the powder to other sites, while thick layers require high saturation for the powders to bind.[145] The liquid binder can be organic or water-based, and binds the powder and hardens the wetted areas.[146–148] The next step in the printing process is to apply a heat treatment to complete the binder reaction and dry the scaffold. Finally, excess loose powder is removed from the scaffold, in a process called depowdering.[149]

Among the materials used for 3D printed bone scaffolds are: calcium phosphate bioceramics, which usually have good osteoconductivity and bioactivity; tricalcium phosphate powders; and biphasic calcium phosphate ceramics composed of a mixture of hydroxyapatite and β-tricalcium phosphate; mesoporous bioactive glasses, among others (reviewed in ref. 150). Seitz *et al.* used 3D printing with a modified powdered hydroxyapatite substrate to generate a porous ceramic scaffold that is customisable using patient imaging data.[151] Yao and colleagues furthermore generated cylindrical porous scaffolds designed using femoral and spinal specimens from rabbits using a similar technique (Figure 5.11a and b), and found good confirmation of scaffolds to design parameters such as porosity and pore connectivity.[152] This approach was validated using computer-aided design/computer-aided manufacturing (CAD/CAM) based on CT data in a mandibular lesion in a porcine model.[153] Hydroxyapatite-based 3D printed scaffolds (Figure 5.11c–f) have been shown to display good ostcoconductivity, with vascularisation in the porous architecture.[154] Osteoblasts seeded on sintered 3D scaffolds made of β-tricalcium phosphate exhibited good proliferation and growth inside the porous structures, where pore size was shown to influence cell density.[145] Additionally, these scaffolds were also demonstrated to promote new bone formation when implanted in rat distal femur model (Figure 5.11g and i). Doping of these scaffolds with SrO–MgO improved osteoid formation[155] (Figure 5.11h and j), while the addition of SiO_2-ZnO increased cell viability in different pore size ranges.[156] A 3D-printed scaffold made of $Ca_7MgSi_4O_{16}$ with a hollow-pipe-packed structure exhibited good compressive strength, which was enough to mechanically support the damaged tissue after implantation, and promoted vascularised bone regeneration in rabbit radius segmental defects, including bone marrow cavity reconnection and bone marrow formation.[157] It is clear that the regenerative capabilities of this scaffold derive from both the composition of the material and the structure, which can be built under precise control through 3D printing. A final point is the possibility to functionalise the 3D-printed scaffolds with other macromolecules, to improve their osoteoinductity and osteonectivity. This was shown by Pati *et al.*, where they used polymer–mineral blends composed of polycaprolactone–poly(lactic-*co*-glycolic acid)–β-TCP that were decorated with extracellular matrix proteins and were able to promote osteoblastic differentiation and bone formation in rat models.[158]

One important limitation of the 3D-printed scaffolds is that the high temperature needed for sintering limits the use of biomolecules during the fabrication of the material. Advances have already been made in this area, with the development of cold-printing techniques, in particular for calcium phosphate-based scaffolds.[159–161] Such methods rely on the use of aqueous acidic binders that are delivered from the inkjet through a dissolution–reprecipitation mechanism. The feasibility of this technique has already been demonstrated for calcium phosphate cements, and can be exploited to produce composite scaffolds that incorporate biopolymers. Following this approach, Inzana *et al.* produced 3D scaffolds combining both collagen and

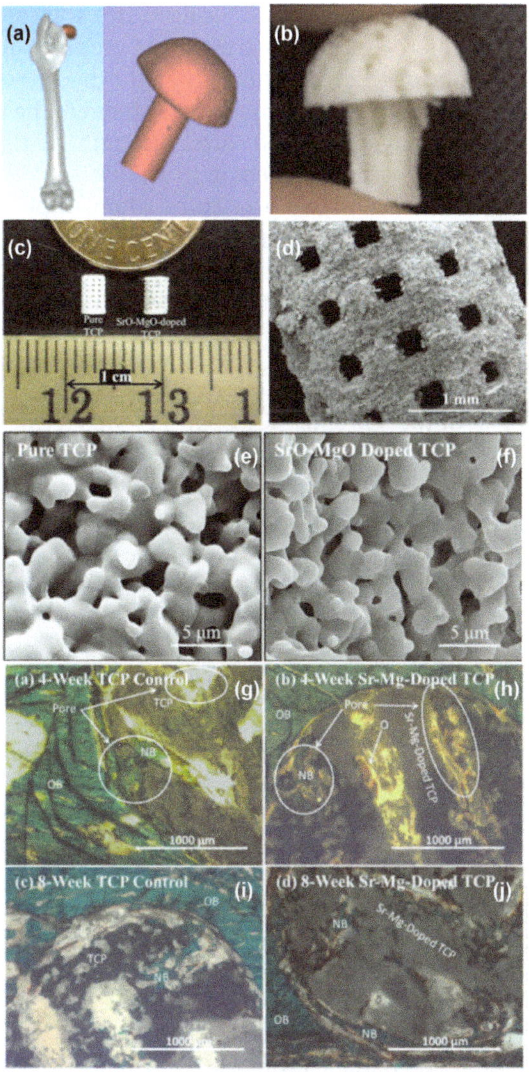

Figure 5.11 (a) CT data-based computer design of rabbit femoral scaffold. (b) Rabbit femoral scaffold mimic printed based on the design shown in (a). Reproduced from ref. 152 with permission from Springer Nature, Copyright 2015. (c) Photograph of microwave-sintered 3D-printed TCP and SrO and MgO-doped TCP scaffolds. (d) SEM image of the 3D-printed TCP scaffold. (e), (f) Surface morphology of the 3D-printed and microwave-sintered TCP (e) and SrO and MgO-doped TCP scaffolds (f). (g)–(j) Histological analysis showing the development of new bone and bone remodelling after 4 and 8 weeks of implantation of the TCP (g and i) and SrO and MgO-doped TCP scaffolds in rats (h and j). (c)–(j) Reproduced from ref. 155 with permission from the Royal Society of Chemistry.

hydroxyapatite, where collagen was dissolved in the phosphoric acid binder.[162] While the micro- and nano-structure of the final material was not characterised, it is reasonable to assume that the hydroxyapatite crystals were intermixed with the collagen fibril network. Indeed, the presence of the collagen significantly increased the flexural strength of the scaffold and improved the viability of the cells seeded on it. Additionally, the 3D construct showed promising osteoconductive properties when implanted in murine femoral defect models. Thus, while this type of scaffold combining both hydroxyapatite and collagen still needs further development and optimisation, it has very good potential as an implant material. This is unsurprising, given that similar scaffolds produced by conventional methods also display similar properties. The advantage of the 3D-printing method is that it allows precise control over the structure and shape of the material. On the other hand, to date it is still not possible to produce similar 3D-printed scaffolds made of collagen-containing intrafibrillar mineralisation, which would be a better mimic of bone structure. Nevertheless, using cold printing opens new possibilities in terms of combining different biopolymers into the scaffold, or incorporating different biomolecules, such as growth factors, during the manufacturing process. These possibilities provide new routes in which the osteoinductive and osteoconductive properties of the scaffold can be controlled and optimised.

One further limitation of 3D printing is that most common biodegradable materials are not compatible with the technique. The reason is that organic solvent binders will readily dissolve most commercially available print heads. One approach to overcome this issue is the use of indirect 3D printing, in which inverse moulds of the desired structures are printed and the biodegradable polymer solution is cast into the replica.[163,164] This method has been successfully used to produce 3D scaffolds made of chitosan and polycaprolactone, which were then coated with hydroxyapatite to improve their osteoinductive properties.[163] It must be noted, however, that the major challenge in this type of approach is the complete removal of the replica material from the internal pores.

Another challenge in the fabrication of 3D-printed scaffolds is that they generally exhibit poor mechanical properties. The main approach to improve the mechanical properties is sintering. However, during this process the scaffold might shrink, causing cracks and rendering the material unusable. The high temperature needed for sintering also limits the use of biomolecules during the fabrication of the material, as discussed above. Removing loose powders from interconnected pores is another challenge, and any leftover might sinter with the porous part, decreasing pore size or even causing complete blockage. Finally, there must be a balance between the mechanical properties and the size/density of the pores – the more porous the structure, the poorer the mechanical properties will be.

Stereolithography is another form of 3D printing where a UV laser selectively irradiates the surface of the UV-curable photopolymer following a shape or pattern determined by the CAD/CAM, causing it to harden and generating a cross-sectional layer of the final object.[165] New layers are built subsequently and stacked together until the 3D object is complete. Over the years stereolithography has given rise to microstereolithography, where a focusing lens is used to polymerise micrometre-sized areas of the polymer with a resolution of up to 1 μm, the highest resolution among solid free-form fabrication techniques.[166] Several different scaffolds with promising biodegradable and bone-regenerating properties have been produced using this method, including materials based on poly(propylene fumarate),[166,167] gelatine and trimethylene carbonate.[166] One limitation of this technique, however, is that photocurable polymers tend to have poor mechanical properties and in many cases do not promote cell adhesion or proliferation. One strategy to tackle these issues is to blend the polymers with mineral, such as hydroxyapatite. A further challenge with microstereolithography is that it requires polymers that are photocurable, which restricts the types of materials that can be used.

Fused-deposition modelling is a technique that uses the melt extrusion of a thermoplastic, layer-by-layer, to produce a 3D scaffold with controllable porosity. Scaffolds based on polycaprolactone[168,169] or polycaprolactone coated with calcium phosphate[170] seem promising in supporting osteoblasts and chondrocytes cultures, leading to the deposition of new extracellular matrix. Most notably, Xu *et al.* used computer tomography-guided fused-deposition modelling to produce a polycaprolactone–hydroxyapatite-based 3D scaffold mimicking goat femurs, with cortical bone-like and cancellous bone-like features.[171] The material was shown to have good mechanical properties; good biocompatibility *in vitro* and *in vivo*; and it promoted the formation of new bone when implanted in goats, while it underwent degradation over time. The main limitations of fused-deposition modelling are that it requires the use of preformed fibres with consistent size and material properties and it has limited application to biodegradable polymers.[166]

Selective laser sintering also proved effective in producing anatomically relevant scaffolds, which also conform to standard scaffold design principles. This technique involves using a high-power laser to selectively sinter powders following the information provided by the CAD/CAM manufacturing based on a CT scan. This approach allows the use of a wide range of materials, including polymers such as poly(lactic acid), polycaprolactone and bioceramics.[144] It was used experimentally to produce a construct mimicking the complex morphology of the temporomandibular joint (TMJ) in a pig model.[172] *In vivo* implantation of these scaffolds demonstrated significant osseointegration after 1 and 3 months, which was able to support normal function.[172] The disadvantage of this method is that it requires high operating pressures, which limits the use of biopolymers. Other methodologies that have been employed with success include using perfusion bioreactors. Decellularised trabecular bone was shaped using image-guided micromilling and seeded with adipose- or bone marrow-derived stem cells before culture in perfusion chambers designed to mimic the TMJ morphology.[173,174] These scaffolds were validated in a porcine model and demonstrated a greater degree of bone formation and vascularisation than those without seeded cells, or untreated lesions.[174]

Clearly, rapid-prototyping techniques are very promising. In combination with input from CT scans of patients, they allow the production bone implant materials with well-controlled shapes, external and internal architectures and mechanical properties that are tailored to the patient, precisely matching the clinical need.[151,152] Optimisation, however, is still needed to expand the range of materials that can be used as powders and binders to produce a 3D scaffold. Given that materials like biomimetic-mineralised collagen, wood-based templates and polymer–mineral composites have demonstrated great potential in promoting bone healing, the next logical step is to develop a range of techniques to work on such materials. This will lead to the production, for instance, of 3D scaffolds made of collagen-containing intrafibrillar mineral and organised in hierarchical structures that closely mimic that of bone, from the nano- to the micro-scales. The organisation of the collagen fibrils, along with the mineral content and material properties in such material, could be precisely tailored to match the type of bone at the injury site. Considering the current limitations of 3D printing, developing this capability is still a challenge.

5.7 Conclusions and Outlook

The limited ability that bone has to self-repair makes it necessary to develop synthetic scaffolds that promote regeneration and restore tissue function. To achieve that, such scaffolds must be able to recruit cells around the implantation site and promote their migration, adhesion, the secretion of new extracellular matrix and eventually remodelling, together with the formation of new tissue. In the last years, many advances have been made, with the development of scaffolds based on collagen–mineral composites; calcium phosphate cements; polymer–mineral blends; and the use of wood as a template to produce hierarchical scaffolds. All these materials have shown, to different extents, very promising osteoconductive, osteoinductive, osseointegration and resorbable capabilities. So far, all these materials have been produced on a laboratory scale. If they are to be used by doctors and surgeons, methods for their medium- and large-scale production at an accessible cost will need to be developed. Furthermore, the scaffolds must also display mechanical properties that are comparable to those at the implantation site. Given the structural, mechanical and biological complexity of bone, producing implant materials that are specifically tailored according to different functional needs is still a challenge. It demands a deep understanding of the structure and properties of different types of bone with distinct functions, and the capability to produce scaffolds with highly controlled architectures and properties. Together with this, controlling the internal architecture of the scaffolds is highly important, as pore size and connectivity are essential for integration of the material into the tissue and the promotion of cellular proliferation, differentiation and vascularisation.

One of the difficulties in this respect is that increasing pore size or pore density will lead to poorer mechanical properties, so a compromise will need to be found between these two parameters, which will depend on the mechanical requirements of the implantation site. Nevertheless, designing and fabricating scaffolds with such controllable structures is still challenging. Here, rapid prototyping techniques, including 3D printing and selective laser sintering, among others, offer a potential solution. The shape, size, and internal and external architecture of the scaffold can be designed and matched to the lesion site, both functionally and anatomically. However, further development is still necessary to allow the use of materials like biopolymers or biopolymer–mineral composites, such as collagen–hydroxyapatite, to be printed into an implant.

Our understanding of tissue anatomy from the nano- to the macro-scopic scale, of cell adhesion and differentiation, and of cell–extracellular matrix interactions is advancing rapidly, and feeding into the design and development of materials capable of promoting bone healing by stimulating osteoblast activity. These advances are going hand in hand with the development of advanced techniques for the fabrication of 3D materials, and ultimately will lead to the production of bone-replacement materials that are incorporated into the clinical setting.

Acknowledgements

This work was supported by Reminova and by the Biotechnology and Biological Sciences Research Council (BBSRC) [grant number BB/M010996/1].

References

1. W. J. Landis, M. J. Song, A. Leith, L. Mcewen and B. F. Mcewen, *J. Struct. Biol.*, 1993, **110**, 39–54.
2. W. Traub, T. Arad and S. Weiner, *Matrix*, 1992, **12**, 251–255.
3. M. J. Olszta, X. G. Cheng, S. S. Jee, R. Kumar, Y. Y. Kim, M. J. Kaufman, E. P. Douglas and L. B. Gower, *Mater. Sci. Eng., R.*, 2007, **58**, 77–116.
4. J. P. R. O. Orgel, T. C. Irving, A. Miller and T. J. Wess, *Proc. Natl. Acad. Sci. U. S. A.*, 2006, **103**, 9001–9005.
5. J. P. R. O. Orgel, A. Miller, T. C. Irving, R. F. Fischetti, A. P. Hammersley and T. J. Wess, *Structure*, 2001, **9**, 1061–1069.
6. A. J. Hodge and J. A. Petruska, in *Aspects of Protein Structure*, ed. G. N. Ramachandran, Academic Press, New York, 1963, pp. 289–300.
7. A. Miller, *Philos. Trans. R. Soc., A*, 1984, **304**, 455–477.
8. D. J. S. Hulmes, T. J. Wess, D. J. Prockop and P. Fratzl, *Biophys. J.*, 1995, **68**, 1661–1670.
9. W. Traub, T. Arad and S. Weiner, *Proc. Natl. Acad. Sci. U. S. A.*, 1989, **86**, 9822–9826.
10. M. J. Glimcher and H. Muir, *Philos. Trans. R. Soc., A*, 1984, **304**, 479–508.

11. E. McNally, F. Nan, G. A. Botton and H. P. Schwarcz, *Micron*, 2013, **49**, 46–53.
12. E. A. McNally, H. P. Schwarcz, G. A. Botton and A. L. Arsenault, *PLoS One*, 2012, 7, e29258.
13. H. P. Schwarcz, *Semin. Cell Dev. Biol.*, 2015, **46**, 44–50.
14. H. P. Schwarcz, E. A. McNally and G. A. Botton, *J. Struct. Biol.*, 2014, **188**, 240–248.
15. N. Reznikov, R. Almany-Magal, R. Shahar and S. Weiner, *Bone*, 2013, **52**, 676–683.
16. N. Reznikov, R. Shahar and S. Weiner, *Bone*, 2014, **59**, 93–104.
17. N. Reznikov, R. Shahar and S. Weiner, *Acta Biomater.*, 2014, **10**, 3815–3826.
18. X. Su, K. Sun, F. Cui and W. Landis, *Bone*, 2003, **32**, 150–162.
19. S. Weiner and H. D. Wagner, *Annu. Rev. Mater. Sci.*, 1998, **28**, 271–298.
20. J. W. C. Dunlop and P. Fratzl, *Annu. Rev. Mater. Res.*, 2010, **40**, 1–24.
21. J. D. Currey, *Bones. Structure and Mechanics*, Princeton University Press, Princeton, 2002.
22. H. J. Gao, B. H. Ji, I. L. Jager, E. Arzt and P. Fratzl, *Proc. Natl. Acad. Sci. U. S. A.*, 2003, **100**, 5597–5600.
23. P. Fratzl, *Nat. Mater.*, 2008, **7**, 610–612.
24. P. Fratzl and R. Weinkamer, *Prog. Mater. Sci.*, 2007, **52**, 1263–1334.
25. J. D. Currey, *J. Mech. Behav. Biomed.*, 2010, **3**, 357–372.
26. H. R. Wenk and F. Heidelbach, *Bone*, 1999, **24**, 361–369.
27. M. B. Bennett and J. A. Stafford, *J. Zool.*, 1988, **214**, 343–351.
28. H. Schechtman and D. L. Bader, *J. Biomech.*, 1997, **30**, 829–835.
29. S. Weiner, L. Addadi and H. D. Wagner, *Mat. Sci. Eng. C-Bio. S.*, 2000, **11**, 1–8.
30. J. D. Currey, *J. Exp. Biol.*, 1999, **202**, 3285–3294.
31. F. J. O'Brien, *Mater. Today*, 2011, **14**, 88–95.
32. G. Bouet, D. Marchat, M. Cruel, L. Malaval and L. Vico, *Tissue Eng., Part B*, 2015, **21**, 133–156.
33. J. Henkel, M. A. Woodruff, D. R. Epari, R. Steck, V. Glatt, I. C. Dickinson, P. F. M. Choong, M. A. Schuetz and D. W. Hutmacher, *Bone Res.*, 2013, **1**, 216–248.
34. L. Roseti, V. Parisi, M. Petretta, C. Cavallo, G. Desando, I. Bartolotti and B. Grigolo, *Mater. Sci. Eng., C*, 2017, **78**, 1246–1262.
35. V. Karageorgiou and D. Kaplan, *Biomaterials*, 2005, **26**, 5474–5491.
36. E. A. Phelps and A. J. Garcia, *Regener. Med.*, 2009, **4**, 65–80.
37. G. X. Pei, Y. X. Qin, D. W. Hutmacher and Z. Y. Zhang, *Biomed. Res. Int.*, 2014, 286978.
38. E. A. Abou Neel, W. Chrzanowski, V. M. Salih, H. W. Kim and J. C. Knowles, *J. Dent.*, 2014, **42**, 915–928.
39. A. Atala, F. K. Kasper and A. G. Mikos, *Sci. Transl. Med.*, 2012, **4**, 160rv12.
40. J. H. Bradt, M. Mertig, A. Teresiak and W. Pompe, *Chem. Mater.*, 1999, **11**, 2694–2701.

41. A. Tampieri, G. Celotti, E. Landi, M. Sandri, N. Roveri and G. Falini, *J. Biomed. Mater. Res. A*, 2003, **67a**, 618–625.
42. A. Tampieri, M. Sandri, E. Landi, D. Pressato, S. Francioli, R. Quarto and I. Martin, *Biomaterials*, 2008, **29**, 3539–3546.
43. A. Tampieri, S. Sprio, M. Sandri and F. Valentini, *Trends Biotechnol.*, 2011, **29**, 526–535.
44. A. S. Deshpande and E. Beniash, *Cryst. Growth Des.*, 2008, **8**, 3084–3090.
45. F. Nudelman, K. Pieterse, A. George, P. H. H. Bomans, H. Friedrich, L. J. Brylka, P. A. J. Hilbers, G. de With and N. A. J. M. Sommerdijk, *Nat. Mater.*, 2010, **9**, 1004–1009.
46. P. A. Price, D. Toroian and J. E. Lim, *J. Biol. Chem.*, 2009, **284**, 17092–17101.
47. A. George and A. Veis, *Chem. Rev.*, 2008, **108**, 4670–4693.
48. F. Nudelman, P. H. H. Bomans, A. George, G. de With and N. A. J. M. Sommerdijk, *Faraday Discuss.*, 2012, **159**, 357–370.
49. L. N. Niu, S. E. Jee, K. Jiao, L. Tonggu, M. Li, L. Wang, Y. D. Yang, J. H. Bian, L. Breschi, S. S. Jang, J. H. Chen, D. H. Pashley and F. R. Tay, *Nat. Mater.*, 2017, **16**, 370–378.
50. Y. Wang, T. Azais, M. Robin, A. Vallee, C. Catania, P. Legriel, G. Pehau-Arnaudet, F. Babonneau, M. M. Giraud-Guille and N. Nassif, *Nat. Mater.*, 2012, **11**, 724–733.
51. A. K. Nair, A. Gautieri and M. J. Buehler, *Biomacromolecules*, 2014, **15**, 2494–2500.
52. S. S. Jee, T. T. Thula and L. B. Gower, *Acta Biomater.*, 2010, **6**, 3676–3686.
53. S. S. Jee, L. Culver, Y. P. Li, E. P. Douglas and L. B. Gower, *J. Cryst. Growth*, 2010, **312**, 1249–1256.
54. T. T. Thula, D. E. Rodriguez, M. H. Lee, L. Pendi, J. Podschun and L. B. Gower, *Acta Biomater.*, 2011, 7, 3158–3169.
55. B. H. Ye, X. S. Luo, Z. W. Li, C. P. Zhuang, L. H. Li, L. Lu, S. Ding, J. H. Tian and C. R. Zhou, *Mater. Sci. Eng., C.*, 2016, **68**, 43–51.
56. A. J. Lausch, L. C. Chong, H. Uludag and E. D. Sone, *Adv. Funct. Mater.*, 2018, **28**, 1804730.
57. C. N. Elias, Y. Oshida, J. H. C. Lima and C. A. Muller, *J. Mech. Behav. Biomed.*, 2008, **1**, 234–242.
58. F. Rupp, L. Scheideler, N. Olshanska, M. de Wild, M. Wieland and J. Geis-Gerstorfer, *J. Biomed. Mater. Res. A*, 2006, **76a**, 323–334.
59. W. J. Wu and G. H. Nancollas, *Colloid Surf., B*, 1997, **10**, 87–94.
60. G. Zhao, Z. Schwartz, M. Wieland, F. Rupp, J. Geis-Gerstorfer, D. L. Cochran and B. D. Boyan, *J. Biomed. Mater. Res. A*, 2005, **74a**, 49–58.
61. C. H. Park, H. F. Rios, Q. M. Jin, J. V. Sugai, M. Padial-Molina, A. D. Taut, C. L. Flanagan, S. J. Hollister and W. V. Giannobile, *Biomaterials*, 2012, **33**, 137–145.
62. H. Schliephake and D. Scharnweber, *J. Mater. Chem.*, 2008, **18**, 2404–2414.
63. Z. M. Xia, X. H. Yu and M. Wei, *J. Biomed. Mater. Res. B*, 2012, **100b**, 871–881.

64. T. Ling, J. Lin, J. J. Tu, S. Q. Liu, W. J. Weng, K. Cheng, H. M. Wang, P. Y. Du and G. R. Han, *J. Mater. Sci.: Mater. Med.*, 2013, **24**, 2709–2718.
65. S. D. Miao, W. J. Weng, Z. L. Li, K. Cheng, P. Y. Du, G. Shen and G. R. Han, *J. Mater. Sci.: Mater. Med.*, 2009, **20**, 131–134.
66. A. Stoch, A. Brozek, S. Blazewicz, W. Jastrzebski, J. Stoch, A. Adamczyk and E. Roj, *J. Mol. Struct.*, 2003, **651**, 389–396.
67. J. J. Tu, M. F. Yu, Y. Lu, K. Cheng, W. J. Weng, J. Lin, H. M. Wang, P. Y. Du and G. R. Han, *J. Mater. Sci.: Mater. Med.*, 2012, **23**, 2413–2423.
68. T. Albrektsson and C. J. Johansson, *Eur. Spine J.*, 2001, **10**, S96–S101.
69. M. Doblare, J. M. Garcia and M. J. Gomez, *Eng. Fract. Mech.*, 2004, **71**, 1809–1840.
70. V. Campana, G. Milano, E. Pagano, M. Barba, C. Cicione, G. Salonna, W. Lattanzi and G. Logroscino, *J. Mater. Sci.: Mater. Med.*, 2014, **25**, 2445–2461.
71. R. A. Jain, *Biomaterials*, 2000, **21**, 2475–2490.
72. Y. Shikinami and M. Okuno, *Biomaterials*, 1999, **20**, 859–877.
73. H. W. Kim, H. H. Lee and J. C. Knowles, *J. Biomed. Mater. Res. A*, 2006, **79a**, 643–649.
74. F. Peng, X. Yu and M. Wei, *Acta Biomater.*, 2011, **7**, 2585–2592.
75. F. Khan, R. S. Tare, J. M. Kanczler, R. O. Oreffo and M. Bradley, *Biomaterials*, 2010, **31**, 2216–2228.
76. J. O. Smith, E. R. Tayton, F. Khan, A. Aarvold, R. B. Cook, A. Goodship, M. Bradley and R. O. C. Oreffo, *J. Tissue Eng. Regener. Med.*, 2017, **11**, 1065–1076.
77. F. Khan, J. O. Smith, J. M. Kanczler, R. S. Tare, R. O. C. Oreffo and M. Bradley, *Adv. Funct. Mater.*, 2013, **23**, 2850–2862.
78. S. Bhumiratana, W. L. Grayson, A. Castaneda, D. N. Rockwood, E. S. Gil, D. L. Kaplan and G. Vunjak-Novakovic, *Biomaterials*, 2011, **32**, 2812–2820.
79. Z. Z. Ding, Z. H. Fan, X. W. Huang, Q. Lu, W. A. Xu and D. L. Kaplan, *ACS Appl. Mater. Interfaces*, 2016, **8**, 24463–24470.
80. H. Y. Han, H. Y. Ning, S. S. Liu, Q. Lu, Z. H. Fan, H. J. Lu, G. Z. Lu and D. L. Kaplan, *Adv. Funct. Mater.*, 2016, **26**, 421–432.
81. X. W. Huang, S. M. Bai, Q. Lu, X. Liu, S. S. Liu and H. S. Zhu, *J. Biomed. Mater. Res., Part B*, 2015, **103**, 1402–1414.
82. P. F. He, S. Sahoo, K. S. Ng, K. L. Chen, S. L. Toh and J. C. H. Goh, *J. Biomed. Mater. Res., Part A*, 2013, **101**, 555–566.
83. Z. Z. Ding, H. Y. Han, Z. H. Fan, H. J. Lu, Y. H. Sang, Y. L. Yao, Q. Q. Cheng, Q. Lu and D. L. Kaplan, *ACS Appl. Mater. Interfaces*, 2017, **9**, 16914–16922.
84. D. C. J. Cancian, E. Hochuli-Vieira, R. A. C. Marcantonio and I. R. Garcia, *Int. J. Oral Maxillofac. Implants.*, 2004, **19**, 73–79.
85. G. Schneider, K. Blechschmidt, D. Linde, P. Litschko, T. Korbs and E. Beleites, *J. Mater. Sci. Mater. Med.*, 2010, **21**, 2853–2859.
86. U. Thormann, S. Ray, U. Sommer, T. ElKhassawna, T. Rehling, M. Hundgeburth, A. Henss, M. Rohnke, J. Janek, K. S. Lips, C. Heiss,

G. Schlewitz, G. Szalay, M. Schumacher, M. Gelinsky, R. Schnettler and V. Alt, *Biomaterials*, 2013, **34**, 8589–8598.
87. H. Shi, X. Ye, J. Zhang and J. TYe, *ACS Biomater. Sci. Eng.*, 2019, **5**, 262–271.
88. A. Tampieri, S. Sprio, A. Ruffini, G. Celotti, I. G. Lesci and N. Roveri, *J. Mater. Chem.*, 2009, **19**, 4973–4980.
89. A. Tampieri, A. Ruffini, A. Ballardini, M. Montesi, S. Panseri, F. Salamanna, M. Fini and S. Sprio, *Biomater. Sci.*, 2018, 7, 307–321.
90. S. Sprio, E. Campodoni, M. Sandri, L. Preti, T. Keppler, F. A. Muller, N. M. Pugno and A. Tampieri, *Int. J. Mol. Sci.*, 2018, **19**, 3604.
91. S. J. Hollister, *Adv. Mater.*, 2009, **21**, 3330–3342.
92. D. W. Hutmacher, J. T. Schantz, C. X. F. Lam, K. C. Tan and T. C. Lim, *J. Tissue Eng. Regener. Med.*, 2007, **1**, 245–260.
93. X. H. Liu and P. X. Ma, *Ann. Biomed. Eng.*, 2004, **32**, 477–486.
94. S. J. Hollister, *Nat. Mater.*, 2005, **4**, 518–524.
95. T. D. Roy, J. L. Simon, J. L. Ricci, E. D. Rekow, V. P. Thompson and J. R. Parsons, *J. Biomed. Mater. Res. A*, 2003, **66a**, 283–291.
96. K. U. Lewandrowski, J. D. Gresser, S. P. Bondre, A. E. Silva, D. L. Wise and D. J. Trantolo, *J. Biomater. Sci., Polym. Ed.*, 2000, **11**, 879–889.
97. B. J. Story, W. R. Wagner, D. M. Gaisser, S. D. Cook and A. M. Rust-Dawicki, *Int. J. Oral Maxillofac. Implants.*, 1998, **13**, 749–757.
98. M. C. Kruyt, J. D. de Bruijn, C. E. Wilson, F. C. Oner, C. A. van Blitterswijk, A. J. Verbout and W. J. A. Dhert, *Tissue Eng.*, 2003, **9**, 327–336.
99. S. Hulbert, F. Young, R. Mathews, J. Klawitter, C. Talbert and F. Stelling, *J. Biomed. Mater. Res.*, 1970, **4**, 433–456.
100. E. Tsuruga, H. Takita, H. Itoh, Y. Wakisaka and Y. Kuboki, *J. Biochem.*, 1997, **121**, 317–324.
101. Y. Kuboki, Q. M. Jin and H. Takita, *J. Bone Joint Surg. Am.*, 2001, **83a**, S105–S115.
102. Q. M. Jin, H. Takita, T. Kohgo, K. Atsumi, H. Itoh and Y. Kuboki, *J. Biomed. Mater. Res.*, 2000, **52**, 841–851.
103. Y. Kuboki, Q. Jin, M. Kikuchi, J. Mamood and H. Takita, *Connect. Tissue Res.*, 2002, **43**, 529–534.
104. J. C. Robins, N. Akeno, A. Mukherjee, R. R. Dalal, B. J. Aronow, P. Koopman and T. L. Clemens, *Bone*, 2005, **37**, 313–322.
105. M. Kanichai, D. Ferguson, P. J. Prendergast and V. A. Campbell, *J. Cell. Physiol.*, 2008, **216**, 708–715.
106. V. V. Meretoja, R. L. Dahlin, S. Wright, F. K. Kasper and A. G. Mikos, *Biomaterials*, 2013, **34**, 4266–4273.
107. L. A. Turner and M. J. Dalby, *Biomater. Sci.*, 2014, **2**, 1574–1594.
108. C. M. Murphy, F. J. O'Brien, D. G. Little and A. Schindeler, *Eur. Cells Mater.*, 2013, **26**, 120–132.
109. J. Y. Lim and H. J. Donahue, *Tissue Eng.*, 2007, **13**, 1879–1891.
110. M. J. Dalby, D. McCloy, M. Robertson, H. Agheli, D. Sutherland, S. Affrossman and R. O. Oreffo, *Biomaterials*, 2006, **27**, 2980–2987.

111. H. L. Khor, Y. Kuan, H. Kukula, K. Tamada, W. Knoll, M. Moeller and D. W. Hutmacher, *Biomacromolecules*, 2007, **8**, 1530–1540.
112. T. Sjöström, A. S. Brydone, R. D. Meek, M. J. Dalby, B. Su and L. E. McNamara, *Nanomedicine*, 2013, **8**, 89–104.
113. E. K. F. Yim, S. W. Pang and K. W. Leong, *Exp. Cell Res.*, 2007, **313**, 1820–1829.
114. S. Oh, K. S. Brammer, Y. S. J. Li, D. Teng, A. J. Engler, S. Chien and S. Jin, *Proc. Natl. Acad. Sci. U. S. A.*, 2009, **106**, 2130–2135.
115. R. J. McMurray, N. Gadegaard, P. M. Tsimbouri, K. V. Burgess, L. E. McNamara, R. Tare, K. Murawski, E. Kingham, R. O. C. Oreffo and M. J. Dalby, *Nat. Mater.*, 2011, **10**, 637–644.
116. R. K. Das, O. F. Zouani, C. Labrugere, R. Oda and M. C. Durrieu, *ACS Nano*, 2013, **7**, 3351–3361.
117. S. N. Robertson, P. Campsie, P. G. Childs, F. Madsen, H. Donnelly, F. L. Henriquez, W. G. Mackay, M. Salmeron-Sanchez, M. P. Tsimbouri, C. Williams, M. J. Dalby and S. Reid, *Philos. Trans. R. Soc., A*, 2018, **376**, 20170290.
118. A. M. Handorf, Y. X. Zhou, M. A. Halanski and W. J. Li, *Organogenesis*, 2015, **11**, 1–15.
119. P. J. Ehrlich and L. E. Lanyon, *Osteoporosis Int.*, 2002, **13**, 688–700.
120. R. L. De Souza, M. Matsuura, F. Eckstein, S. C. Rawlinson, L. E. Lanyon and A. A. Pitsillides, *Bone*, 2005, **37**, 810–818.
121. A. J. Engler, S. Sen, H. L. Sweeney and D. E. Discher, *Cell*, 2006, **126**, 677–689.
122. L. G. Vincent, Y. S. Choi, B. Alonso-Latorre, J. C. del Alamo and A. J. Engler, *Biotechnol. J.*, 2013, **8**, 472–484.
123. C. M. Murphy, A. Matsiko, M. G. Haugh, J. P. Gleeson and F. J. O'Brien, *J. Mech. Behav. Biomed.*, 2012, **11**, 53–62.
124. H. Sun, F. Zhu, Q. Hu and P. H. Krebsbach, *Biomaterials*, 2014, **35**, 1176–1184.
125. T. J. Qu, J. J. Jing, Y. S. Ren, C. Ma, J. Q. Feng, Q. Yu and X. H. Liu, *Acta Biomater.*, 2015, **16**, 60–70.
126. Y. Wang, Y. Hua, Q. Zhang, J. Yang, H. J. Li, Y. Li, M. Cao, Q. Cai, X. P. Yang, X. Zhang and C. Y. Li, *J. Tissue Eng. Regener. Med.*, 2018, **12**, 1545–1555.
127. E. Ozcivici, Y. K. Luu, B. Adler, Y. X. Qin, J. Rubin, S. Judex and C. T. Rubin, *Nat. Rev. Rheumatol.*, 2010, **6**, 50–59.
128. W. R. Thompson, C. T. Rubin and J. Rubin, *Gene*, 2012, **503**, 179–193.
129. J. H. Chen, C. Liu, L. D. You and C. A. Simmons, *J. Biomech.*, 2010, **43**, 108–118.
130. J. C. Chen and C. R. Jacobs, *Stem Cell Res. Ther.*, 2013, **4**, 107.
131. E. Baas, J. H. Kuiper, Y. Yang, M. A. Wood and A. J. El Haj, *J. Biomech.*, 2010, **43**, 733–739.
132. A. Sittichockechaiwut, A. M. Scutt, A. J. Ryan, L. F. Bonewald and G. C. Reilly, *Bone*, 2009, **44**, 822–829.
133. A. Sittichokechaiwut, J. H. Edwards, A. M. Scutt and G. C. Reilly, *Eur. Cells Mater.*, 2010, **20**, 45–57.

134. J. R. Mauney, S. Sjostorm, J. Blumberg, R. Horan, J. P. O'Leary, G. Vunjak-Novakovic, V. Volloch and D. L. Kaplan, *Calcif. Tissue Int.*, 2004, **74**, 458–468.
135. J. D. Boerckel, Y. M. Kolambkar, H. Y. Stevens, A. S. P. Lin, K. M. Dupont and R. E. Guldberg, *J. Orthop. Res.*, 2012, **30**, 1067–1075.
136. A. Roshan-Ghias, F. M. Lambers, M. Gholam-Rezaee, R. Müller and D. P. Pioletti, *Bone*, 2011, **49**, 1357–1364.
137. R. Garman, G. Gaudette, L. R. Donahue, C. Rubin and S. Judex, *J. Orthop. Res.*, 2007, **25**, 732–740.
138. L. Xie, J. M. Jacobson, E. S. Choi, B. Busa, L. R. Donahue, L. M. Miller, C. T. Rubin and S. Judex, *Bone*, 2006, **39**, 1059–1066.
139. A. S. G. Curtis, S. Reid, I. Martin, R. Vaidyanathan, C. A. Smith, H. Nikukar and M. J. Dalby, *IEEE Trans. Nanobioscience*, 2013, **12**, 247–254.
140. H. Nikukar, S. Reid, P. M. Tsimbouri, M. O. Riehle, A. S. G. Curtis and M. J. Dalby, *ACS Nano*, 2013, **7**, 2758–2767.
141. G. D. Pemberton, P. Childs, S. Reid, H. Nikukar, P. M. Tsimbouri, N. Gadegaard, A. S. G. Curtis and M. J. Dalby, *Nanomedicine*, 2015, **10**, 547–560.
142. G. D. Pemberton, P. Childs, S. Reid, H. Nikukar, P. M. Tsimbouri, N. Gadegaard, A. S. Curtis and M. J. Dalby, *Nanomedicine*, 2015, **10**, 547–560.
143. P. M. Tsimbouri, P. G. Childs, G. D. Pemberton, J. L. Yang, V. Jayawarna, W. Orapiriyakul, K. Burgess, C. Gonzalez-Garcia, G. Blackburn, D. Thomas, C. Vallejo-Giraldo, M. J. P. Biggs, A. S. G. Curtis, M. Salmeron-Sanchez, S. Reid and M. J. Dalby, *Nat. Biomed. Eng.*, 2017, **1**, 758–770.
144. G. Brunello, S. Sivolella, R. Meneghello, L. Ferroni, C. Gardin, A. Piattelli, B. Zavan and E. Bressan, *Biotechnol. Adv.*, 2016, **34**, 740–753.
145. S. Tarafder, V. K. Balla, N. M. Davies, A. Bandyopadhyay and S. Bose, *J. Tissue Eng. Regener. Med.*, 2013, **7**, 631–641.
146. B. Leukers, H. Gulkan, S. H. Irsen, S. Milz, C. Tille, H. Seitz and M. Schieker, *Materialwiss. Werkst.*, 2005, **36**, 781–787.
147. E. Vorndran, M. Klarner, U. Klammert, L. M. Grover, S. Patel, J. E. Barralet and U. Gbureck, *Adv. Eng. Mater.*, 2008, **10**, B67–B71.
148. P. H. Warnke, H. Seitz, F. Warnke, S. T. Becker, S. Sivananthan, E. Sherry, Q. Liu, J. Wiltfang and T. Douglas, *J. Biomed. Mater. Res., Part B*, 2010, **93b**, 212–217.
149. A. Khalyfa, S. Vogt, J. Weisser, G. Grimm, A. Rechtenbach, W. Meyer and M. Schnabelrauch, *J. Mater. Sci. Mater. Med.*, 2007, **18**, 909–916.
150. S. Bose, S. Vahabzadeh and A. Bandyopadhyay, *Mater. Today*, 2013, **16**, 496–504.
151. H. Seitz, W. Rieder, S. Irsen, B. Leukers and C. Tille, *J. Biomed. Mater. Res., Part B*, 2005, **74b**, 782–788.
152. Q. Q. Yao, B. Wei, Y. Guo, C. Z. Jin, X. T. Du, C. Yan, J. W. Yan, W. H. Hu, Y. Xu, Z. Zhou, Y. J. Wang and L. M. Wang, *J. Mater. Sci. Mater. Med.*, 2015, **26**, 51.

153. L. Ciocca, F. De Crescenzio, M. Fantini and R. Scotti, *Comput. Med. Imag. Grap.*, 2009, **33**, 58–62.
154. S. T. Becker, H. Bolte, K. Schunemann, H. Seitz, J. J. Bara, B. E. Beck-Broichsitter, P. A. J. Russo, J. Wiltfang and P. H. Warnke, *Int. J. Oral Maxillofac. Surg.*, 2012, **41**, 1153–1160.
155. S. Tarafder, N. M. Davies, A. Bandyopadhyay and S. Bose, *Biomater. Sci.*, 2013, **1**, 1250–1259.
156. G. A. Fielding, A. Bandyopadhyay and S. Bose, *Dent. Mater.*, 2012, **28**, 113–122.
157. W. Zhang, C. Feng, G. Yang, G. Li, X. Ding, S. Wang, Y. Dou, Z. Zhang, J. Chang, C. Wu and X. Jiang, *Biomaterials*, 2017, **135**, 85–95.
158. F. Pati, T. H. Song, G. Rijal, J. Jang, S. W. Kim and D. W. Cho, *Biomaterials*, 2015, **37**, 230–241.
159. A. Butscher, M. Bohner, C. Roth, A. Ernstberger, R. Heuberger, N. Doebelin, P. R. von Rohr and R. Muller, *Acta Biomater.*, 2012, **8**, 373–385.
160. U. Gbureck, T. Hozel, U. Klammert, K. Wurzler, F. A. Muller and J. E. Barralet, *Adv. Funct. Mater.*, 2007, **17**, 3940–3945.
161. K. Igawa, M. Mochizuki, O. Sugimori, K. Shimizu, K. Yamazawa, H. Kawaguchi, K. Nakamura, T. Takato, R. Nishimura, S. Suzuki, M. Anzai, U. Chung and N. Sasaki, *J. Artif. Organs*, 2006, **9**, 234–240.
162. J. A. Inzana, D. Olvera, S. M. Fuller, J. P. Kelly, O. A. Graeve, E. M. Schwarz, S. L. Kates and H. A. Awad, *Biomaterials*, 2014, **35**, 4026–4034.
163. J. Y. Lee, B. Choi, B. Wu and M. Lee, *Biofabrication*, 2013, **5**, 045003.
164. E. Tamjid, A. Simchi, J. W. C. Dunlop, P. Fratzl, R. Bagheri and M. Vossoughi, *J. Biomed. Mater. Res., Part A*, 2013, **101**, 2796–2807.
165. H. Kodama, *Rev. Sci. Instrum.*, 1981, **52**, 1770–1773.
166. J. W. Lee, J. Y. Kim and D. W. Choo, *Int. J. Stem Cells*, 2010, **3**, 85–95.
167. M. N. Cooke, J. P. Fisher, D. Dean, C. Rimnac and A. G. Mikos, *J. Biomed. Mater. Res., Part B*, 2003, **64b**, 65–69.
168. T. Cao, K. H. Ho and S. H. Teoh, *Tissue Eng.*, 2003, **9**, S103–S112.
169. D. W. Hutmacher, T. Schantz, I. Zein, K. W. Ng, S. H. Teoh and K. C. Tan, *J. Biomed. Mater. Res.*, 2001, **55**, 203–216.
170. C. Choong, J. T. Triffitt and Z. F. Cui, *Food Bioprod. Process.*, 2004, **82**, 117–125.
171. N. Xu, X. J. Ye, D. X. Wei, J. Zhong, Y. Y. Chen, G. H. Xu and D. N. He, *ACS Appl. Mater. Interfaces*, 2014, **6**, 14952–14963.
172. M. H. Smith, C. L. Flanagan, J. M. Kemppainen, J. A. Sack, H. Chung, S. Das, S. J. Hollister and S. E. Feinberg, *Int. J Med. Robot. Comp.*, 2007, **3**, 207–216.
173. W. L. Grayson, M. Frohlich, K. Yeager, S. Bhumiratana, M. E. Chan, C. Cannizzaro, L. Q. Wan, X. S. Liu, X. E. Guo and G. Vunjak-Novakovic, *Proc. Natl. Acad. Sci. U. S. A.*, 2010, **107**, 3299–3304.
174. S. Bhumiratana, J. C. Bernhard, D. M. Alfi, K. Yeager, R. E. Eton, J. Bova, F. Shah, J. M. Gimble, M. J. Lopez, S. B. Eisig and G. Vunjak-Novakovic, *Sci. Transl. Med.*, 2016, **8**, 343ra83.

Subject Index